Springer Texts in Electrical Engineering

Springer Texts in Electrical Engineering

Multivariable Feedback Systems
F.M. Callier/C.A. Desoer

Linear Programming
M. Sakarovitch

Introduction to Random Processes
E. Wong

Stochastic Processes in Engineering Systems
E. Wong/B. Hajek

Introduction to Probability
J.B. Thomas

Elements of Detection and Signal Design
C.L. Weber

Charles L. Weber

Elements of Detection and Signal Design

With 57 Illustrations

Springer-Verlag
New York Berlin Heidelberg
London Paris Tokyo

Charles L. Weber
Electrical Engineering Department
University of Southern California
Los Angeles, California 90089-0272 USA

Library of Congress Cataloging in Publication Data
Weber, Charles L.
Elements of detection and signal design.
(Springer texts in electrical engineering)
Includes index.
1. Statistical communication theory. 2. Signal theory (Telecommunication) I. Title. II. Series.
TK5101.W38 1987 621.38'043 87-12707

Previous edition: C.L. Weber, *Elements of Detection and Signal Design*

Printed and bound by R.R. Donnelley & Sons, Harrisonburg, Virginia.
Printed in the United States of America.

9 8 7 6 5 4 3 2 1

ISBN 0-387-96529-7 Springer-Verlag New York Berlin Heidelberg
ISBN 3-540-96529-7 Springer-Verlag Berlin Heidelberg New York

To My Parents

Preface to the Springer-Verlag Edition

Due to a steady flow of requests over several years, Springer-Verlag now provides a corrected reprint of this text. It is designed to serve as a text for a first semester graduate level course for students in digital communication systems. As a prerequisite, it is presumed that the reader has an understanding of basic probability and stochastic processes.

The treatment of digital communications in this book is intended to serve as an introduction to the subject. Part one is a development of the elements of statistical communication theory and radar detection. The text begins with a general model of a communication system which is extensively developed and the performance analyses of various conventional systems. The first part also serves as introductory material for the second part of the text which is a comprehensive study of the theory of transmitter optimization for coherent and noncoherent digital communication systems, that is, the theory of signal design.

Important features of this text include: (1) The first unified and comprehensive treatment of the theory of signal design for digital communication systems, (2)

The fundamentals of statistical communication theory from a general model which encompasses most of the classical examples encountered in practice, (3) Discussions of the Bayes, minimax, and Neyman-Pearson performance criteria, (4) An extensive treatment of coherent and noncoherent M-ary digital communication systems, (5) A chapter devoted to the detection of a stochastic process in noise and (6) Many examples and problems, thereby making it suitable as a graduate text.

Charles L. Weber

Preface

Part One of this text is intended as an elementary introduction to statistical decision theory as applied to radar detection and statistical communication theory. A general model of a communication system is considered which encompasses most of the classical examples encountered in practice. This model is then used as a basis for presenting the fundamentals of the theory of receiver optimization as esthetically developed by the architects of statistical communication theory in the recent past. Examples are considered in detail and ample problems are provided, so that this first part may adequately serve as a graduate text as well as introductory material for Part Two.

Part Two is the first comprehensive presentation of the theory of signal design for coherent and noncoherent digital communication systems. This unified treatment consists of a theoretical study of transmitter-receiver optimization, based principally on a different variational approach. Since some of the proofs are esoteric, the development is presented in a theorem-proof form, so that the results may be employed without their verification.

The text is directed to the reader who has the background attained from a one-semester course covering the fundamentals of the theory of probability and stochastic processes. Such prerequisite material is presented, for example, in the first part of Davenport and Root, "Introduction to Random Signals and Noise" (McGraw-Hill, New York, 1958), or Papoulis, "Probability, Random Variables, and Stochastic Processes" (McGraw-Hill, New York, 1965). When more advanced concepts are required, necessary introductory background is provided. No previous knowledge of statistical decision theory is assumed; the text was written as a second course for graduate students.

Pertinent references are listed at the end of each chapter, but no attempt has been made to provide a complete bibliography, since adequate ones already exist.

In a text of this nature it is impossible to acknowledge all those who have influenced or contributed to the presentation of the material. The author is particularly indebted, however, to L. Davisson, R. Scholtz, D. Slepian, and A. Viterbi for their many helpful suggestions, which are gratefully acknowledged. A particular appreciation is extended to A. V. Balakrishnan, without whose pioneering work and continuing guidance in this subject area this work would not have developed so rapidly. Also, special thanks are extended to the Technical Reports Groups of the University of Southern California and the University of California at Los Angeles for extensive assistance throughout the preparation of the manuscript.

Charles L. Weber

Contents

PART ONE ELEMENTS OF DETECTION

CHAPTER 1 Introduction 3

CHAPTER 2 A mathematical model 7

CHAPTER 3 General decision-theory concepts 11

CHAPTER 4 Binary detection systems minimizing the average risk 19

4.1 Binary decision functions 20
Example 4.1 Specialization to a signal space of two elements 22
4.2 Vector model 25
4.3 Coherent binary phase modulation 30

CHAPTER 5 Minimax decision-rule concepts 40

CHAPTER 6 Radar detection theory 45

6.1 Radar-system-design philosophy 46
6.2 Vector model 48

CHAPTER 7 Binary composite-hypothesis testing 52

Example 7.1 Detection of one pulse of known arrival time 56
Example 7.2 Detection with the complex-envelope representation 61
Example 7.3 Binary noncoherent communication system 63

CHAPTER 8 Detection and communication in colored noise 73

8.1 Detection in colored noise 74
8.2 Coherent binary communication in colored noise 82
8.3 Noncoherent binary communication in colored noise 84

CHAPTER 9 Detecting a stochastic signal in noise 93

9.1 Detection of a random vector 94
9.2 Detection of a stochastic process in noise 97

CHAPTER 10 M-ary digital communication systems 106

10.1 Coherent M-ary communication 109
10.2 Noncoherent M-ary communication 111

PART TWO SIGNAL DESIGN

CHAPTER 11 Introduction 121

CHAPTER 12 Problem statement for coherent channels 124

12.1 Description in the time domain 124
12.2 Reduction to finite-dimensional euclidean space 127
12.3 Bandwidth considerations 132

CHAPTER 13 Signal design when the dimensionality of the signal set is restricted to 2 134

13.1 Optimal signal selection in two dimensions 135
13.2 Communication efficiency and channel capacity for two-dimensional signal sets 141
13.3 Partial ordering of the class of two-dimensional signal sets 144
13.4 The dependence of some suboptimal signal sets on the signal-to-noise ratio 145

CHAPTER 14 General theory 149

14.1 Introduction 149
14.2 Convex-body considerations: small signal-to-noise ratios 152
14.3 Linearly dependent versus linearly independent signal sets 157
14.4 Gradient of the probability of detection 164
14.5 Signal sets whose convex hull does not include the origin 167
14.6 The admissible α space 175
14.7 Series expansions and asymptotic approximations 178

CHAPTER 15 Optimality for coherent systems when dimensionality is not specified: regular simplex coding 189

15.1 Necessary (first-order) considerations for optimality 189
15.2 Uniqueness of the regular simplex satisfying necessary conditions for all signal-to-noise ratios 194

15.3 Global optimality of the regular simplex for large signal-to-noise ratios 199
15.4 Sufficient (second-order) conditions for optimality 200
15.5 Maximizing the minimum distance 214

CHAPTER 16 Optimality for coherent systems when the dimensionality is restricted to $D \leq M - 2$ **218**

16.1 Necessary (first-order) conditions 220
16.2 Sufficient (second-order) conditions 229
16.3 Choosing the largest of several local maxima 235
16.4 Five signal vectors in three dimensions 237

CHAPTER 17 Optimality for coherent systems when the dimensionality is restricted to $D \leq M - K$, where $K \leq M/2$ **241**

17.1 Necessary (first-order) conditions 243
17.2 Choosing the largest of several local extrema 248
17.3 The effect of dimensionality on system performance 249

CHAPTER 18 Additional solutions for three-dimensional signal structures **252**

CHAPTER 19 Signal-design concepts for noncoherent channels **255**

19.1 Necessary (first-order) conditions for noncoherent optimality 263
19.2 Evaluation of probability of error for the orthogonal noncoherent signal structure 265
19.3 Sufficient (second-order) conditions for noncoherent optimality 267
19.4 Global optimality when $M = 2$ 277

APPENDIX A Summary of conditional gaussian probability density functions **281**

APPENDIX B Karhunen-Loeve expansion **281**

APPENDIX C **Modified Bessel function of the first kind** 282

APPENDIX D **Marcum's Q function** 283

APPENDIX E **Summary of tetrachoric series** 284

APPENDIX F **Chi-squared distribution** 285

INDEX 287

part one

Elements of Detection

1
Introduction

Statistical communication theory encompasses the application of the theory of probability to mathematical models which represent communication systems, that is, systems whose purpose is to transfer information from a source to a destination. This theory may be conveniently divided into three overlapping parts: the source or generation of information, its transmission, and the reception of the signals containing the desired information. The purpose of the reception operation is either estimation, an attempt to determine or estimate some parameter or quantity associated with the received signal, or detection, an attempt to describe which of a predetermined set (usually finite) of possible messages the signal represents.

Information is a measure of the increase in knowledge as a result of the reception of a signal. In particular, if the receiver knows in advance the message being transferred by the system, no information is conveyed. The *channel* is the medium through which the messages are transmitted. The disturbances which the signal waveform encounters during its transmission through the channel are called *noises*.

The theory, as it will be developed here, is directed toward a better understanding of several types of communication, telemetry, and radar-system problems. In Part One we shall examine the reception of signals in the most efficient manner and in Part Two the most efficient manner of transmitting signals. A few of the important examples follow, and as the theory is developed, we shall see how it is applicable to these and other examples.

Example 1.1 Digital communication system A *digital communication system* is one whose transmitter must convey to its receiver a sequence of messages. The set of messages is from the finite alphabet, $m_1, m_2, \ldots, m_M$. The transmitter conveys a message in this sequence by transmitting one of M different waveforms during a specified interval of time, say, $[0,T]$. The messages might correspond to a set of numbers or a digital representation of an analog variable, or to the letters of some nonnumerical alphabet. The physical medium through which the waveform must travel between the transmitter and the receiver and those subsystem components over which the systems engineer has no control is the *channel*.[1] The atmosphere through which electromagnetic radiation travels, for either straight-line transmission or communication links that employ ionospheric reflection, is a common example of a communication channel. Other examples are *electromagnetic waveguides*, and media which transmit acoustical waveforms. For instance, water is the transmission medium for sonar systems, which detect the presence of submarines, underwater mines, or the like by means of inaudible high-frequency vibrations originating at or reflecting from these objects. These systems are designated *passive* and *active* sonar systems, respectively. Wires, such as telephone lines, are also channels. If the channel were noiseless, so that the transmitted waveform could reach the receiver undistorted, then the receiver would be perfectly able to determine which message had been transmitted. Unfortunately, physical channels do not behave so cooperatively; they distort the waveform in a variety of ways.

The distortion in a radio-communication channel is characterized by the type of interference that it places on the reception of electromagnetic energy radiated from the transmitter. These disturbances can conveniently be classified into three categories. One form of interference which is always present is thermal noise in the antenna and front-end components of the receiver. Although modern technological advances in low-temperature receivers have reduced this form of interference by at

[1] The channel is essentially that part of the system which is fixed, and generally includes subsystems such as transmitter and receiver front-end components, antennas, or a phase-locked loop, as well as the transmission medium.

least an order of magnitude, there are still many applications where thermal noise dominates. In electromagnetic communication via the ionosphere, where the channel employs tropospheric propagation and ionospheric reflection, a form of interference that is often more significant is fading and multipath propagation of the transmitted waveforms. This can usually be modeled by introducing random parameters in the transmitted signal in addition to the additive thermal disturbances, or more generally, by introducing a random transformation on the transmitted waveform. The third type of disturbance is electromagnetic radiation at frequencies within the received band. This is of least interest in all but certain military applications (such as jamming), since in communication situations it can most often be avoided through preliminary system-design considerations and proper choices of frequency bands.

Because of the increasing emphasis on line-of-sight communication brought on by the development of space vehicles and satellite relay systems, the study of communication systems perturbed only by additive noise has extensive application. Since this form of radio-frequency interference can never be completely avoided, we shall consider it in some detail. The performance analysis of some of the analog as well as digital communication systems we shall examine is relatively simple when the disturbances are additive.

Because of these various types of disturbances, or noise, a distorted version of the transmitted signal arrives at the receiver, and as a result, the receiver may not always be able to decide correctly every time which message the received waveform represents. Such a system thus has a *nonzero probability of error;* the amount of this probability of error is the logical criterion for judging the performance of a digital communication system. If we are considering a certain family of systems, the preferred one is that which has the lowest probability of error. In some digital systems, criteria other than the probability of error are of interest. For example, the *mean square error* and the *signal-to-noise ratio* are of interest in continuous estimation problems, or in systems where analog signals are quantized.

Example 1.2 Estimation: filtering and prediction In many situations, such as radio and television transmission or certain control system applications, we should like the receiver to reproduce the transmitted signal in the best possible way from the distorted waveform that it receives, again based on some predesignated criterion, which usually is the minimum *mean square error*. If the receiver is required to estimate the present value of the transmitted signal, given part or all of the past values and/or the present value of the received waveform, the receiver is classified as a *causal*[1] *filter*. If it is required to estimate the value of the transmitted waveform at some future time, given part or all of the past observable waveform, the receiver is termed a *predictor*. The receiver may also be required to *interpolate*, that is, to obtain the best possible estimate of a signal in the presence of interference at some time within the interval of observation. All these types of estimators may be classified under the more general classification of *analog communication systems*.

It should be noted that a filter and an interpolator can be perfect if the received waveform is undistorted, but the predictor will generally have errors even when there is no interference in the received wave.

Example 1.3 Radar systems The purpose of a radar system is to discover the presence of distant objects, such as airplanes, missiles, or ships, by transmitting pulses of electromagnetic energy and detecting the presence or absence of the reflected echo. A known waveform of electromagnetic energy is transmitted into the region where the target is suspected. If there is an object in this region, a portion of the electromagnetic energy will be reflected from the object to the receiver. The received waveform is not only stochastic, inasmuch as the receiver will generally not know the amplitude, time of arrival, phase, and doppler frequency shift of the reflected energy, but there may also be reflections both from other objects and from the background. Given all these interferences, the receiver must determine the presence or absence of a target. This situation is similar to that in Example 1.1 if we set $M = 2$, where the waveform corresponding to *no target present* is exactly zero. When we study radar systems, however, we shall see that the logical criterion for system evaluation will differ from that for digital communication systems. Aspects of Example 1.2 can also be applied to radar systems if, given a decision of *signal present*, the receiver is also to estimate unknown parameters such as signal amplitude or time of arrival. Estimating the echo's time of arrival is equivalent to estimating the target's range, and estimating the echo's doppler frequency shift corresponds to estimating the rate of change of the range, or the *range rate*.

[1] A filter, in general, may use past, present, and future data. A special case is the causal filter, which cannot utilize future data.

2
A Mathematical Model

The mathematical model presented in this chapter is intended to be sufficiently general to include the examples of Chapter 1 as well as many others, and yet sufficiently specific to enable us to apply the concepts of statistical-decision theory as well. We shall examine the general reception problem as a statistical-decision problem and hence shall assume certain statistics of the messages and channel disturbances depending on the criterion for system evaluation prescribed and the set of decisions permitted by the receiver. We shall also have to make certain assumptions concerning the nature of the available data, such as whether it resulted from discrete or continuous sampling of the observed waveform. Finally, we must specify the performance criterion by which we are seeking the optimal system.

The first entity present in virtually all communication situations is the set of elements which we shall designate as the *signal space* or *message space*. These elements contain the information which the system is to transfer to the receiver. In Example 1.1, for instance, there are M waveforms which comprise the signal space, one of which the system is to communicate from its origin to its destination every T sec. Here the signal space consists of a finite number of elements. In estimation problems the parameters which the receiver is attempting to determine make up the signal space. In some cases the parameters take on only a finite set of values, but more generally these parameters take on a continuum of values. The signal set then consists of infinitely many elements.

In a radar system whose sole purpose is to detect the presence or absence of a target, the signal set consists of only two elements, $s_1(t)$, corresponding to *target present,* and $s_0(t)$ (which normally is identically zero), corresponding to *target not present.* If, in addition to detecting the presence of the target, the system is asked to estimate unknown parameters of the received echo, the signal space must be appropriately enlarged and in general will have to be infinite-dimensional.

The general model consists of an abstract space of sufficient dimensionality, known as the signal space $\mathbf{\Omega}$ with elements s. The occurrence of elements in this space will be assumed to be governed by the *probability density function* $\pi(s)$, known as the *a priori distribution* of the signals. It specifies the probabilities or the probability density with which the possible signals may be transmitted.

In each of the examples cited note that the observer or the receiver is able to observe only distorted versions of the transmitted signal waveforms, and not the elements of the signal space themselves. We are therefore led to introduce as part of the model a second abstract space called the *observation space* $\mathbf{\Gamma}$, with elements y which correspond to observations or possible inputs to the receiver. As the examples have indicated, because of factors such as the randomness of the disturbances in the channel, there is in general no deterministic mapping from the elements of the signal space to the observation space. In addition, it is normally impossible to make a deterministic association of a given observation with a given element of the signal space. This randomness in the observation is dictated by the statistics of the channel noise. Hence for a given element $s \in \mathbf{\Omega}$ the values that y may assume can at best be represented probabilistically. Thus for each $s \in \mathbf{\Omega}$ there is a *conditional probability density function* $f(y|s)$ defined on $\mathbf{\Gamma}$. This density function is governed by the statistical characteristics of the channel. That is, the channel provides the probability density under which the different observations may occur, given that the signal s has been transmitted.

As the examples indicate, and as is the case in most communication situations, the receiver must make a decision of some kind on the basis of the available observation. It may have to determine which element of a finite set of elements in the signal space has been transmitted or make an estimate of some parameter of the transmitted signal. In any event, it is clearly necessary to define for the model a *decision space* $\mathbf{A}$, with elements a which represent the choices or decisions available to the receiver.

There may be a one-to-one correspondence between the elements of the signal space and those of the decision space, as in Example 1.1. However, other decisions may also be allowed by the receiver. For instance, in Example 1.1, if the decisions $a_1, \ldots, a_M$ correspond, respectively, to the messages $m_1, \ldots, m_M$ in the signal space, an additional decision a_0 might be introduced for the case where the receiver is uncertain about the transmitted signal, that is, where the received observation was sufficiently distorted that any choice from among the decisions $a_1, \ldots, a_M$ would entail substantial uncertainty. With the addition of a decision a_0, the receiver is essentially allowed to conclude *I don't know*. Or, what is sometimes even more meaningful, the additional decision a_0 may correspond to making a decision *continue observation;* that is, as the observations are being received, the receiver may at any time halt the observation and make a decision or, alternatively, continue to observe until the decision can be made with sufficient certainty. In this case there must be a feedback link to the transmitter to tell it when to stop sending; that is, T is random. The study of this class of decisions is entitled *sequential-decision theory*. The nature of the observation space may be such that the concept of a sequential-decision device is meaningless. However, in most cases where a sequential device is applicable, an acceptable model can be developed.

We have not yet mentioned anything specific about the nature of the different spaces in the model—whether they are to be considered discrete, continuous vectors, etc. As we develop specific examples, we shall indicate the nature of the spaces involved. For the present this will not be necessary.

The crucial point in arriving at an optimal receiver is to specify the *decision function*, which is the mapping or assigning of decisions to the possible observations. Hence the receiver is a *mapping* from the observation space to the decision space, and optimal receiver design reduces to determining this map on the basis of the designated criterion; this mapping must be made without knowledge of or independently of the par-

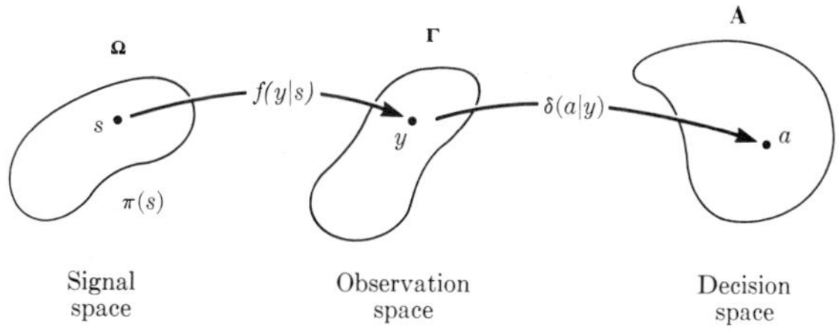

Fig. 2.1 Diagram of the mathematical model.

ticular s that results in the observation y. We shall use the notation $\delta(a|y)$ to designate this mapping, noting that it need not be a deterministic map. In general, it could be probabilistic, in which case for each $y \in \mathbf{\Gamma}$ there will be the density function $\delta(a|y)$ over the decision space $\mathbf{A}$. We shall see, however, that one of the fundamental theorems of statistical-decision theory will allow us to restrict attention only to deterministic decision functions for most problems of practical significance.

The model that has been presented is schematically drawn in Fig. 2.1. This model can be generalized, and we shall do so to some extent later on. For the next several chapters we shall study this model and apply it to specific examples.

REFERENCES

2.1 Middleton, D.: "Introduction to Statistical Communication Theory," McGraw-Hill, New York, 1960.

2.2 Helstrom, C. W.: "Statistical Theory of Signal Detection," Pergamon Press, New York, 1960.

2.3 Ferguson, T. S.: "Mathematical Statistics: A Decision Theoretic Approach," Academic, New York, 1967.

3
General Decision-theory Concepts

Let us begin with a few fundamental aspects of statistical-decision theory so that we may apply them directly to problems in communication theory. Summarizing from Chapter 2, we have the following definitions:

1. $\mathbf{\Gamma}$ is the *observation space* which has elements y.
2. $\mathbf{A}$ is the *decision space* with elements a, called actions or decisions.
3. $\mathbf{\Omega}$ is the *signal space* with elements s, called the messages or signals.

The transmission of information from one point to another has been characterized in the following way: At the transmitter a point s in the signal space $\mathbf{\Omega}$ is chosen randomly according to an a priori probability distribution $\pi(s)$ over $\mathbf{\Omega}$. This signal is sent from the transmitter to the receiver through a medium called the channel, which produces random fluctuations on s such that at the point of reception s is not the observed variable; instead, an element $y \in \mathbf{\Gamma}$ is observed, which is a random variable governed by a distribution conditioned on the transmitted signal s. That is to say, for each $s \in \mathbf{\Omega}$ the observed random variable behaves ac-

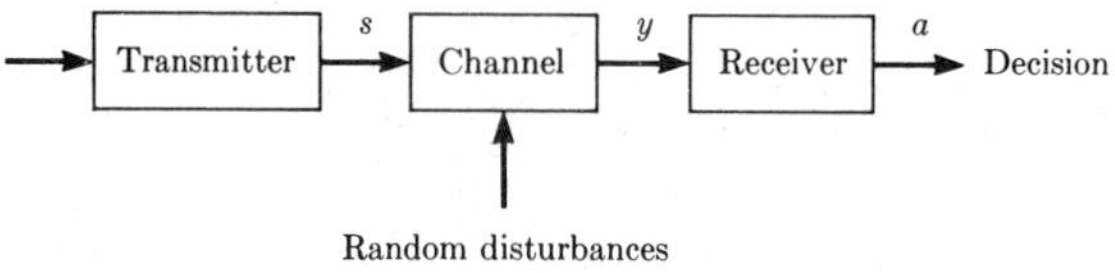

Fig. 3.1 A general communication system.

cording to the conditional probability distribution function $f(y|s)$, whose form depends on the type of disturbances in the channel. The receiver observes y and then chooses an element $a \in \mathbf{A}$ on the basis of a probability distribution over $\mathbf{A}$ which is conditioned on the observed variable y. We designate this conditional distribution as $\delta(a|y)$ and call it the *decision function*. Note again that the receiver must base its choice of a only on y; that is, the receiver cannot discern the actual signal chosen at the transmitter. A block diagram of such a communication system is shown in Fig. 3.1.

With this overall configuration in mind we now define the pertinent functions more precisely.

DEFINITION 3.1 *A decision function $\delta(a|y)$ is defined as the conditional probability of deciding on a point $a \in \mathbf{A}$ given a point $y \in \mathbf{\Gamma}$.*

Thus

$$0 \leq \delta(a|y) \leq 1 \qquad \text{for each } a \text{ and } y \tag{3.1}$$

and

$$\sum_{\mathbf{A}} \delta(a|y) = 1 \qquad \text{for each } y \in \mathbf{\Gamma} \tag{3.2}$$

DEFINITION 3.2 *A nonrandomized decision function, also called a deterministic function, is of the form*

$$\delta(a|y) \triangleq \begin{cases} 1 & \text{for } a = a_0 \\ 0 & \text{for } a \neq a_0 \end{cases} \tag{3.3}$$

That is, for some $y \in \mathbf{\Gamma}$ the probability of choosing a_0 is 1 and the probability of choosing any $a \neq a_0$ is 0.

With the restriction in (3.1) and the summation in (3.2) we are implying that the number of elements in the decision space is at most countably infinite. As Examples 1.2 to 1.3 indicate, this need not be the case. More generally, then, $\delta(a|y)$ is defined as a conditional probability density function, and

$$\int_{\mathbf{A}} \delta(a|y)\, da = 1 \qquad \text{for each } y \in \mathbf{\Gamma} \tag{3.4}$$

Taken in this sense, it includes the cases described by Definition 3.1.

DEFINITION 3.3 *A binary system is characterized by* **A** *consisting of only two elements,* a_1 *and* a_2.

A binary system, however, places no restriction on the possible number of signals in the signal space. In a binary system, then,

$$\delta(a_1|y) = 1 - \delta(a_2|y) \qquad \text{for each } y \in \mathbf{\Gamma} \tag{3.5}$$

An example of a *randomized decision function* in the binary case is

$$\delta(a_1|y') = \epsilon \qquad 0 < \epsilon < 1$$
$$\delta(a_2|y') = 1 - \epsilon$$

so whenever y' is observed, a_1 is decided with probability ϵ and a_2 is decided with probability $1 - \epsilon$. In general, ϵ will depend on y. Implementation of such decision functions would be through a random table, a random-number generator, or such.

DEFINITION 3.4 *The cost function specifies the penalty or loss or cost that is assessed for choosing a when s was transmitted.*

DEFINITION 3.5 *A simple cost function, denoted as* $C(s,a)$, *is the cost of deciding a given that s is the signal.*

Simple cost functions do not depend on the probability distribution of the variables involved. An example of a simple cost function is as follows: let

$$\mathbf{\Omega} = (s_1,s_2)$$
$$\mathbf{A} = (a_1,a_2)$$

and

$$C(s_i,a_j) = c_{ij} \qquad i, j = 1, 2$$

such that c_{ij} does not depend on the probability distribution of s_i and a_j.

With $\mathbf{\Omega}$ and **A** defined as above, an example of a cost function which is not simple is

$$c_{ij} = -p_{ij} \log p_{ij}$$

where p_{ij} is the probability of the joint occurrence of s_i and a_j. This particular cost function is of fundamental importance in *information theory,* where

$$H = -\sum_i \sum_j p_{ij} \log p_{ij}$$

is defined as the *entropy,* or *uncertainty,* of the random variables s_i and a_j.

For a given s there may be certain elements a in $\mathbf{A}$ that are preferable to others, and their costs will be appropriately less. For a given s the most preferred elements a are those with the lowest cost; choosing such an a corresponds to having made a correct decision, while choosing an a with a larger cost should be taken as an incorrect decision.

We shall be concerned here primarily with simple cost functions. The cost function must be decided upon at the outset by the systems analyst.

DEFINITION 3.6 *The average risk, known as the Bayes risk, is the expected value of the cost function, averaged over all $s \in \mathbf{\Omega}$ and $a \in \mathbf{A}$.*

The average risk R depends on the decision function $\delta(a|y)$ and the a priori signal distribution $\pi(s)$. Thus

$$R(\pi,\delta) \triangleq E(C(s,a)) \tag{3.6}$$

The evaluation of a system such as that in Fig. 3.1 consists of determining $R(\pi,\delta)$ for a given $\pi(s)$ and $\delta(a|y)$. This is not the only type of system evaluation or method of evaluating a system; others will be discussed later.

DEFINITION 3.7 *A decision rule is the criterion used to determine the best decision function; that is, a system is optimized according to a specified decision rule.*

DEFINITION 3.8 BAYES DECISION RULE *A decision function $\delta_B(a|y)$ is optimum in the Bayes sense with respect to a given a priori signal distribution $\pi(s)$ if*

$$R(\pi,\delta_B) \leq R(\pi,\delta) \qquad \textit{for all } \delta \tag{3.7}$$

Since $\delta(a|y)$ corresponds to the receiver operation, determining the optimum $\delta(a|y)$ corresponds to finding the best receiver for a given transmitter. Thus we define the *best receiver* (Bayes receiver) as that receiver which performs the operation indicated by the Bayes decision rule.

If the transmitter subsystem is also to be determined, we have control over the signals used, and possibly control over $\pi(s)$. For the present we shall restrict ourselves to receiver optimization for a given transmitter, whether or not we know what these transmitter characteristics are. Given the transmitter and channel characteristics, along with a specified cost function and decision rule, we can determine the best receiver, that is, that $\delta(a|y)$ which minimizes $R(\pi,\delta)$. In subsequent chapters we shall find the best receiver under different conditions and evaluate the resulting systems performance. In Part Two, we shall take up transmitter optimization.

In general, from the considerations thus far we have

$$R(\pi,\delta) = E(C(s,a)) = \int_{\Omega} ds \int_{\mathbf{A}} da\, C(s,a)f(s,a) \tag{3.8}$$

where $f(s,a)$ is the joint distribution of s and a.

However (see Fig. 3.1), between s and a there appears the observed variable y, so that

$$f(s,a) = \int_{\Gamma} f(s,y,a)\, dy$$

Now the joint density $f(s,y,a)$ can be expressed as

$$f(s,y,a) = \delta(a|s,\, y)f(y,s)$$

but $\delta(a|s,\, y)$ corresponds to the receiver operation and, as described above, cannot depend directly on s. Thus

$$\delta(a|s,\, y) = \delta(a|y)$$

and if we express $f(y,s)$ as

$$f(y,s) = f(y|s)\pi(s)$$

the average risk can equivalently be expressed as

$$R(\pi,\delta) = \int_{\Omega} ds\, \pi(s) \int_{\mathbf{A}} da\, C(s,a) \int_{\Gamma} dy\, \delta(a|y)f(y|s) \tag{3.9}$$

When the probability density functions are completely given, the problem falls into the classification of simple hypothesis testing, to use statistical terminology. For the present, at any rate, we shall assume that all necessary density functions are known. This assumption will be relaxed later.

Note that no restrictions have been placed on the type of elements which comprise $\mathbf{\Omega}$, $\mathbf{\Gamma}$, and $\mathbf{A}$. They may be one-dimensional, k-dimensional vectors, functions of time, or more abstract entities. Regardless of the form the different variables take, the Bayes decision function minimizes the average risk, and this optimization always requires knowledge of the a priori distribution $\pi(s)$. In practical situations this information is not always available.

The question immediately arises, then, of what criterion or optimization technique can be employed when $\pi(s)$ is not known. With $\pi(s)$ unknown the Bayes solution cannot be found. Alternatively, consider the following.

DEFINITION 3.9 *The conditional risk $r(s,\delta)$ is the risk associated with a given signal s and decision function δ; that is,*

$$r(s,\delta) \triangleq E_s(C(s,a)) \tag{3.10}$$

By this notation we mean that s is fixed and the average is taken over a only. Equivalently,

$$r(s,\delta) = \int_{\mathbf{A}} da \int_{\Gamma} dy\, C(s,a)\delta(a|y)f(y|s) \tag{3.11}$$

and

$$R(\pi,\delta) = E(r(s,\delta)) = \int_{\Omega} r(s,\delta)\pi(s)\, ds \tag{3.12}$$

DEFINITION 3.10 MINIMAX DECISION RULE *A decision function δ_m is said to be minimax if*

$$\max_s r(s,\delta_m) = \min_\delta \max_s r(s,\delta) \leq \max_s r(s,\delta) \qquad \textit{for all } \delta \tag{3.13}$$

This is known as minimizing the set of worst possible cases and does not require knowledge of $\pi(s)$. Thus, if $\pi(s)$ is unknown, one possible decision rule is the minimax decision rule. In this case system evaluation consists in determining $r(s,\delta_m)$ for the different $s \in \Omega$ after δ_m has been determined.

DEFINITION 3.11 NATURAL ORDERING *A decision function $\delta_1(a|y)$ is as good as $\delta_2(a|y)$ if*

$$r(s,\delta_1) \leq r(s,\delta_2) \qquad \textit{for all } s \in \Omega$$

DEFINITION 3.12 *A decision function $\delta_1(a|y)$ is better than $\delta_2(a|y)$ if*

$$r(s,\delta_1) \leq r(s,\delta_2) \qquad \textit{for all } s \in \Omega$$

and

$$r(s,\delta_1) < r(s,\delta_2) \qquad \textit{for at least one } s \in \Omega$$

DEFINITION 3.13 *Two decision functions, δ_1 and δ_2, are equivalent if*

$$r(s,\delta_1) = r(s,\delta_2) \qquad \textit{for all } s \in \Omega$$

DEFINITION 3.14 *A decision function $\delta(a|y)$ is said to be admissible if there exists no decision function better than δ. A decision function is said to be inadmissible if it is not admissible.*

DEFINITION 3.15 *A class C of decision functions is said to be complete if for any $\delta \notin C$ there exists a $\delta' \in C$ which is better than δ.*

DEFINITION 3.16 *A class of decision functions C_0 is essentially complete if for any $\delta \notin C_0$ there exists a $\delta' \in C_0$ which is as good as δ.*

THEOREM 3.1 *If $\mathbf{A}$ is a convex set in E_k, a k-dimensional euclidean space, if*

$C(s,a)$ is a convex function of $a \in \mathbf{A}$ for each $s \in \mathbf{\Omega}$, and if for every $s \in \mathbf{\Omega}$ there exists an $\epsilon > 0$ and a B such that

$$C(s,a) \geq \epsilon|a| + B$$

then the class of nonrandomized decision functions is essentially complete.

For the proof of this fundamental theorem see Blackwell and Girshick [3.4] or Ferguson [3.5].

Therefore, when the above hypotheses are satisfied, we immediately know that the optimal decision function is *nonrandomized,* and we can restrict attention to functions which map the observation space $\mathbf{\Gamma}$ into the decision space $\mathbf{A}$. Most problems that have any practical significance for which the signal and decision spaces consist of a continuum of elements satisfy the above hypotheses.

When the decision space is discrete and consists of a finite or countably infinite set of elements, $\mathbf{A}$ is not a convex set and hence does not satisfy the hypotheses of Theorem 3.1. We shall demonstrate subsequently, however, that the optimal decision function for such cases is also deterministic. The optimal-decision-function problem is then equivalent to optimally partitioning the observation space into disjoint regions, say $\mathbf{\Gamma}_i$, called *decision regions,* which are such that when y falls in the region $\mathbf{\Gamma}_i$, the receiver will arrive at the decision a_i.

In most cases it is extremely difficult to determine these optimal decision regions. We shall consider later several cases which are of practical significance where the decision regions can be found.

As an example of a cost function and the risk associated with it, consider again Example 1.1. $\mathbf{A}$ now consists of M points, with a_i corresponding to the decision s_i *was sent,* where $i = 1, \ldots, M$. If we let the cost function be the simple cost function

$$C(s_i,a_j) = \begin{cases} 0 & i = j \\ 1 & i \neq j \end{cases} \tag{3.14}$$

then the average risk is just

$$R(\pi,\delta) = E(C(s_i,a_j)) = \sum_{i \neq j}\sum P(s_i,a_j)$$

where $P(s_i,a_j)$ is the probability that s_i is sent *and* a_j is decided.

It is evident that $R(\pi,\delta)$ is just the probability that an error is made, and minimization of the average risk is equivalent in this case to minimization of the probability of error. That is, the probability of error is defined as the average risk with the cost function in (3.14). In other words,

$$P_e \triangleq R(\pi,\delta) = \sum_{i \neq j}\sum P(s_i,a_j) = 1 - \sum_{i=1}^{M} P(s_i,a_i) = 1 - P_d \tag{3.15}$$

where P_d is the *probability of detection.*[1] Equivalently, the probability of detection is defined as

$$P_d \triangleq \sum_{i=1}^{M} P(s_i,a_i) \tag{3.16}$$

In Example 1.3 we may take **A** to be the real line segment representing possible estimates of the range and let $a = s$ be the correct decision, where s is the *true target range.* If we take

$$C(s,a) = (s - a)^2 \tag{3.17}$$

errors are penalized according to the square of their magnitude, and the associated average risk is then the mean square error.

REFERENCES

3.1 Wald, A.: "Statistical Decision Functions," Wiley, New York, 1950.

3.2 Lehmann, E. L.: "Testing Statistical Hypotheses," Wiley, New York, 1959.

3.3 Middleton, D.: "Introduction to Statistical Communication Theory," McGraw-Hill, New York, 1960.

3.4 Blackwell, D., and M. A. Girshick: "Theory of Games and Statistical Decisions," Wiley, New York, 1954.

3.5 Ferguson, T. S.: "Mathematical Statistics: A Decision Theoretic Approach," Academic, New York, 1967.

[1] Also called *probability of correct decision* in a communications environment.

4
Binary Detection Systems Minimizing the Average Risk

We can now apply the concepts of the previous chapter to the special case of a binary system and determine the optimal decision function as well as the resulting minimum average risk. We shall see that the optimal decision function is the *likelihood ratio*, regardless of the cost function and regardless of the a priori probabilities, and that these parameters affect only a threshold level.

An example of a binary system in which both the signal space and the decision space consist of only two points is the *return-to-zero binary system*. In the return-to-zero system each signal is zero at $t = 0$ and $t = T$, where T is the signal duration time. Two possible sets of signals for such a system are shown in Fig. 4.1, where for the first set

$$\mathcal{E} \triangleq \int_0^T [s_0(t)]^2\, dt = \int_0^T [s_1(t)]^2\, dt$$

implying signals with the same average power; this clearly is not the case for the second set.

An example of a binary system with a signal space of more than

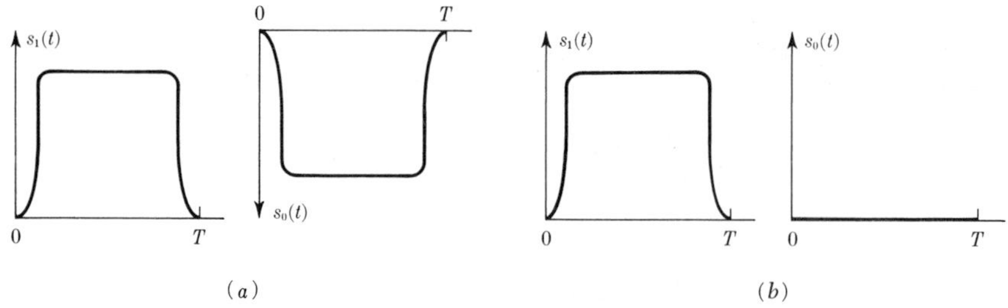

Fig. 4.1 Examples of return-to-zero binary signal sets: (*a*) equal average power, (*b*) unequal average power.

two elements is the *non-return-to-zero binary system*, which consists of the four signal waveforms $s_1(t), \ldots, s_4(t)$, as shown in Fig. 4.2. The mathematical model of a binary signal set takes this form when the rise and fall times of the signals are taken into account. If $s_1(t)$ or $s_2(t)$ is transmitted, a_0 is the correct decision; if $s_3(t)$ or $s_4(t)$ is transmitted, a_1 is the correct decision.

4.1 BINARY DECISION FUNCTIONS

Since **A** consists of only two elements, the integration over **A** in (3.9) can be carried out, resulting in

$$R(\pi,\delta) = \int_{\Omega} \pi(s)\, ds \int_{\Gamma} f(y|s)\, dy\, [C(s,a_0)\delta(a_0|y) + C(s,a_1)\delta(a_1|y)] \tag{4.1}$$

Using (3.5) in (4.1), we have that

$$R(\pi,\delta) = \int_{\Omega} \pi(s)\, ds \int_{\Gamma} dy\, f(y|s)C(s,a_1) + \int_{\Omega} \pi(s)\, ds \int_{\Gamma} dy\, f(y|s)[C(s,a_0) - C(s,a_1)]\, \delta(a_0|y) \tag{4.2}$$

We wish to find that $\delta(a|y)$ which minimizes $R(\pi,\delta)$ in (4.2). The first integral in (4.2), which we define as I_1, does not depend on $\delta(a|y)$. Thus minimizing $R(\pi,\delta)$ consists in determining that $\delta(a|y)$ which minimizes

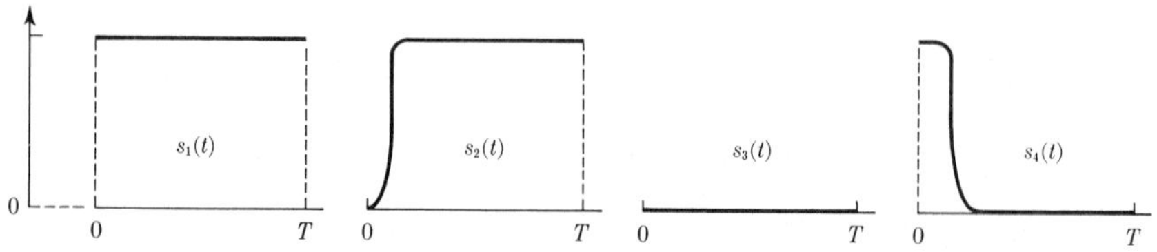

Fig. 4.2 Signal waveforms for a non-return-to-zero binary system.

the second integral, I_2. For this we define

$$g(y) \triangleq \int_{\Omega} ds\, \pi(s) f(y|s)[C(s,a_0) - C(s,a_1)] \tag{4.3}$$

Then $R(\pi,\delta) = I_1 + I_2$, and

$$I_2 \triangleq \int_{\Gamma} dy\, g(y)\delta(a_0|y) \tag{4.4}$$

We now partition $\mathbf{\Gamma}$ into three disjoint sets, $\mathbf{\Gamma}^+$, $\mathbf{\Gamma}^0$, and $\mathbf{\Gamma}^-$, such that

$$\begin{aligned} \mathbf{\Gamma}^+ &= \{y|g(y) > 0\} \\ \mathbf{\Gamma}^0 &= \{y|g(y) = 0\} \\ \mathbf{\Gamma}^- &= \{y|g(y) < 0\} \end{aligned} \tag{4.5}$$

then

$$I_2 = \int_{\mathbf{\Gamma}^+} \delta(a_0|y)g(y)\, dy - \int_{\mathbf{\Gamma}^-} \delta(a_0|y)|g(y)|\, dy + \int_{\mathbf{\Gamma}^0} g(y)\delta(a_0|y)\, dy \tag{4.6}$$

The third integral in (4.6) is always zero, so $\delta(a_0|y)$ can be anything when $g(y) = 0$ without affecting the value of $R(\pi,\delta)$. The integrands in both the remaining integrals are nonnegative for all y. Thus to make I_2 as small as possible, we must make the integral over $\mathbf{\Gamma}^+$ as small as possible and the integral over $\mathbf{\Gamma}^-$ as large as possible. Since

$$0 \leq \delta(a_0|y) \leq 1$$

the optimum δ in the Bayes sense is

$$\delta_B(a_0|y) = \begin{cases} 0 & \text{if } y \in \mathbf{\Gamma}^+ \\ 1 & \text{if } y \in \mathbf{\Gamma}^- \end{cases} \tag{4.7a}$$

or, equivalently,

$$\delta_B(a_1|y) = \begin{cases} 1 & \text{if } y \in \mathbf{\Gamma}^+ \\ 0 & \text{if } y \in \mathbf{\Gamma}^- \end{cases} \tag{4.7b}$$

With this necessary restriction imposed on δ, δ_B minimizes the first integral and maximizes the second integral. The corresponding minimum average risk is

$$R(\pi,\delta_B) = I_1 + \int_{\mathbf{\Gamma}^-} g(y)\, dy \tag{4.8}$$

To keep the system from being unstable when $g(y) = 0$, an equality sign might be attached to the definition of either $\mathbf{\Gamma}^+$ or $\mathbf{\Gamma}^-$. Figure 4.3 is a block diagram of the optimal system.

We have, then, that in a binary system the optimal decision function δ_B based on the Bayes decision rule can be chosen to be nonrandom-

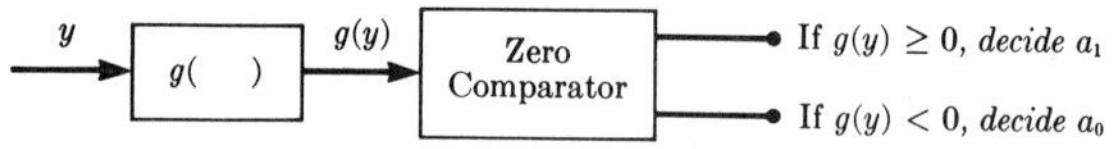

Fig. 4.3 Optimal receiver for a binary system.

ized [it could be random where $g(y) = 0$ or on sets with measure zero with respect to $g(y)$].

The optimal receiver can be expressed somewhat differently when we divide the signal space into disjoint sets according to

$$\begin{aligned} \mathbf{\Omega}_1 &= \{s | C(s,a_1) \leq C(s,a_0)\} \\ \mathbf{\Omega}_0 &= \{s | C(s,a_1) > C(s,a_0)\} \end{aligned} \tag{4.9}$$

Then

$$g(y) = \int_{\mathbf{\Omega}_1} ds\, \pi(s) f(y|s)[C(s,a_0) - C(s,a_1)] - \int_{\mathbf{\Omega}_0} ds\, \pi(s) f(y|s)[C(s,a_1) - C(s,a_0)] \tag{4.10}$$

The Bayes decision can then be expressed as

$$\delta_B(a_1|y) = \begin{cases} 1 & \text{when } \int_{\mathbf{\Omega}_1} ds\, \pi(s) f(y|s)[C(s,a_0) - C(s,a_1)] \\ & \qquad \geq \int_{\mathbf{\Omega}_0} ds\, \pi(s) f(y|s)[C(s,a_1) - C(s,a_0)] \\ 0 & \text{when } \int_{\mathbf{\Omega}_1} ds\, \pi(s) f(y|s)[C(s,a_0) - C(s,a_1)] \\ & \qquad < \int_{\mathbf{\Omega}_0} ds\, \pi(s) f(y|s)[C(s,a_1) - C(s,a_0)] \end{cases} \tag{4.11}$$

This is the most general form of a binary system, inasmuch as no specific form has been assumed for the a priori probabilities, the channel characteristics, or costs. We now make the following specialization.

Example 4.1 Specialization to a signal space of two elements For $\mathbf{A} = (a_0,a_1)$ and $\mathbf{\Omega} = (s_0,s_1)$ we define a *cost matrix* as

$$\mathbf{C} \triangleq \begin{bmatrix} C(s_0,a_0) & C(s_0,a_1) \\ C(s_1,a_0) & C(s_1,a_1) \end{bmatrix} \triangleq \begin{bmatrix} C_{00} & C_{01} \\ C_{10} & C_{11} \end{bmatrix} \tag{4.12}$$

The C_{ij} may take on any values, but here we make the following realistic restrictions. We associate a_i with the correct decision when s_i was transmitted. Since the cost of correct decisions is naturally taken to be less than that for incorrect decisions, we make the assumption

$$C_{ii} < C_{ij} \qquad \text{for all } i \neq j \tag{4.13}$$

Then from (4.3) we can write for the special case $\mathbf{\Omega} = (s_0,s_1)$

$$g(y) = \pi_0(C_{00} - C_{01}) f(y|s_0) + \pi_1(C_{10} - C_{11}) f(y|s_1) \tag{4.14}$$

where

$$\pi_0 \triangleq \text{probability of transmitting } s_0 = \pi(s_0)$$
$$\pi_1 \triangleq \text{probability of transmitting } s_1 = \pi(s_1) = 1 - \pi(s_0)$$

The optimal decision function then becomes

$$\delta_B(a_1|y) = \begin{cases} 1 & \text{when } \dfrac{\pi_1 C_{10} f(y|s_1) + \pi_0 C_{00} f(y|s_0)}{\pi_0 C_{01} f(y|s_0) + \pi_1 C_{11} f(y|s_1)} \geq 1 \\ 0 & \text{when } \dfrac{\pi_1 C_{10} f(y|s_1) + \pi_0 C_{00} f(y|s_0)}{\pi_0 C_{01} f(y|s_0) + \pi_1 C_{11} f(y|s_1)} < 1 \end{cases} \tag{4.15}$$

The decision a_1 has been arbitrarily assigned when equality occurs in the decision function. This does not alter the optimality of the result, since we are speaking of events of measure zero.

DEFINITION 4.1 *The likelihood ratio is defined as*

$$L(y) \triangleq \frac{f(y|s_1)}{f(y|s_0)} \tag{4.16}$$

In defining the likelihood ratio we have assumed that $f(y|s_0)$ does not vanish at the y where $f(y|s_1)$ vanishes. In practical applications this will always be so. This mathematical detail can be easily overcome, however, in examples where this assumption is invalid. In terms of $L(y)$ the decision function becomes

$$\delta_B(a_1|y) = \begin{cases} 1 & \text{if } L(y) \geq \dfrac{\pi_0(C_{01} - C_{00})}{\pi_1(C_{10} - C_{11})} \\ 0 & \text{if } L(y) < \dfrac{\pi_0(C_{01} - C_{00})}{\pi_1(C_{10} - C_{11})} \end{cases} \tag{4.17}$$

Thus the optimal decision function is a threshold device. Note that the magnitude of the threshold depends only on the difference of the entries in the cost matrix, and not on their magnitude. Also, the costs and a priori probabilities affect only the threshold level of the decision function.

The corresponding minimum average risk is

$$R(\pi,\delta_B) = \pi_0 C_{01} \int_{\Gamma^+} f(y|s_0)\, dy + \pi_1 C_{11} \int_{\Gamma^+} f(y|s_1)\, dy + \pi_0 C_{00} \int_{\Gamma^-} f(y|s_0)\, dy + \pi_1 C_{10} \int_{\Gamma^-} f(y|s_1)\, dy$$

We introduce the notation

$$\begin{aligned} P_{01} &\triangleq \int_{\Gamma^+} f(y|s_0)\, dy = \text{Pr (deciding } a_1|s_0 \text{ was transmitted)} \\ P_{11} &\triangleq \int_{\Gamma^+} f(y|s_1)\, dy = \text{Pr (deciding } a_1|s_1 \text{ was transmitted)} \\ P_{00} &\triangleq \int_{\Gamma^-} f(y|s_0)\, dy = \text{Pr (deciding } a_0|s_0 \text{ was transmitted)} \\ P_{10} &\triangleq \int_{\Gamma^-} f(y|s_1)\, dy = \text{Pr (deciding } a_0|s_1 \text{ was transmitted)} \end{aligned} \tag{4.18}$$

Then

$$R(\pi,\delta_B) = \pi_0 C_{01} P_{01} + \pi_1 C_{11} P_{11} + \pi_0 C_{00} P_{00} + \pi_1 C_{10} P_{10} \tag{4.19}$$

Since

$$P_{01} + P_{00} = 1$$

and

$$P_{10} + P_{11} = 1$$

$R(\pi,\delta_B)$ can be equivalently expressed as

$$R(\pi,\delta_B) = \pi_0 C_{00} + \pi_1 C_{11} + \pi_0 (C_{01} - C_{00}) P_{01} + \pi_1 (C_{10} - C_{11}) P_{10}$$

DEFINITION 4.2 *The cost matrix* **C** *is the set of costs in matrix form; that is,*

$$\mathbf{C} \triangleq [C_{ij}]$$

The cost matrix most often used in communication telemetry systems is the one defined by Eq. (3.14), which in a binary system is

$$\mathbf{C} = \begin{bmatrix} 0 & 1 \\ 1 & 0 \end{bmatrix} \tag{4.20}$$

It assesses no loss to correct decisions and the same loss to all incorrect decisions. In this case the decision function becomes

$$\delta_B(a_0|y) = \begin{cases} 0 & \text{if } L(y) \geq \dfrac{\pi_0}{\pi_1} \\ 1 & \text{if } L(y) < \dfrac{\pi_0}{\pi_1} \end{cases} \tag{4.21}$$

and the corresponding average risk is

$$R(\pi,\delta_B) = \pi_1 P_{10} + \pi_0 P_{01}$$

Hence our definition of the probability of error in (3.15) becomes

$$\begin{aligned} P_e &= \Pr(s_1) \Pr(\text{deciding } a_0|s_1 \text{ was transmitted}) \\ &\qquad + \Pr(s_0) \Pr(\text{deciding } a_1|s_0 \text{ was transmitted}) \\ &= \pi_0 P_{01} + \pi_1 P_{10} \end{aligned} \tag{4.22}$$

In other words, for this special cost matrix the probability of error and the average risk are synonymous. An observer who makes his decision on the basis of minimization of the probability of error is called an *ideal observer* or *ideal receiver*.

For binary systems the probability of detection is defined as

$$P_d \triangleq 1 - P_e = \pi_0 P_{00} + \pi_1 P_{11} \tag{4.23}$$

Note that since the optimal receiver is a threshold device, any strictly monotonic function of the likelihood ratio is an equivalent device. As we shall see, one that is often used is $\ln L(y)$. Also, minimizing the probability of error or average risk is equivalent to maximizing the probability of detection, an alternative which, as we shall see in later chapters, is often preferable.

4.2 VECTOR MODEL

Let us now apply these results to a specific problem. The method most often used in analyzing communication systems with time-varying transmittable waveforms is to replace all waveforms by finite-dimensional vectors, the advantage of which will become apparent in subsequent chapters. Initially, then, we shall analyze situations under the assumption that the system is already represented by a vector model.

For this, suppose we have a receiver that is able to observe an m-dimensional random vector

$$\mathbf{Y} = \begin{bmatrix} y_1 \\ \cdot \\ \cdot \\ \cdot \\ y_m \end{bmatrix}$$

which is of the form

$$\mathbf{Y} = \mathbf{S} + \mathbf{N} \tag{4.24}$$

where $\mathbf{S}$ and $\mathbf{N}$ are independent, $\mathbf{S}$ is either $\mathbf{S}_0$ or $\mathbf{S}_1$, with a priori probabilities π_0 and π_1, respectively, and $\mathbf{N}$ is an m-dimensional gaussian random vector with zero mean and covariance matrix equal to

$$\sigma^2\mathbf{I} \tag{4.25}$$

where $\mathbf{I}$ is the m-by-m identity matrix. Assume that both signals have the same energy; that is,

$$\|\mathbf{S}_0\|^2 = \|\mathbf{S}_1\|^2 = \mathcal{E}$$

Under these conditions we are asked to find the receiver which best determines which signal has been transmitted, where by "best" we mean that which minimizes the probability of error P_e. This is a binary communication problem in the presence of additive gaussian noise.

We have seen that the probability of error is the same as the average risk when the cost matrix is

$$\mathbf{C} = \begin{bmatrix} 0 & 1 \\ 1 & 0 \end{bmatrix}$$

We also have that for this cost matrix the optimal receiver is the one which forms the likelihood ratio

$$L(\mathbf{Y}) = \frac{f(\mathbf{Y}|\mathbf{S}_1)}{f(\mathbf{Y}|\mathbf{S}_0)}$$

and decides that $\mathbf{S}_1$ or $\mathbf{S}_0$ was transmitted if $L(\mathbf{Y})$ is greater or less, respectively, than the threshold π_0/π_1.

In order to determine the form of $L(\mathbf{Y})$, we note first that[1]

$$f(\mathbf{Y}|\mathbf{S}) = \frac{f(\mathbf{Y},\mathbf{S})}{f(\mathbf{S})}$$

Since $\mathbf{S}$ and $\mathbf{N}$ are assumed independent, we note that

$$f_{SN}(\mathbf{S},\mathbf{N}) = f_S(\mathbf{S})f_N(\mathbf{N})$$

If we make the transformation

$$\mathbf{Y} = \mathbf{S} + \mathbf{N}$$
$$\mathbf{S} = \mathbf{S}$$

Then

$$f_{YS}(\mathbf{Y},\mathbf{S}) = f_{SN}(\mathbf{S}, \mathbf{Y} - \mathbf{S})|J| = f_S(\mathbf{S})f_N(\mathbf{Y} - \mathbf{S})|J|$$

where J is the jacobian between $(\mathbf{Y},\mathbf{S})$ and $(\mathbf{S},\mathbf{N})$, which clearly has absolute value of unity.

Thus

$$f(\mathbf{Y}|\mathbf{S}) = \frac{f_S(\mathbf{S})f_N(\mathbf{Y} - \mathbf{S})}{f_S(\mathbf{S})} = f_N(\mathbf{Y} - \mathbf{S})$$

Therefore, from the hypothesis that the signal and noise are additive and independent, we find that the likelihood ratio can be expressed as

$$L(\mathbf{Y}) = \frac{f_N(\mathbf{Y} - \mathbf{S}_1)}{f_N(\mathbf{Y} - \mathbf{S}_0)} \tag{4.26}$$

Now,

$$f_N(\mathbf{N}) = \frac{1}{(\sqrt{2\pi}\,\sigma)^m} \exp\left(-\frac{1}{2\sigma^2}\|\mathbf{N}\|^2\right)$$

so that the likelihood ratio may be written explicitly as

$$L(\mathbf{Y}) = \frac{\exp\left(-\frac{1}{2\sigma^2}\|\mathbf{Y} - \mathbf{S}_1\|^2\right)}{\exp\left(-\frac{1}{2\sigma^2}\|\mathbf{Y} - \mathbf{S}_0\|^2\right)} \tag{4.27}$$

[1] It is assumed that $L(\mathbf{Y})$ is strictly positive, which is the case for gaussian densities as well as for π_0/π_1.

or, equivalently,

$$L(\mathbf{Y}) = \frac{\exp\left[-\frac{1}{2\sigma^2}(\|\mathbf{S}_1\|^2 - 2\mathbf{Y}^T\mathbf{S}_1)\right]}{\exp\left[-\frac{1}{2\sigma^2}(\|\mathbf{S}_0\|^2 - 2\mathbf{Y}^T\mathbf{S}_0)\right]}$$

where T means *the transpose of*. Since $\|\mathbf{S}_1\|^2 = \|\mathbf{S}_0\|^2$, the expression for $L(\mathbf{Y})$ can be simplified to

$$L(\mathbf{Y}) = \exp\left[\frac{1}{\sigma^2}(\mathbf{Y}^T\mathbf{S}_1 - \mathbf{Y}^T\mathbf{S}_0)\right] \tag{4.28}$$

Since the logarithm is a (strictly) monotonically increasing function, the optimal decision function in (4.28) can also be expressed as

$$\begin{aligned} &\text{If } \ln L(\mathbf{Y}) \geq \ln \frac{\pi_0}{\pi_1} \qquad \text{decide } a_1 \\ &\text{If } \ln L(\mathbf{Y}) < \ln \frac{\pi_0}{\pi_1} \qquad \text{decide } a_0 \end{aligned} \tag{4.29}$$

For this problem the optimal decision function reduces to

$$\begin{aligned} &\text{If } \mathbf{Y}^T\mathbf{S}_1 - \mathbf{Y}^T\mathbf{S}_0 \geq \sigma^2 \ln \frac{\pi_0}{\pi_1} \qquad \text{choose } a_1 \\ &\text{If } \mathbf{Y}^T\mathbf{S}_1 - \mathbf{Y}^T\mathbf{S}_0 < \sigma^2 \ln \frac{\pi_0}{\pi_1} \qquad \text{choose } a_0 \end{aligned} \tag{4.30}$$

Thus the optimal receiver correlates the incoming signal $\mathbf{Y}$ with $\mathbf{S}_1$ and $\mathbf{S}_0$ and compares the difference with a threshold. This is the simplest form of what is known as the *matched filter;* it is a partitioning of the observation space.

Note that if $\pi_0 = \pi_1$ the threshold level is zero, and the decision function is then independent of the noise power σ^2. Note also that if $\mathbf{N} = \mathbf{0}$, then $\mathbf{Y} = \mathbf{S}$. If, for example, $\mathbf{S} = \mathbf{S}_1$, then, since $\mathbf{S}_1{}^T\mathbf{S}_1 > \mathbf{S}_1{}^T\mathbf{S}_0$ (provided, of course, that $\mathbf{S}_1 \neq \mathbf{S}_0$), the receiver will always decide that $\mathbf{S}_1$ was transmitted when $\mathbf{S}_1$ actually was transmitted, and similarly for $\mathbf{S}_0$. Hence the probability of error is clearly zero. Such a case is termed the *singular case.* We have the *nonsingular case* whenever $P_e > 0$.

Now let us determine the probability of detection for the optimal decision function in (4.30). For this we have

$$\begin{aligned} P_d &= \pi_1 P_{11} + \pi_0 P_{00} \\ &= \pi_1 \Pr\left[\mathbf{Y}^T(\mathbf{S}_1 - \mathbf{S}_0) \geq \sigma^2 \ln \frac{\pi_0}{\pi_1} \,\middle|\, \mathbf{Y} = \mathbf{S}_1 + \mathbf{N}\right] \\ &\qquad + \pi_0 \Pr\left[\mathbf{Y}^T(\mathbf{S}_1 - \mathbf{S}_0) < \sigma^2 \ln \frac{\pi_0}{\pi_1} \,\middle|\, \mathbf{Y} = \mathbf{S}_0 + \mathbf{N}\right] \end{aligned}$$

First we evaluate P_{11}, which can be expressed in terms of the random-noise vector $\mathbf{N}$ as

$$P_{11} = \Pr\left[(\mathbf{S}_1 + \mathbf{N})^T(\mathbf{S}_1 - \mathbf{S}_0) \geq \sigma^2 \ln \frac{\pi_0}{\pi_1}\right]$$

$$= \Pr\left[\mathbf{N}^T(\mathbf{S}_1 - \mathbf{S}_0) \geq \sigma^2 \ln \frac{\pi_0}{\pi_1} - \mathbf{S}_1{}^T(\mathbf{S}_1 - \mathbf{S}_0)\right]$$

We set $\mathbf{S}_1{}^T\mathbf{S}_0 \triangleq \mathcal{E} \cos \theta$, where θ is the angle between the m-dimensional vectors $\mathbf{S}_0$ and $\mathbf{S}_1$. If we set

$$\eta \triangleq \mathbf{N}^T(\mathbf{S}_1 - \mathbf{S}_0)$$

then η is a gaussian random variable (since linear functions of gaussian random variables are gaussian random variables), with

$$E(\eta) = 0 \tag{4.31}$$

and

$$E(\eta^2) = E((\mathbf{S}_1 - \mathbf{S}_0)^T\mathbf{N}\mathbf{N}^T(\mathbf{S}_1 - \mathbf{S}_0)) = 2\sigma^2\mathcal{E}(1 - \cos \theta) \tag{4.32}$$

Hence

$$P_{11} = \Pr\left[\eta \geq \sigma^2 \ln \frac{\pi_0}{\pi_1} - \mathcal{E}(1 - \cos \theta)\right] \tag{4.33}$$

If we define

$$\zeta \triangleq \frac{\eta}{\sqrt{E(\eta^2)}}$$

then ζ is a gaussian random variable with zero mean and unit variance. Upon substitution into Eq. (4.33) we obtain

$$P_{11} = \Pr\left[\zeta > \frac{\ln (\pi_0/\pi_1) - (\mathcal{E}/\sigma^2)(1 - \cos \theta)}{\sqrt{(2\mathcal{E}/\sigma^2)(1 - \cos \theta)}}\right]$$

$$= \int_{K_1}^{\infty} \frac{1}{\sqrt{2\pi}} \exp\left(-\tfrac{1}{2}\zeta^2\right) d\zeta \triangleq \operatorname{erfc} K_1 \tag{4.34}$$

where

$$K_1 \triangleq \frac{\ln (\pi_0/\pi_1) - (\mathcal{E}/\sigma^2)(1 - \cos \theta)}{\sqrt{(2\mathcal{E}/\sigma^2)(1 - \cos \theta)}} \tag{4.35}$$

erf is known as the *error function* and erfc is its *complement* on $(-\infty, \infty)$. Similarly,

$$P_{00} = \operatorname{erfc} K_0 \tag{4.36}$$

where

$$K_0 \triangleq \frac{-\ln (\pi_0/\pi_1) - (\mathcal{E}/\sigma^2)(1 - \cos \theta)}{\sqrt{(2\mathcal{E}/\sigma^2)(1 - \cos \theta)}} \tag{4.37}$$

and $\mathcal{E}/\sigma^2$ is the signal-to-noise ratio.

In the special case when the transmitted vectors are equally likely ($\pi_0 = \pi_1 = \frac{1}{2}$), then

$$K_0 = K_1 = -\sqrt{\frac{\mathcal{E}}{2\sigma^2}(1 - \cos\theta)} \triangleq K$$

and

$$P_{11} = P_{00} = \text{erfc } K$$

Finally, we have

$$P_d = \text{erfc}\left(-\sqrt{\frac{\mathcal{E}}{2\sigma^2}(1 - \cos\theta)}\right) \tag{4.38}$$

which demonstrates that when the signals are equally likely, the performance of a binary communication system depends only on the signal-to-noise ratio $\mathcal{E}/\sigma^2$ and the normalized inner product of the two signals, $\cos\theta$.

We can make some further comments. For any given $\mathcal{E}/\sigma^2$, P_d is maximized with respect to $\cos\theta$ by

$$\cos\theta = -1$$

or, equivalently, when

$$\mathbf{S}_0 = -\mathbf{S}_1 \tag{4.39}$$

This is true for any a priori distribution and corresponds to placing the tips of the two vectors as far apart as possible. In the equally-likely case, from (4.38) the probability of detection becomes

$$P_d = \text{erfc}\left(-\sqrt{\frac{\mathcal{E}}{\sigma^2}}\right) \tag{4.40}$$

In (4.40) the only significant signal parameter is $\mathcal{E}$, showing that the total signal energy is all that affects system performance, and that for this binary system signal shape is not of significance as long as it has energy $\mathcal{E}$ and (4.39) is satisfied. This is of fundamental importance in practical system design.

The only reason for using $\ln L(\mathbf{Y})$ instead of $L(\mathbf{Y})$ is the simplicity obtained in the decision function when gaussian random variables are assumed. When $\ln L(\mathbf{Y})$ is used, the signal probabilities, noise variance, and the costs enter only in the setting of the threshold and do not alter the form of the actual receiver structure.

It can be shown that the only requirement necessary for $P_{00} = P_{11}$ is $\pi_0 = \pi_1$. This is the case even if the signal powers are unequal.

From (4.38) it follows immediately that

$$P_e = \operatorname{erfc} \sqrt{\frac{\mathcal{E}}{2\sigma^2}(1 - \cos\theta)} \tag{4.41}$$

and from (4.40), when $\mathbf{S}_0 = -\mathbf{S}_1$, we obtain

$$P_e = \operatorname{erfc} \sqrt{\frac{\mathcal{E}}{\sigma^2}} \tag{4.42}$$

Therefore, if a binary system requires the probability of error to be less than some specified level, we have immediately the minimal allowable signal-to-noise ratio that can be tolerated (the error function is tabulated in tables). Having $\mathcal{E}/\sigma^2 = 0$ is equivalent to just guessing. If $\sigma^2 \to 0$, or if $\mathcal{E} \to \infty$, then $\mathcal{E}/\sigma^2 \to \infty$ and $P_e \to 0$, giving the singular case.

We have emphasized equally likely signals, as this is the most meaningful in a communication environment. This is not the case, however, in radar problems, where a different design philosophy is used (see Chap. 6).

4.3 COHERENT BINARY PHASE MODULATION

The following is a binary digital communication system which was described in a general manner in Example 1.1. In particular, we shall study here the special case where the transmitter may choose between one of two possible waveforms over each T-sec interval. The consideration here is the extreme case in which the carrier phase is known exactly to the receiver. Such systems are termed *coherent*. In Chapter 7 we shall consider the other extreme, termed *noncoherent*, in which the carrier phase is totally unknown to the receiver.

In a coherent binary phase-modulation system the two transmittable signals are assumed to be of the form

$$s_i(t) = A \cos [\omega_c t + \phi_i(t)] \qquad i = 0, 1$$

over each interval of time $0 \le t \le T$, where T is the observation time or bit time. The overall waveform actually consists of a time-multiplexed sequence of such waveforms. The analysis and performance of this system can be determined from examination of the single observation interval $[0,T]$.

The transmittable signals have the same average power

$$P_{\text{av}} = \frac{1}{T}\int_0^T s_i^2(t)\, dt = \frac{A^2}{2} \qquad i = 0, 1$$

and energy $\mathcal{E} = A^2T/2$, where we have made the assumption that the phase variations $\{\phi_i(t)\}$ are slowly varying with respect to the carrier frequency ω_c. The digital set $\mathbf{\Omega}$ may be interpreted either as $\mathbf{\Omega} = \{\phi_0(t),\phi_1(t)\}$

or $\mathbf{\Omega} = \{s_0(t), s_1(t)\}$. In either case the a priori probabilities are π_0 and π_1, respectively.

The channel will be such that it adds white gaussian noise with one-sided spectral density N_0 watts/Hz. The white-noise disturbances are usually caused by thermal noise[1] in the front end of the receiver, or from galactic disturbances in the channel. The receiver is further assumed to be synchronous; that is, it has knowledge of the starting time of each observation interval, thus eliminating the possibility that $s_i(t)$ will change to $s_j(t)$ (for $j \neq i$) during any observation interval. Let us now use the Bayes decision rule to determine the optimal receiver for this binary communication system and then evaluate the resulting probability of error versus signal-to-noise ratio.

Determining the optimal binary communication receiver has been shown to be the same as evaluating the ratio

$$\ln L(\mathbf{Y}) = \ln \frac{f(\mathbf{Y}|\mathbf{S}_1)}{f(\mathbf{Y}|\mathbf{S}_0)}$$

and the comparison of this ratio to a threshold. For simplicity, let the costs and a priori probabilities be related by

$$\pi_1 C_{10} = \pi_0 C_{01}$$

and

$$C_{11} = C_{00} = 0$$

so that the decision function reduces to

$$\ln L(\mathbf{Y}) \lesseqgtr 0$$

There are essentially two approaches that can be used to evaluate this likelihood ratio:

1. We can sample the received waveform $y(t)$ m times during the time interval $[0,T]$. This results in an m-dimensional vector with an m-dimensional probability density function, for which the likelihood ratio may be determined as in the previous example. Then we allow the number of samples to increase and the time interval between samples to decrease in such a manner that in the limit we obtain an integral, or continuous sampling over $[0,T]$. This is called the *sampling method.*

[1] The thermal-noise one-sided power spectral density in an R-ohm resistor is kT_0R, where $k = 1.38 \times 10^{-16}$ ergs/deg is *Boltzmann's constant* and T_0 is the temperature in degrees Kelvin. When normalized to a 1-ohm resistor, the spectral density is $N_0 = kT_0$ watts/Hz. We shall employ the two-sided spectral density, in which case the spectral density becomes $N_0/2$ watts/Hz. In either case the total noise power at the receiver with noise bandwidth of W Hz is N_0W watts.

2. We assume that we are observing $y(t)$ continuously at the outset, in which case m-dimensional probability density functions have no meaning. This is called the *continuous model.*

In the first approach we are attempting to interpret continuous operations on sampling results, which can lead to mathematical complications. The second is the more rigorous method and involves use of the Karhunen-Loeve expansion (described in Appendix B), which rigorously converts the continuous problem into the form of a vector. Here we shall use the sampling method, deferring to Chap. 8 the proof that the continuous method provides the same result.

Since the signal and noise are independent, we may write

$$L(\mathbf{Y}) = \frac{f_N(\mathbf{Y} - \mathbf{S}_1)}{f_N(\mathbf{Y} - \mathbf{S}_0)}$$

where the vectors $\mathbf{Y}$, $\mathbf{S}_0$, and $\mathbf{S}_1$ represent the samples of $y(t)$, $s_0(t)$, and $s_1(t)$, respectively. If we first assume that the additive noise has flat spectral density over the frequency interval $[-W,W]$, as in Fig. 4.4, the noise samples will be mutually independent when the sampling rate is $2W$ samples per second. Hence set

$$m = 2WT \tag{4.43}$$

Because of the independent noise sample, we can write $L(\mathbf{Y})$ as

$$L(\mathbf{Y}) = \frac{\exp\left[-(1/2\sigma^2)(\|\mathbf{Y}\|^2 + \|\mathbf{S}_1\|^2 - 2\mathbf{Y}^T\mathbf{S}_1)\right]}{\exp\left[-(1/2\sigma^2)(\|\mathbf{Y}\|^2 + \|\mathbf{S}_0\|^2 - 2\mathbf{Y}^T\mathbf{S}_0)\right]}$$

or, equivalently,

$$\ln L(\mathbf{Y}) = \frac{1}{\sigma^2}\mathbf{Y}^T(\mathbf{S}_1 - \mathbf{S}_0) \tag{4.44}$$

where

$$\sigma^2 = N_0 W \tag{4.45}$$

is the variance of each noise sample. In terms of the samples themselves,

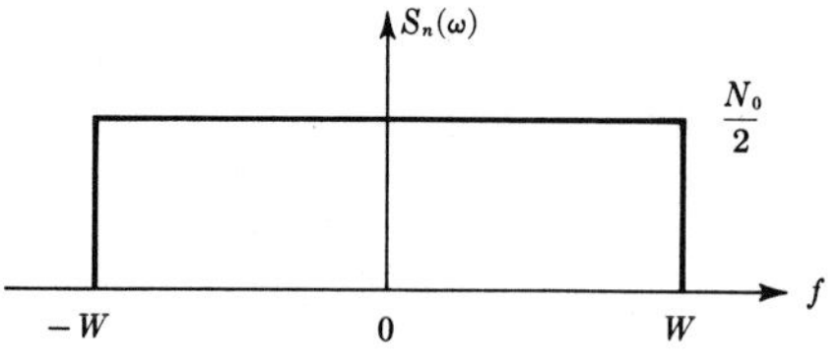

Fig. 4.4 Band-limited flat noise spectral density.

(4.44) can be expressed as

$$\ln L(\mathbf{Y}) = \frac{2}{N_0} \sum_{k=1}^{m} y\left(\frac{k}{2W}\right)\left[s_1\left(\frac{k}{2W}\right) - s_0\left(\frac{k}{2W}\right)\right]\Delta t \tag{4.46}$$

where $\Delta t = 1/2W$.

The crucial point in the sampling method is as follows. In (4.46) let $m \to \infty$ and $W \to \infty$ simultaneously, so that (4.43) remains valid. Then the sum in (4.46), by definition of an integral in the Reimann sense, is equal, in the limit as $m \to \infty$, to

$$\ln L(y(t); 0 \leq t \leq T) = \frac{2}{N_0} \int_0^T y(t)[s_1(t) - s_0(t)]\, dt \tag{4.47}$$

This is the optimal receiver for the binary coherent communication system. The limiting procedure is valid only in the case where $n(t)$ is white gaussian noise. Without this prerequisite, successive samples would inevitably become correlated, thus nullifying the optimability of the vector form of the decision function.

This optimal-receiver operation can be interpreted either as selection of that waveform whose cross correlation with $y(t)$ is larger, leading to the name *correlation detection*, or as the passing of $y(t)$ through time-varying linear filters which are matched to the signals $s_1(t)$ and $s_0(t)$ and selection of the signal corresponding to the filter whose output at time $t = T$ is larger, leading to the name *matched-filter detection*. Mathematically, correlation detection and matched-filter detection are synonymous. The weighting functions for the matched filters would be

$$h_1(t) = s_1(T - t) \qquad h_0(t) = s_0(T - t)$$

or, equivalently,

$$\ln L(y(t)) = \frac{2}{N_0} \int_0^T y(t) h(T - t)\, dt$$

where

$$h(t) = s_1(T - t) - s_0(T - t)$$

With our choice of costs and a priori probabilities, the threshold is zero. In general, of course, this is not the case. The structure of this receiver detector is shown in Fig. 4.5.

With the probability of error as the performance criterion, we have

$$P_e = 1 - P_d$$

where the probability of detection is clearly

$$P_d = \pi_0 P_{00} + \pi_1 P_{11}$$

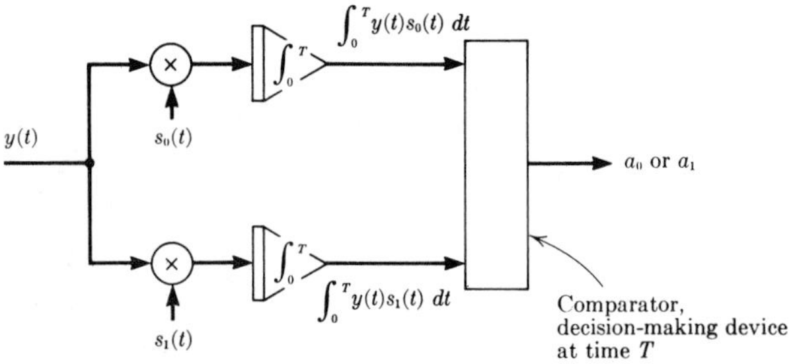

Fig. 4.5 Optimal detector for the coherent binary communication system for equally likely, equal energy signals.

When $s_0(t)$ is transmitted, $y(t) = s_0(t) + n(t)$ and

$$P_{00} = \Pr\left[\int_0^T y(t)s_0(t)\,dt \geq \int_0^T y(t)s_1(t)\,dt \middle| y(t) = s_0(t) + n(t)\right] \tag{4.48}$$

where we have assumed that the costs and a priori probabilities are such that the threshold is zero. Equivalently, P_{00} can be written as

$$P_{00} = \Pr\left\{\int_0^T n(t)[s_1(t) - s_0(t)]\,dt \leq \int_0^T s_0(t)[s_0(t) - s_1(t)]\,dt\right\} \tag{4.49}$$

The right side of the inequality in (4.49) can be written in terms of the phase variations $\theta_1(t)$ and $\theta_0(t)$ as

$$\frac{\mathcal{E}}{T}\int_0^T \{1 - \cos[\theta_1(t) - \theta_0(t)]\}\,dt$$

where the narrowband assumption of $\theta_1(t)$ and $\theta_0(t)$ with respect to ω_c has been employed.

If we define the gaussian random variable

$$\eta \triangleq \int_0^T n(t)[s_1(t) - s_0(t)]\,dt \tag{4.50}$$

which can be shown to have statistics

$$\begin{aligned} E(\eta) &= 0 \\ E(\eta^2) &= \frac{\mathcal{E}N_0}{T}\int_0^T \{1 - \cos[\theta_1(t) - \theta_0(t)]\}\,dt \end{aligned} \tag{4.51}$$

then P_{00} can be expressed as

$$P_{00} = \Pr\left(\eta > -\frac{\mathcal{E}}{T}\int_0^T \{1 - \cos[\theta_1(t) - \theta_0(t)]\}\,dt\right)$$

In terms of the normalized gaussian random variable ζ, defined as

$$\zeta \triangleq \frac{\eta}{\sqrt{E(\eta^2)}}$$

which has zero mean and unit variance, then

$$P_{00} = \Pr\left(\zeta > -\sqrt{\frac{\mathcal{E}}{N_0T}\int_0^T \{1 - \cos\,[\theta_1(t) - \theta_0(t)]\}\,dt}\right)$$

$$= \operatorname{erfc}\left(-\sqrt{\frac{\mathcal{E}}{N_0T}\int_0^T \{1 - \cos\,[\theta_1(t) - \theta_0(t)]\}\,dt}\right)$$

With the assumptions we have made

$$P_{00} = P_{11}$$

and hence

$$P_e = \operatorname{erfc}\sqrt{\frac{\mathcal{E}}{N_0T}\int_0^T \{1 - \cos\,[\theta_1(t) - \theta_0(t)]\}\,dt} \tag{4.52}$$

Again, since erfc is a monotonically decreasing function of its argument, the probability of error in (4.52) is minimized by maximizing the argument. It is immediate that the best choice is

$$\theta_1(t) - \theta_0(t) = \pi \tag{4.53}$$

or, equivalently,

$$s_1(t) = -s_0(t)$$

often referred to as *antipodal signals.*

This minimum value of the probability of error is

$$P_e = \operatorname{erfc}\sqrt{\frac{2\mathcal{E}}{N_0}} \tag{4.54}$$

Therefore the performance of the optimal binary coherent communication systems in the presence of additive white gaussian noise depends only on the ratio of signal-energy-to-noise-spectral-density $\mathcal{E}/N_0$. This performance curve is plotted along with certain other binary-system-performance curves to be discussed in Chap. 7.

PROBLEMS

4.1 In the example in Sec. 4.2, for an arbitrary cost matrix, noise power σ^2, and a priori probabilities (π_0,π_1), show that the optimal Bayes decision function is

$$\text{If } \mathbf{Y}^T(\mathbf{S}_1 - \mathbf{S}_0) \geq \sigma^2\left(\ln\frac{\pi_0}{\pi_1} + \ln\frac{C_{01} - C_{00}}{C_{10} - C_{11}}\right) \qquad \text{decide } a_1$$

otherwise choose a_0, where we have assumed $\|\mathbf{S}_0\|^2 = \|\mathbf{S}_1\|^2$. Determine the corresponding system performance characteristics.

4.2 If in Prob. 4.1 the signal powers are unequal, show that the optimal Bayes decision function is

$$\text{If } \mathbf{Y}^T(\mathbf{S}_1 - \mathbf{S}_0) \geq \frac{\|\mathbf{S}_1\|^2 - \|\mathbf{S}_0\|^2}{2} + \sigma^2 \ln \frac{\pi_0(C_{01} - C_{00})}{\pi_1(C_{10} - C_{11})} \qquad \text{choose } a_1$$

otherwise choose a_0. Determine the corresponding system performance characteristics.

4.3 In a binary decision problem the signal consists of the two signal vectors $\mathbf{S}_0$ and $\mathbf{S}_1$, which are equally likely. The received vector is

$$\mathbf{Y} = \mathbf{S} + \mathbf{N}$$

where the additive-noise vector $\mathbf{N}$ has zero mean and a probability density function depending only on its magnitude; that is,

$$f(n_1, n_2, \ldots, n_m) = f(\|\mathbf{N}\|)$$

where f is a monotonically decreasing function of its argument and independent of $\mathbf{S}$. Determine the optimal decision function for this problem.

4.4 Prove that any choice of $\delta(a|y)$ other than δ_B in (4.7) will result in a greater average risk (neglecting those δ which differ from δ_B on a set of measure zero).

4.5 Suppose y is a random variable which has one of two possible probability density functions, either a gaussian distribution with zero mean and variance σ^2, or a gaussian with mean μ and variance σ^2. Assume the probability π_0 that y comes from the zero mean density and π_1 that it comes from the density with mean μ. Assume that y is sampled m times, with the samples designated as $y_1, \ldots, y_m$, and assume mutually independent samples. Determine the Bayes decision function for deciding whether the $\{y_i\}$ were chosen from the zero-mean distribution or from the distribution with mean μ. Evaluate the type I and type II errors (see Chap. 6).

4.6 In Prob. 4.2 (*a*) Express the optimal receiver operation and corresponding system performance characteristics in terms of

$$\|\mathbf{S}_1\|^2 = \mathcal{E}_1 \qquad \|\mathbf{S}_0\|^2 = \mathcal{E}_0 \qquad \mathcal{E}_1 \neq \mathcal{E}_0$$

for an arbitrary cost matrix.

(*b*) Define $\mathbf{S}_1{}^T\mathbf{S}_0 = \sqrt{\mathcal{E}_1\mathcal{E}_0} \cos\theta$ and determine the optimal θ in this case.

(*c*) Set

$$\mathbf{C} = \begin{bmatrix} 0 & 1 \\ 1 & 0 \end{bmatrix}$$

and write the resulting probability of error.

Note: For binary coherent systems, regardless of the costs, a priori distribution, noise variance, and signal powers, the relative direction of the two signals should be opposite.

4.7 The a posteriori receiver is defined as that receiver which concludes that $\mathbf{S}_i$ was transmitted if the a posteriori probability of $\mathbf{S}_i$, given the observation vector $\mathbf{Y}$, is greater than that for $\mathbf{S}_j$ for all $i \neq j$. In particular, decide $\mathbf{S}_1$ if $\Pr(\mathbf{S}_1|\mathbf{Y}) > \Pr(\mathbf{S}_0|\mathbf{Y})$. Show that the optimal receiver in Eq. (4.21) coincides with the a posteriori receiver.

4.8 Show that the only requirement necessary for $P_{00} = P_{11}$ is that $\pi_0 = \pi_1$, and that this is the case even if the signal energies are unequal.

4.9 Compare the performance of the following two binary communication systems: system A consists of two equilikely antipodal signal vectors, each with norm P; system B has one signal vector with norm $2P$, the second vector being the zero vector. Assume that the noise is additive, white, and gaussian. Which system is better? Do both systems have the same average power?

4.10 What signal-to-noise ratio is required of a binary antipodal communication system with equally likely a priori signals to maintain a probability of error of 10^{-2}, 10^{-4}, and 10^{-6}? If $N_0 = 1$ watt/Hz, what signal energy is required in each case?

4.11 Consider the following binary vector communication system. Let the observed data be represented by the m-dimensional vector

$$\mathbf{Y} = \mathbf{S}_i + \mathbf{N} \qquad i = 0, 1$$

where the $\mathbf{S}_i$ have a priori probabilities π_0 and π_1, respectively. Assume that the additive noise vector is gaussian with zero mean and covariance matrix $\mathbf{\Lambda}_N$. (*a*) Determine the structure of the optimal Bayes receiver for deciding which signal was transmitted and determine the expression for the resulting probability of error in terms of error functions. (*b*) Show that when

$$\mathbf{\Lambda}_N = \sigma^2 \begin{bmatrix} 1 & e^{-\Delta} & e^{-2\Delta} & \cdots & e^{-(m-1)\Delta} \\ e^{-\Delta} & 1 & e^{-\Delta} & \cdots & e^{-(m-2)\Delta} \\ \cdots & \cdots & \cdots & \cdots & \cdots \\ e^{-(m-1)\Delta} & e^{-(m-2)\Delta} & \cdots & \cdots & 1 \end{bmatrix}$$

the optimal test statistic can be written as

$$L(\mathbf{Y}) = L_1(\mathbf{Y}) - L_2(\mathbf{Y})$$

where

$$L_j(\mathbf{Y}) \triangleq \frac{1}{\sigma^2(1-b^2)} \left[b^2 \sum_{k=2}^{m-1} (x_k^j)^2 - 2b \sum_{k=1}^{m-1} x_k^j x_{k+1}^j + \sum_{k=1}^{m} (x_k^j)^2 \right]$$

with $x_k^j = y_k - s_k^j \qquad k = 1, \ldots, m$ and $j = 0, 1$

and $b \triangleq e^{-\Delta}$

Reference: Schwartz [4.7].

4.12 In a binary communication problem there are two signal waveforms corresponding to each correct decision. Determine the Bayes receiver for a generalized cost matrix. Determine the Bayes receiver if the noise is additive and independent of the signal.

4.13 Find the Bayes decision function for deciding between hypotheses H_0 and H_1, where H_0 is the hypothesis that an observed random variable x is distributed according to the distribution

$$p_0(x) = \frac{1}{\sqrt{2\pi}} \exp\left(-\tfrac{1}{2}x^2\right)$$

and H_1 is the hypothesis that x is distributed according to the rectangular distribution

$$p_1(x) = \begin{cases} \frac{1}{5} & 0 \le x \le 5 \\ 0 & \text{otherwise} \end{cases}$$

The a priori probabilities are

$$\pi_0 = \tfrac{3}{4} \qquad \pi_1 = \tfrac{1}{4}$$

The cost matrix is

$$\mathbf{C} = \begin{bmatrix} 0 & 1 \\ 1 & 0 \end{bmatrix}$$

Find the resulting probability of error.

4.14 (*a*) Consider the problem that y is a random variable with density function

$$f(y) = \begin{cases} K \exp(-Ky) & y \geq 0 \\ 0 & y < 0 \end{cases}$$

where K is unknown but is hypothesized to be one of two different values, K_0 or K_1. Assume that y is sampled m times with independent samples. What is the Bayes decision function for this decision problem?

(*b*) Assume that $\mathbf{K}$, with elements $K_1, \ldots, K_m$, is an m-dimensional vector and that we must choose between two m-dimensional vectors $\mathbf{K}_0$ or $\mathbf{K}_1$, where y_i has density

$$K_i^\circ \exp(-K_i^\circ y_i) \qquad \text{or} \qquad K_i' \exp(-K_i' y_i)$$

What is the optimal decision function?

(*c*) Assume that the m-dimensional random vector $\mathbf{Y}$ is of the form

$$\mathbf{Y} = \mathbf{S} + \mathbf{N}$$

where $\mathbf{S}$ is a zero mean gaussian random vector independent of $\mathbf{N}$, and $\mathbf{N}$ is a zero mean gaussian random vector with covariance matrix $\mathbf{\Lambda}_N$. $\mathbf{S}$ has covariance matrix either $\mathbf{\Lambda}_1$ or $\mathbf{\Lambda}_2$. We are asked to decide optimally in the Bayes sense, after one observation of the random vector $\mathbf{Y}$, which covariance matrix the signal $\mathbf{S}$ has. What is this optimal decision function?

4.15 Consider the following sampling model of a binary communication system. The observed signal has the form

$$\mathbf{Y} = \lambda\mathbf{S} + \mathbf{N}$$

where the choice of the signal $\mathbf{S}$ depends on the a priori signal probabilities π_0 and π_1; that is, for the signal with probability π_0 the transmitter chooses

$$\pi_1 \mathbf{e}$$

and for the other signal chooses

$$-\pi_0 \mathbf{e}$$

where $\mathbf{e}$ is a unit vector in m dimensions. λ is the signal-to-noise ratio and $\mathbf{N}$ is an m-dimensional gaussian random vector with covariance matrix equal to the m-by-m identity matrix.

(*a*) Find the receiver that gives the smallest probability of error (on the assumption that the receiver knows the a priori probabilities), and find this minimum as a function of λ.

(*b*) Determine for (*a*) whether the threshold depends on λ.

4.16 A communication system transmits binary information by on-off modulation in the presence of noise. The receiver input y for any bit has the conditional probability densities

$$f_0(y) = y \exp \frac{-y^2}{2} \qquad \text{if the transmitter is off}$$

$$f_1(y) = y \exp\left[-\tfrac{1}{2}(y^2 + A^2)\right] I_0(Ay) \qquad \text{if the transmitter is on}$$

where $y \geq 0$. If the on and off states have equal a priori probabilities, obtain a transcendental equation for the threshold T_0 which minimizes the probability of error. Find the resulting probability of error.

REFERENCES

4.1 Viterbi, A. J.: "Principles of Coherent Communications," McGraw-Hill, New York, 1966.

4.2 Hancock, J. C., and P. A. Wintz: "Signal Detection Theory," McGraw-Hill, New York, 1966.

4.3 Wozencraft, J. M., and I. M. Jacobs: "Principles of Communication Engineering," Wiley, New York, 1965.

4.4 Helstrom, C. W.: "Statistical Theory of Signal Detection," Pergamon Press, New York, 1960.

4.5 Middleton, D.: "Introduction to Statistical Communication Theory," McGraw-Hill, New York, 1960.

4.6 Davenport, W. B., and W. L. Root: "An Introduction to the Theory of Random Signals and Noise," McGraw-Hill, New York, 1958.

4.7 Schwartz, M. I.: On the Detection of Known Binary Signals in Gaussian Noise of Exponential Covariance, *IEEE Trans. Inform. Theory,* vol. IT-11, no. 3, pp. 330–335, 1965.

5

Minimax Decision-rule Concepts

In real situations parameters such as a priori probabilities are rarely known exactly. In some problems a priori probabilities can be estimated with a high degree of confidence. In situations where this cannot be done the *minimax* criterion is often an acceptable alternative for determining the best receiver operation. The minimax decision function corresponds to the Bayes decision function for the a priori probabilities which makes the Bayes risk a maximum. The minimax criterion has the additional characteristic of being independent of the actual a priori probabilities.

Let us consider two methods of determining minimax decision functions for binary problems, both of which are directly extendable to multiple decision functions. We shall restrict ourselves initially to binary decisions and also restrict consideration to signal spaces of two elements, s_0 and s_1.

To start with, certain relationships between average risk and conditional risks will be of value. We have noted that the conditional risks are linear functions of P_{01} and P_{10}. We assume, as before, that $C_{01} > C_{00}$ and $C_{10} > C_{11}$. Then

$$\begin{aligned} C_{00} \le r(s_0,\delta) \le C_{01} \\ C_{11} \le r(s_1,\delta) \le C_{10} \end{aligned} \tag{5.1}$$

Also,

$$R(\pi,\delta) = E(r(s,\delta)) = \pi_0 r(s_0,\delta) + \pi_1 r(s_1,\delta)$$

or, equivalently,

$$R(\pi,\delta) = r(s_0,\delta) + [r(s_1,\delta) - r(s_0,\delta)]\pi_1 \tag{5.2}$$

Hence $R(\pi,\delta)$, for any δ, is a linear function of the a priori probabilities, as indicated in Fig. 5.1. It is immediate that for any δ

$$\max_{\pi} R(\pi,\delta) = \max_{s} r(s,\delta) \tag{5.3}$$

Thus

$$\min_{\delta} \max_{\pi} R(\pi,\delta) = \min_{\delta} \max_{s} r(s,\delta) \tag{5.4}$$

We define the δ which satisfies Eq. (5.4) as the *minimax decision function*, which we designate as δ_m.

The *minimax theorem*, one of the fundamental theorems of decision theory, states that under very general conditions [5.4] (much more general than binary decision problems)

$$\min_{\delta} \max_{\pi} R(\pi,\delta) = \max_{\pi} \min_{\delta} R(\pi,\delta) \tag{5.5}$$

So, from Eqs. (5.4) and (5.5),

$$\min_{\delta} \max_{s} r(s,\delta) = \max_{\pi} \min_{\delta} R(\pi,\delta) \tag{5.6}$$

We define the *least favorable a priori distribution* π_L, as that one which maximizes $R(\pi,\delta_B)$, where δ_B is the Bayes decision rule for π.

Thus, by (5.6), the minimax decision rule can be found by finding the Bayes decision function corresponding to the least favorable a priori distribution.

From (5.2) we note that $R(\pi,\delta)$ is independent of the a priori distribution if and only if the conditional risks are equal.

We can now prove the following theorem.

THEOREM 5.1 *For the binary decision problem assume that there exists a* δ_m *such that*

$$r(s_0,\delta_m) = r(s_1,\delta_m) \tag{5.7}$$

Assume also that δ_m is a Bayes decision function for some a priori distribution. Then δ_m is the minimax decision function.

Proof: In contradiction, assume that δ_m is not minimax; then there exists a δ' such that

$$\max_s r(s,\delta') < \max_s r(s,\delta_m)$$

Then, by assumption,

$$\max_s r(s,\delta') < r(s_1,\delta_m)$$

and

$$\max_s r(s,\delta') < r(s_0,\delta_m)$$

Hence

$$r(s,\delta') < r(s,\delta_m) \qquad \text{for all } s$$

So, for any a priori distribution,

$$R(\boldsymbol{\pi},\delta') = \pi_0 r(s_0,\delta') + \pi_1 r(s_1,\delta') < \pi_0 r(s_0,\delta_m) + \pi_1 r(s_1,\delta_m) = R(\boldsymbol{\pi},\delta_m)$$

Thus δ_m cannot be the Bayes decision function for any a priori distribution, which is a contradiction.

This is proof that there cannot be a horizontal line in Fig. 5.1 below that corresponding to δ_m.

A decision function that satisfies (5.7) in Theorem 5.1 is called an *equalizer decision function.*

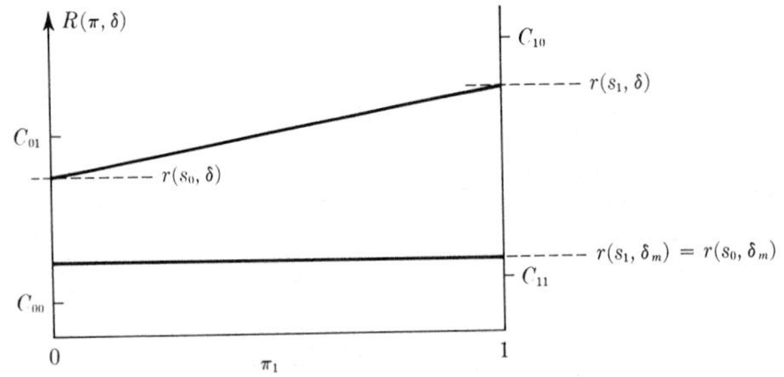

Fig. 5.1 Average risk versus a priori probabilities for arbitrary decision function δ and minimax decision function δ_m.

In summary, we have two methods of finding minimax decision functions:

1. Find the Bayes decision function corresponding to the *least favorable a priori distribution.*
2. Find an *equalizer decision function* which is Bayes for some a priori distribution.

Following are some more general versions of this theorem, stated without proof, which are applicable to the M-ary decision-function problem.

THEOREM 5.2 *Assume that* Ω *has a finite number of elements (say,* M*); assume that there exists a* δ_m *such that*

$$r(s_i,\delta_m) = r \qquad \textit{for all } s_i \in \Omega$$

and assume that δ_m *is admissible* (as in Chap. 3). *Then* δ_m *is the minimax decision function for the* M*-ary decision-rule problems.*

THEOREM 5.3 *Assume that* Ω *has a finite number of elements (say,* M*); assume that there exists a Bayes solution* δ_B *with respect to the a priori distribution* $(\pi_1, \ldots, \pi_M)$*; and assume* $\pi_j > 0$ for $j = 1, \ldots, M$. *Then* δ_B *is admissible.*

For the proofs of these theorems, see Ferguson [5.4].

These two theorems provide one method of finding the minimax decision function in M dimensions. The second method in M dimensions is also the same as in the binary case, to find the Bayes decision function for the least favorable a priori distribution. In general, however, minimax decision functions are difficult to find. The curves in Fig. 5.2 exemplify

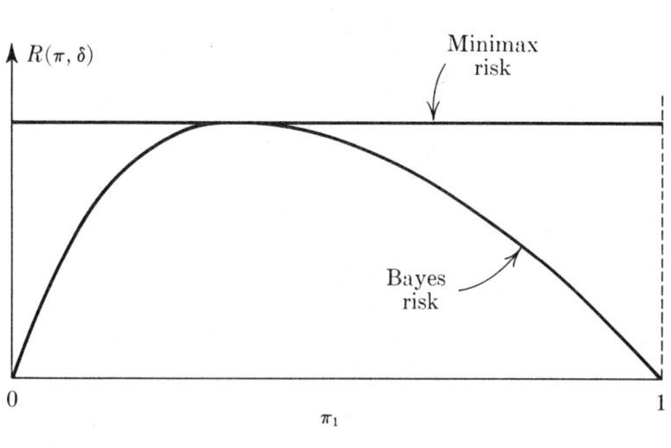

Fig. 5.2 Bayes and minimax risk functions.

the Bayes and minimax risks geometrically for the binary-decision-rule problems.

PROBLEMS

5.1 Consider the following sampling model for a binary pulse-code-modulation communication system:

$$\mathbf{Y} = \lambda\mathbf{S} + \mathbf{N}$$

where the choice of $\mathbf{S}$ depends on the a priori signal probabilities π_0 and π_1. That is, for the signal vector which has probability π_0 the transmitter chooses

$$\mathbf{S}_0 = \pi_1\mathbf{e}$$

and for the other signal chooses

$$\mathbf{S}_1 = -\pi_0\mathbf{e}$$

where

$\mathbf{e}$ = a vector of unit magnitude
λ = signal-to-noise ratio
$\mathbf{N}$ = a k-dimensional gaussian random vector which has zero mean and covariance matrix equal to the k-by-k identity matrix

(*a*) First find the receiver that minimizes the probability of error, with the assumption that the receiver knows the signal probabilities. Find this minimum as a function of λ. Determine the resulting probability of error.

(*b*) On the assumption that the receiver does not know the a priori probabilities, determine the minimax receiver. Again evaluate the probability of error as a function of λ.

5.2 In a binary communication problem there are two signal waveforms corresponding to each correct decision. Determine the minimax decision function for a generalized cost function.

REFERENCES

5.1 Helstrom, C. W.: "Statistical Theory of Signal Detection," Pergamon Press, New York, 1960.
5.2 Middleton, D.: "Introduction to Statistical Communication Theory," McGraw-Hill, New York, 1960.
5.3 Selin, I.: "Detection Theory," Princeton University Press, Princeton, N.J., 1965.
5.4 Ferguson, T. S.: "Mathematical Statistics: A Decision Theoretic Approach," Academic, New York, 1967.

6
Radar Detection Theory

Radar detection is a particular kind of binary decision problem. Initially we shall make the simplifying assumption that the signal space also consists of only two elements and require the receiver to determine in the presence of channel disturbances whether s_1 or s_0 has been transmitted (to choose a_1 or a_0, respectively), where s_1 corresponds to a completely known signal and s_0 normally corresponds to *signal absent;* that is

$$s_1 = s \qquad \text{for } \textit{signal present}$$
$$s_0 = 0 \qquad \text{for } \textit{signal absent}$$

This simple hypothesized assumption will later be relaxed so that s is a sample function from certain types of stochastic processes. The resulting mathematical model will then be a composite hypothesis (see Chap. 7) and will then have practical significance.

In the radar problem the a priori probabilities π_0 and π_1 are unknown and are very difficult to estimate; hence we can speak only of conditional risks. For an arbitrary decision function

$$r(s_0,\delta) = C_{00} + (C_{01} - C_{00})P_{01}$$
$$r(s_1,\delta) = C_{11} + (C_{10} - C_{11})P_{10}$$

In problems where the a priori probabilities are hard to establish, but where the costs can be set, we have seen that the minimax criterion is the most logical choice of performance criterion. When the costs are also difficult to establish, which is particularly true in radar detection problems, we adopt the following system-design philosophy.

6.1 RADAR-SYSTEM-DESIGN PHILOSOPHY

There are two types of errors. The first is a decision that a signal is present when no signal has been sent. The probability of this error is P_{01}. It is called the *false-alarm probability* α and corresponds to a *type I* error in statistical terminology.

The second is a decision that no signal is present when one is actually there. The probability of this error is $P_{10} = 1 - P_{11}$. It is denoted by β and corresponds to a *type II error*. P_{11} is the *detection probability* of the signal given that it is actually present.

The optimization problem for radar detection is usually stated as follows: for a given false-alarm probability α maximize P_{11}, the probability of detecting the signal given that it is present. We shall see that there is a unique solution to this problem, that the solution is again a threshold device, and that the threshold level is a function only of the false-alarm probability α.

In a radar problem the cost associated with a type II error is generally set much higher than that for a type I error. For example, a decision that an enemy bomber is coming when it actually is not is obviously less costly than deciding no bomber is present when one is in fact approaching.

If the cost matrix is specified as in the preceeding chapter, then the criterion stated is equivalent to specifying a value for $r(s_0,\delta)$ and then minimizing $r(s_1,\delta)$ subject to this restriction. Stating the problem in terms of P_{01} and P_{11} has the advantage of not requiring specification of a cost matrix. In both cases the results are the same. In either case the statement of the problem is equivalent to a simplified version of the *Neyman-Pearson lemma.*

Let us now develop the optimal receiver. From Chap. 4 we have

$$P_{01} = \int_\Gamma \delta(a_1|y)f(y|s_0)\,dy = \alpha \tag{6.1}$$

where α is a preset constant with $0 < \alpha < 1$. Under this restriction we want to find the $\delta(a_1|y)$ which maximizes

$$P_{11} = \int_\Gamma \delta(a_1|y)f(y|s_1)\,dy \tag{6.2}$$

For each $K > 0$ let $\mathbf{\Gamma}_K$ be that subset of $\mathbf{\Gamma}$ where the ratio

$$\frac{f(y|s_1)}{f(y|s_0)} > K \tag{6.3}$$

Assume for the present that $f(y|s_1)$ and $f(y|s_0)$ are density functions with no discrete probabilities.[1] Let us write

$$P_{11} = P_{11} - K\alpha + K\alpha \tag{6.4}$$

and choose K so that

$$\int_{\mathbf{\Gamma}_K} f(y|s_0)\, dy = \alpha \tag{6.5}$$

This is always possible, since we have eliminated discrete probabilities. If $\alpha = 0$, then $\mathbf{\Gamma}_K$ is the empty set,[2] which implies that K is unbounded. If $\alpha = 1$, then $\mathbf{\Gamma}_K = \mathbf{\Gamma}$, which implies $K = 0$. $\mathbf{\Gamma}_K$ varies continuously with K at all interior points; therefore a K can be found for every α. Substituting (6.1) and (6.2) into (6.4), we may write

$$\begin{aligned} P_{11} &= \int_{\mathbf{\Gamma}} \delta(a_1|y) f(y|s_1)\, dy - K \int_{\mathbf{\Gamma}} \delta(a_1|y) f(y|s_0)\, dy + K\alpha \\ &= \int_{\mathbf{\Gamma}} \delta(a_1|y) \left[\frac{f(y|s_1)}{f(y|s_0)} - K\right] f(y|s_0)\, dy + K\alpha \end{aligned} \tag{6.6}$$

All functions in (6.6) are always nonnegative. Thus P_{11} is maximum when $\delta(a_1|y)$ is chosen to be

$$\delta_R(a_1|y) = \begin{cases} 1 & \text{if } \dfrac{f(y|s_1)}{f(y|s_0)} \geq K \\ 0 & \text{if } \dfrac{f(y|s_1)}{f(y|s_0)} < K \end{cases} \tag{6.7a}$$

or

$$\delta_R(a_1|y) = \begin{cases} 1 & \text{if } y \in \mathbf{\Gamma}_K \\ 0 & \text{if } y \notin \mathbf{\Gamma}_K \end{cases} \tag{6.7b}$$

From (6.3) and (6.5) it follows immediately that K depends only on α. Note also that $\delta_R(a_1|y)$ is the Bayes solution for the a priori probabilities which satisfy

$$\frac{\pi_0}{\pi_1} = K \qquad \pi_0 + \pi_1 = 1 \qquad \mathbf{C} = \begin{bmatrix} 0 & 1 \\ 1 & 0 \end{bmatrix} \tag{6.8}$$

This proves all the claims and therefore completes the development.[3] It

[1] With discrete probabilities permitted, certain intricate problems arise, which can be overcome with greater mathematical rigor. The results are the same as presented, and this is an ample presentation for engineering applications.

[2] Or has measure equal to zero.

[3] This is not an obvious development of the Neyman-Pearson criterion but is taken on account of its simplicity.

can be shown that any other choice of a decision function results in a lower P_{11}, thus establishing uniqueness.

The threshold level K can be found directly from the false-alarm probability:

$$P_{01} = \Pr\left[\frac{f(y|s_1)}{f(y|s_0)} \geq K|s_0 \text{ was transmitted}\right] = \alpha \tag{6.9}$$

The corresponding maximum detection probability is

$$P_{11} = \Pr\left[\frac{f(y|s_1)}{f(y|s_0)} \geq K|s_1 \text{ was transmitted}\right] \tag{6.10}$$

One slight generalization of this problem is the situation in which the transmitted signal is one from a class of signals with a specified probability distribution over this class and the receiver is to decide whether a member of this class has been transmitted or not. This is the model that results in incoherent radar detection, which will be treated later.

Under the assumption that this optimal system can be implemented, the performance of the system reduces to deciding the value to be used for the threshold. This, as we have indicated, involves a tradeoff between the false-alarm probability and the detection probability.

6.2 VECTOR MODEL

In this section we consider a vector model of a radar detection problem. In particular, the observation is assumed to consist of a signal vector $\mathbf{S}$ and additive gaussian noise,

$$\mathbf{Y} = \mathbf{S} + \mathbf{N}$$

where the m-dimensional vector $\mathbf{Y}$ could, for example, represent a sequence of samples of an observed waveform, and the additive noise vector $\mathbf{N}$ is assumed to be gaussian with zero mean and covariance matrix given by $\sigma^2\mathbf{I}$, with $\mathbf{I}$ the m-by-m identity matrix. For a given false-alarm probability α this receiver is to optimally determine the presence or absence of the signal vector $\mathbf{S}$ after observing the received vector $\mathbf{Y}$.

The receiver structure can initially be expressed directly in terms of the likelihood ratio in (6.3):

$$\text{If } \frac{f(\mathbf{Y}|\mathbf{S} = \mathbf{S})}{f(\mathbf{Y}|\mathbf{S} = \mathbf{0})} > K \qquad \text{decide } \textit{signal present} \tag{6.11}$$

where K will be determined from knowledge of α. This can be expressed in terms of the noise density as

$$\frac{f(\mathbf{Y}|\mathbf{S} = \mathbf{S})}{f(\mathbf{Y}|\mathbf{S} = \mathbf{0})} = \frac{f_N(\mathbf{Y} - \mathbf{S})}{f_N(\mathbf{Y} - \mathbf{0})} = \frac{\exp\left(\frac{-1}{2\sigma^2}\|\mathbf{Y} - \mathbf{S}\|^2\right)}{\exp\left(\frac{-1}{2\sigma^2}\|\mathbf{Y}\|^2\right)}$$

and after we have taken logarithms and simplified the optimal detector is

$$\text{If } \mathbf{Y}^T\mathbf{S} > \frac{\|\mathbf{S}\|^2}{2} + \sigma^2 \ln K \qquad \text{decide } \textit{signal present}$$

If we define a normalized signal vector $\mathbf{S}_n$ as

$$\mathbf{S}_n \triangleq \frac{\mathbf{S}}{\|\mathbf{S}\|}$$

we can express the detector in terms of $\mathbf{S}_n$ as

$$\mathbf{Y}^T\mathbf{S}_n > \frac{\|\mathbf{S}\|}{2} + \frac{\sigma^2 \ln K}{\|\mathbf{S}\|} \triangleq K_0 \tag{6.12}$$

where K_0 is to be determined in terms of the false-alarm probability. For a given false-alarm probability, K_0 must be chosen so that

$$\Pr\,(\mathbf{Y}^T\mathbf{S}_n \geq K_0|\textit{signal absent}) = \alpha \tag{6.13}$$

That is, the probability of deciding *target present* when there is no target is equal to α, or equivalently, K_0 is to be chosen so that

$$\Pr\,(\mathbf{N}^T\mathbf{S}_n \geq K_0) = \alpha \tag{6.14}$$

Normalizing (6.13) by defining

$$\zeta \triangleq \frac{\mathbf{N}^T\mathbf{S}_n}{\sqrt{\operatorname{Var} \mathbf{N}^T\mathbf{S}_n}} = \frac{\mathbf{N}^T\mathbf{S}_n}{\sigma} \tag{6.15}$$

the condition on K_0 can be expressed as

$$\Pr\left(\zeta \geq \frac{K_0}{\sigma}\right) = \operatorname{erfc} \frac{K_0}{\sigma} = \alpha \tag{6.16}$$

where ζ is a gaussian random variable with zero mean and unit variance. Equation (6.16) is plotted in Fig. 6.1. For a given false-alarm probability there is a unique threshold which can be determined from this gaussian curve. With this, the resultant detection probability can be expressed as

$$\begin{aligned} P_{11} &= \Pr\,(\mathbf{Y}^T\mathbf{S}_n > K_0|\mathbf{Y} = \mathbf{S} + \mathbf{N}) \\ &= \Pr\,(\mathbf{N}^T\mathbf{S}_n > K_0 - \|\mathbf{S}\|) \end{aligned}$$

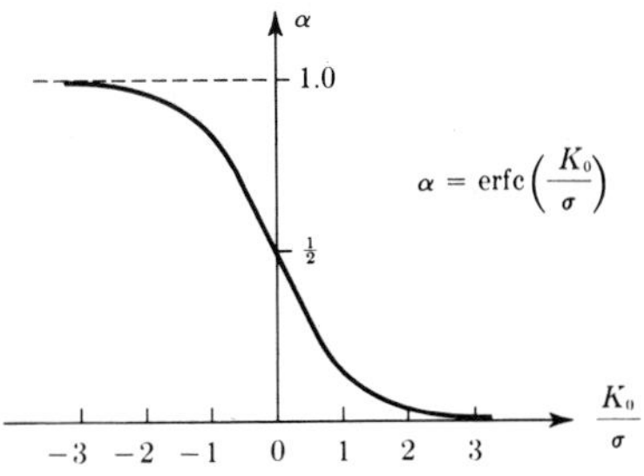

Fig. 6.1 Threshold versus false-alarm probability.

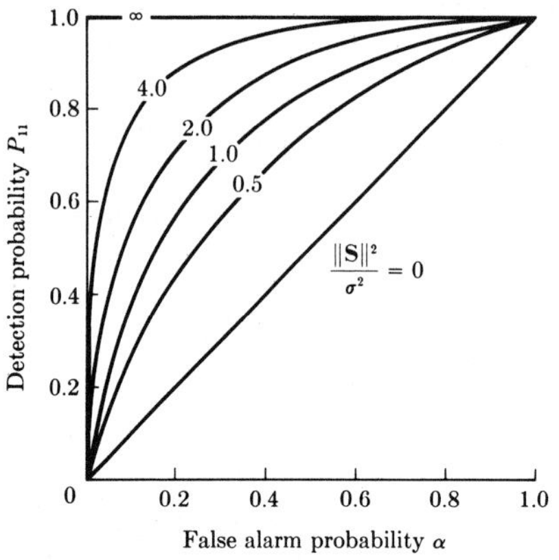

Fig. 6.2 Detection probability versus false-alarm probability for various signal-to-noise ratios.

Normalizing as we did above, we have

$$P_{11} = \Pr\left(\zeta > \frac{K_0}{\sigma} - \frac{\|S\|}{\sigma}\right) = \operatorname{erfc}\left(\operatorname{erfc}^{-1}\alpha - \frac{\|S\|}{\sigma}\right) \tag{6.17}$$

Therefore the performance of the detector is a function of the two parameters false-alarm probability α and signal-to-noise ratio $\|S\|^2/\sigma^2$.

The detection probability is plotted as a function of the false-alarm probability for various signal-to-noise ratios in Fig. 6.2.

PROBLEMS

6.1 For a given false-alarm probability α show that any decision function other than δ_R in (6.7) results in a smaller detection probability.

6.2 *Determining the presence or absence of a deterministic signal in additive white gaussian noise* Assume that the observed waveform is either $y(t) = s(t) + n(t)$ for $0 \le t \le T$ or $y(t) = n(t)$ for $0 \le t \le T$, where $s(t)$ is a signal waveform which is completely known to the receiver. The additive noise is gaussian with two-sided spectral density $N_0/2$.

(*a*) Using the Neyman-Pearson criterion, show that the optimal detector is the matched filter

$$\int_0^T y(t)s(t)\,dt$$

whose output at time T is compared to the threshold K_0 and decides *signal present* if the output is greater than K_0, where K_0 is determined by

$$\operatorname{erfc}\sqrt{\frac{2}{N_0\mathcal{E}}}\,K_0 = \alpha$$

where α is the false-alarm probability and

$$\mathcal{E} = \int_0^T s^2(t)\,dt$$

(*b*) Show that the resulting detection probability is

$$P_{11} = \operatorname{erfc}\left(\operatorname{erfc}^{-1}\alpha - \sqrt{\frac{2\mathcal{E}}{N_0}}\right)$$

(*c*) Show that the optimal detector can be specified so that it does not require knowledge of the received signal energy.

REFERENCES

6.1 Helstrom, C. W.: "Statistical Theory of Signal Detection," Pergamon Press, New York, 1960.
6.2 Middleton, D.: "Introduction to Statistical Communication Theory," McGraw-Hill, New York, 1960.
6.3 Wainstein, L. A., and V. D. Zubakov, "Extraction of Signals from Noise," Prentice-Hall, Englewood Cliffs, N.J., 1962.
6.4 Selin, I.: "Detection Theory," Princeton University Press, Princeton, N.J., 1965.

7 Binary Composite-hypothesis Testing

The communication detection systems discussed thus far have been describable in terms of deciding between a simple hypothesis and a simple alternative. In this section we shall consider binary decision rules which are describable in terms of a composite hypothesis versus a composite alternative. More precisely, by simple hypothesis is meant that under the assumed hypothesis the probability distribution of the test data is completely specified. When the assumed hypothesis does not specify completely the probability distribution of the obtained data, the hypothesis is called *composite*. Composite-hypothesis testing for binary systems results in a generality in that the signal space is partitioned into two disjoint subsets $\mathbf{\Omega}_0$ and $\mathbf{\Omega}_1$, each of which may contain an arbitrarily large number of signals. For example, $\mathbf{\Omega}_0$ may consist of a class of time waveforms characterized by an unknown parameter, say $\{s_0(t;m)\}$, where m is an unknown random variable or vector; the same is true for $\mathbf{\Omega}_1$.

For example, consider the case where

$$\mathbf{\Omega}_0 = \{A \cos (\omega_0 t + \theta); 0 \le t \le T\}$$
$$\mathbf{\Omega}_1 = \{A \cos (\omega_1 t + \theta); 0 \le t \le T\}$$

where θ is an unknown random variable.

When the receiver is to determine the class in which the transmitted waveform is located, this system is called a *frequency-shift-keying* binary communication system (discussed in Sec. 7.2). If, instead, $\mathbf{\Omega}_0$ consists of the single signal

$$s_0(t) = 0 \qquad 0 \le t \le T$$

and $\mathbf{\Omega}_1$ is as described above, and the receiver is now to distinguish between *signal present* and *signal absent*, we have one form of a radar detection model. We now determine the configuration of the Bayes optimal decision function. Recall from Sec. 4.1 that minimizing the average risk for binary decision functions is equivalent to finding the decision function $\delta(a_0|y)$ which minimizes the integral

$$I_2 = \int_{\mathbf{\Gamma}} \delta(a_0|y)\, dy \int_{\mathbf{\Omega}} f(y|s)\pi(s)[C(s,a_0) - C(s,a_1)]\, ds \tag{7.1}$$

We partition the signal space as described in Fig. 7.1, where π_i is defined as the a priori probability that S is contained in the subspace $\mathbf{\Omega}_i$ for $i = 0,1$ and $p_i(s)$ is defined as the conditional density of s in $\mathbf{\Omega}_i$, given that $s \in \mathbf{\Omega}_i$ for $i = 0,1$. Then $\pi(s)$ can be expressed as

$$\pi(s) = \pi_0 p_0(s) + \pi_1 p_1(s) \tag{7.2}$$

Upon substitution of this term in (7.1), I_2 becomes

$$I_2 = \int_{\mathbf{\Gamma}} \delta(a_0|y)\, dy \left\{\pi_0 \int_{\mathbf{\Omega}_0} f(y|s)p_0(s)[C(s,a_0) - C(s,a_1)]\, ds \right.$$
$$\left. + \pi_1 \int_{\mathbf{\Omega}_1} f(y|s)p_1(s)[C(s,a_0) - C(s,a_1)]\, ds\right\} \tag{7.3}$$

$$\int_{\mathbf{\Omega}_i} p_i(s)\, ds = 1 \qquad \text{for } i = 0, 1$$

If the cost matrix

$$\mathbf{C} \triangleq \begin{bmatrix} 0 & 1 \\ 1 & 0 \end{bmatrix}$$

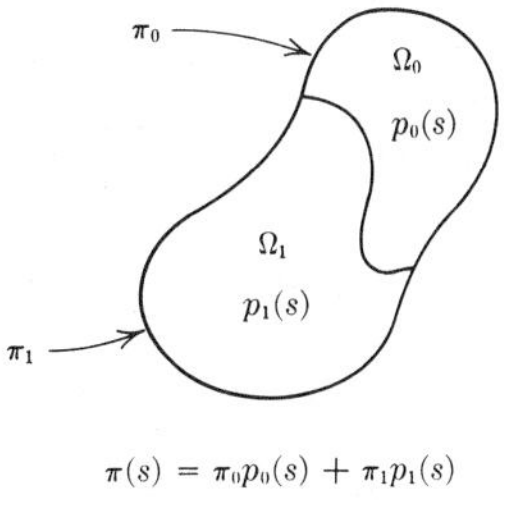

Fig. 7.1 Partitioning of the signal space for binary composite-hypothesis decision functions.

is employed, so that minimizing the average risk corresponds to minimizing the probability of error, we find that the optimal decision function becomes the *generalized likelihood ratio* given by

$$\text{If } \frac{\int_{\Omega_1} f(y|s)p_1(s)\,ds}{\int_{\Omega_0} f(y|s)p_0(s)\,ds} > \frac{\pi_0}{\pi_1} \qquad \text{decide } a_1 \tag{7.4}$$

Equivalently, if the occurrence of signals in one or both of the subspaces of $\mathbf{\Omega}$ is dependent on the random vector $\mathbf{m}$ with density $p_0(\mathbf{m})$ in $\mathbf{\Omega}_0$ and $p_1(\mathbf{m})$ in $\mathbf{\Omega}_1$, the generalized likelihood ratio can be expressed as

$$\text{If } \frac{\int_{\Omega_{1_m}} f(y|s_1(\mathbf{m}))p_1(\mathbf{m})\,d\mathbf{m}}{\int_{\Omega_{0_m}} f(y|s_0(\mathbf{m}))p_0(\mathbf{m})\,d\mathbf{m}} > \frac{\pi_0}{\pi_1} \qquad \text{decide } a_1 \tag{7.5}$$

where $\mathbf{\Omega}_{1_m}$ is the space of values that $\mathbf{m}$ can take on given $s \in \mathbf{\Omega}_1$ and $\mathbf{\Omega}_{0_m}$ is the space of values that $\mathbf{m}$ can take on given $s \in \mathbf{\Omega}_0$. The vector $\mathbf{m}$ is sometimes called a *nuisance vector*, or a vector representing a set of *nuisance parameters*. $s_1(\mathbf{m})$ is a signal from $\mathbf{\Omega}_1$ dependent on the nuisance vector; $s_0(\mathbf{m})$ is similarly defined.

In this development $s_0(\mathbf{m})$ and $s_1(\mathbf{m})$ are dependent on the same random vector. In general this need not be the case. In many radar applications the hypothesis, *signal present*, is a composite hypothesis, while the alternative, *signal absent* or *signal identically zero*, is a simple hypothesis.

Let us now turn to the detection in the presence of additive white gaussian noise of signal waveforms of the form[1]

$$s(t) = A(t)\cos(\omega_c t + \theta(t) + \phi) \qquad 0 \le t \le T \tag{7.6}$$

where the amplitude modulation $A(t)$ and the phase modulation $\theta(t)$ are assumed to be narrowband with respect to ω_c, and ϕ is assumed to be a random variable uniformly distributed between $-\pi$ and π.

From the previous discussion we see that the optimal detector is the one that forms the generalized likelihood ratio

$$\frac{\int_{\Omega_{1_\phi}} f(\mathbf{Y}|\mathbf{S}(\psi))p_1(\psi)\,d\psi}{f(\mathbf{Y}|\mathbf{S} = \mathbf{0})}$$

[1] This is a simplified version of many radar detection problems inasmuch as it is assumed that the interval of time $[0,T]$ during which the signal arrives as well as the carrier frequency ω_c and the modulation waveform are known to the receiver. This will serve to exemplify, however, the manipulations that are pertinent in radar detection analysis. More general situations would take into account unknown arrival time, unknown frequency shifts due to doppler effects, unknown amplitude and phase, etc. Later examples will account for some of these unknowns.

and decides *signal present* if this ratio is greater than a threshold, say, K_0, which is to be determined from the false-alarm probability.

As in the continuous-sampling problem treated thus far, the optimal detector will be determined by first assuming that the observation consists of k uniformly spaced samples of the observed waveform over the interval $[0,T]$.

On the assumption that the signal and noise are independent, the sampled version of the decision function may be expressed as

$$\frac{(1/2\pi)\int_{-\pi}^{\pi} f_n(\mathbf{Y} - \mathbf{S}(\psi))\,d\psi}{f_n(\mathbf{Y})}$$

where the vectors $\mathbf{Y}$ and $\mathbf{S}(\psi)$ represent successive samples of $y(t)$ and $s(t,\psi) = A(t)\cos(\omega_c t + \theta(t) + \psi)$, respectively. We proceed now as we did in Sec. 4.2 to obtain the continuous version of the above decision function. If we allow the number of samples of the observation signal to increase, the noise bandwidth increases simultaneously, becoming white noise in the limit. With the resulting *Reimann sum* expressed in integral form, the test statistic becomes[1]

$$L(y(t);\,0 \le t \le T) = \frac{(1/2\pi)\int_{-\pi}^{\pi} \exp\left\{(-1/N_0)\int_0^T [y(t) - s(t,\psi)]^2\,dt\right\} d\psi}{\exp\left\{(-1/N_0)\int_0^T [y(t)]^2\,dt\right\}} \tag{7.7}$$

where $N_0/2$ is the two-sided spectral density of the additive white gaussian noise. Since $y(t)$ is independent of the integration with respect to ψ, the integral $\int_0^T y^2(t)\,dt$ may be removed, resulting in the simplification

$$L(y(t);\,0 \le t \le T) = \frac{1}{2\pi}\int_{-\pi}^{\pi} \exp\left[\frac{-1}{N_0}\int_0^T s^2(t;\psi)\,dt + \frac{2}{N_0}\int_0^T y(t)s(t;\psi)\,dt\right] d\psi \tag{7.8}$$

In all the problems of practical interest the integral

$$\int_0^T s^2(t;\psi)\,dt$$

will not depend upon ψ because of the narrowband assumption made on the amplitude and phase of the transmitted signal. Therefore the optimal

[1] The variable of integration ψ here should not be confused with the phase of the signal in the observed waveform. Equivalently, for each given observed waveform $y(t)$ over $[0,T]$ there is a value of the functional $L(y(t))$ for $0 \le t \le T$, and the integration with respect to ψ is performed without knowledge of the actual phase ϕ of the signal in the observed waveform $y(t)$.

decision function can be written in the form

$$\text{If } \frac{1}{2\pi}\int_{\pi}^{\pi} \exp\left[\frac{2}{N_0}\int_0^T y(t)s(t;\psi)\,dt\right]d\psi > K \qquad \text{decide } \textit{signal present} \tag{7.9}$$

where the threshold K is to be set by the false-alarm probability α.

We have a completely general result thus far, with the exception that the signal was assumed narrowband with respect to the rf carrier ω_c. The physical result is that we take $y(t)$, correlate it with $s(t,\psi)$, take the exponential of the resultant, and average that over all possible phase angles. It is hopeless, however, to expect to be able to implement such a complex functional. Let us now consider certain specific examples of interest where the above functional can be greatly simplified.

Example 7.1 Detection of one pulse of known arrival time This is an oversimplified version of the radar detection problem, presented in order to introduce the kind of analysis necessary for radar detection problems. Let us assume that the observed waveform is of the form

$$y(t) = A\cos(\omega_c t + \phi) + n(t) \qquad 0 \le t \le T \tag{7.10}$$

where ϕ is a random variable uniformly distributed over the interval $(-\pi,\pi)$, A is a constant, knowledge of which we shall see is not necessary to specify the optimal receiver structure, and $n(t)$ is white gaussian noise with two-sided spectral density $N_0/2$.

Using the Neyman-Pearson criterion to decide the presence or absence of the signal $A\cos(\omega_c t + \phi)$ for $0 \le t \le T$, we form the optimal decision function in (7.9),

$$\frac{1}{2\pi}\int_{-\pi}^{\pi} \exp\left[\frac{2}{N_0}\int_0^T y(t)A\cos(\omega_c t + \psi)\,dt\right]d\psi \gtrless K$$

We then compare this test statistic to a threshold K. If the test statistic exceeds the threshold, *signal present* is decided.

Equivalently, we may express this test statistic as

$$\begin{aligned} L(y) &= \frac{1}{2\pi}\int_{-\pi}^{\pi} \exp\left[\frac{2A}{N_0}(e_c\cos\psi - e_s\sin\psi)\right]d\psi \\ &= \frac{1}{2\pi}\int_{-\pi}^{\pi} \exp\left[\frac{2A}{N_0}\sqrt{e_c^2 + e_s^2}\cos\left(\psi + \tan^{-1}\frac{e_s}{e_c}\right)\right]d\psi \end{aligned} \tag{7.11}$$

where

$$e_c \triangleq \int_0^T y(t)\cos\omega_c t\,dt \tag{7.12}$$

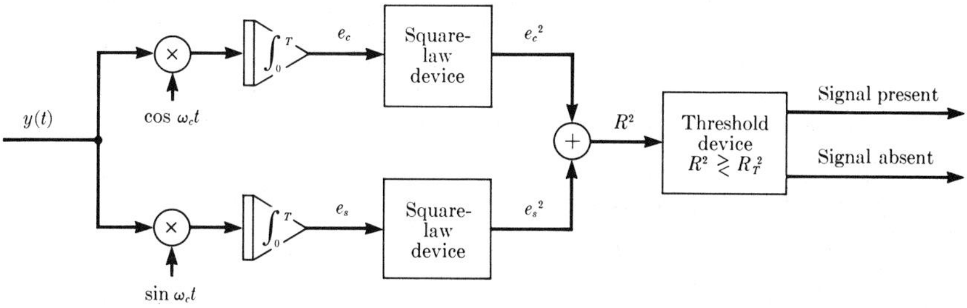

Fig. 7.2 Diagram of an envelope correlation detector for radar detection.

and

$$e_s \triangleq \int_0^T y(t) \sin \omega_c t \, dt \tag{7.13}$$

Representation (7.11) is recognized as

$$L(y) = I_0\left(\frac{2A}{N_0} R\right) \tag{7.14}$$

where

$$R \triangleq \sqrt{e_c{}^2 + e_s{}^2} \tag{7.15}$$

and I_0 is the modified Bessel function of the first kind of order zero (see Appendix C). The physical device that generated R is known as the *envelope correlation detector.*

Equivalently, we may write

$$R^2 = \left[\int_0^T y(t) \cos \omega_c t \, dt\right]^2 + \left[\int_0^T y(t) \sin \omega_c t \, dt\right]^2 \tag{7.16}$$

a mechanization of which appears in Fig. 7.2.

In view of $L(y)$ as expressed in (7.14), it appears that a complete mechanization of the optimal detector would involve the addition of a square-root device followed by an amplifier whose gain is dependent on the signal-to-noise ratio, followed by a nonlinear function generator whose output is the modified Bessel function of the input. This additional equipment is not necessary, however, since I_0 is a strictly monotonically increasing function of its argument. Therefore its inverse exists, and the optimal detector

$$I_0\left(\frac{2A}{N_0} R\right) \gtrless K \tag{7.17}$$

can equivalently be represented as

$$R \gtrless \frac{N_0}{2A} I_0^{-1}(K) \triangleq R_T \tag{7.18}$$

or

$$R^2 \gtrless R_T{}^2$$

inasmuch as the envelope[1] R is nonnegative.

The threshold R_T in (7.18) is determined from the false-alarm probability α; for a given α we must choose R_T such that

$$P_{01} = \Pr\,[R > R_T | y(t) = n(t), 0 \leq t \leq T] = \alpha \tag{7.19}$$

It is readily shown that e_c and e_s are gaussian random variables with mean values

$$E(e_c | y(t) = n(t)) = E(e_s | y(t) = n(t)) = 0 \tag{7.20}$$

and second-order moments[2]

$$E(e_c{}^2 | y(t) = n(t)) = E(e_s{}^2 | y(t) = n(t)) = \frac{N_0 T}{4} \tag{7.21}$$

$$E(e_c e_s | y(t) = n(t)) = 0$$

when the observed waveform is noise only. P_{01}, then, has the representation

$$P_{01} = 2 \iint\limits_C \frac{\exp\,[(-2/N_0 T)(e_c{}^2 + e_s{}^2)]}{\pi N_0 T}\, de_c\, de_s \tag{7.22}$$

where C is that region where $R = \sqrt{e_c{}^2 + e_s{}^2} > R_T$.

If we transform (7.22) to polar coordinates with the transformation

$$\begin{aligned} e_c &\triangleq R \cos \Delta \\ e_s &\triangleq R \sin \Delta \end{aligned} \tag{7.23}$$

we obtain P_{01} in terms of an integral of the *Rayleigh probability distribution* $f_0(R)$,

$$P_{01} = \int_{R_T}^{\infty} f_0(R)\, dR = \int_{R_T}^{\infty} \frac{4R}{N_0 T} \exp \frac{-2R^2}{N_0 T}\, dR = \exp \frac{-2R_T{}^2}{N_0 T} = \alpha \tag{7.24}$$

from which R_T is determined as

$$R_T = \sqrt{\frac{N_0 T}{2} \ln \frac{1}{\alpha}} \tag{7.25}$$

We see, therefore, that the threshold is dependent on the noise spectral density, the observation time, and the false-alarm probability. Significantly, with the Neyman-Pearson radar-design philosophy, the

[1] In basing detection on only one pulse, any strictly monotonically increasing function of the envelope detector serves as an equivalent optimal detector. This is not true, however, for detection based upon more than one pulse.

[2] Neglecting double frequency terms.

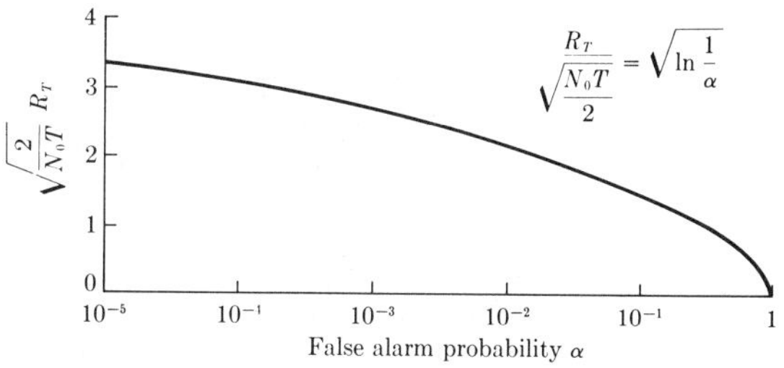

Fig. 7.3 Threshold setting versus false-alarm probability.

threshold setting is independent of the signal amplitude. The threshold R_T is plotted against the false-alarm probability in Fig. 7.3.

With this threshold setting we can determine the detection probability for the optimal receiver. The probability that the envelope correlation detector exceeds the threshold under the hypothesis *signal present* can be written as

$$\begin{aligned} P_{11} &= \Pr\,[R > R_T | y(t) = s(t,\phi) + n(t)] \\ &= \Pr\left[\sqrt{e_c^2 + e_s^2} > \sqrt{\frac{N_0 T}{2}} \ln \frac{1}{\alpha} \,\middle|\, y(t) \right. \\ &\qquad\qquad\qquad \left. = A \cos\,(\omega_c t + \phi) + n(t)\right] \end{aligned} \tag{7.26}$$

To evaluate P_{11} we first fix the phase ϕ of the signal and then determine the conditional density $f(e_c,e_s|\phi)$. For a given ϕ this conditional density is a gaussian with conditional means

$$E(e_c|\phi) = \frac{AT}{2} \cos \phi \qquad E(e_s|\phi) = \frac{AT}{2} \sin \phi \tag{7.27}$$

and conditional second-order statistics given by

$$\sigma^2_{(e_c|\phi)} = \sigma^2_{(e_s|\phi)} = \frac{N_0 T}{4} \tag{7.28}$$

and covariance zero. The conditional density is therefore

$$\begin{aligned} f(e_c,e_s|\phi) = \frac{2}{\pi N_0 T} \exp \left\{ \frac{-2}{N_0 T} \left[\left(e_c - \frac{AT \cos \phi}{2} \right)^2 \right. \right. \\ \left. \left. + \left(e_s - \frac{AT \sin \phi}{2} \right)^2 \right] \right\} \end{aligned}$$

from which we obtain

$$\begin{aligned} f(e_c,e_s) &= \frac{1}{2\pi} \int_0^{2\pi} f(e_c,e_s|\phi)\, d\phi \\ &= \frac{2}{\pi N_0 T} \exp \frac{-2(e_c^2 + e_s^2)}{N_0 T} \exp \frac{-A^2 T}{2N_0} I_0 \left(\frac{2A}{N_0} \sqrt{e_c^2 + e_s^2} \right) \end{aligned}$$

Transforming to obtain R, the test statistic in which we are interested, and again using the transformation in (7.23), we obtain the joint density of R and Δ under the hypothesis *signal present*. Integration with respect to Δ then results in the density $f_1(R)$ of R under the hypothesis *signal present*:

$$f_1(R) \triangleq \frac{4R}{N_0T} \exp \frac{-2R^2}{N_0T} \exp \frac{-\mathcal{E}}{N_0} I_0\left(\frac{2A}{N_0} R\right) \tag{7.29}$$

when $\mathcal{E} \triangleq A^2T/2$ is the total receiver signal energy. The density of the detector output with signal present in (7.29) is the *rician probability density function*. Finally,

$$\begin{aligned} P_{11} &= \int_{R_T}^{\infty} f_1(R)\, dR \\ &= \int_{\sqrt{\frac{N_0T}{2}\ln\frac{1}{\alpha}}}^{\infty} \frac{4R}{N_0T} \exp \frac{-2R^2}{N_0T} \exp \frac{-\mathcal{E}}{N_0} I_0\left(\frac{2A}{N_0} R\right) dR \\ &= \int_{\sqrt{2\ln\frac{1}{\alpha}}}^{\infty} x \exp\left[-\frac{1}{2}\left(x^2 + \frac{2\mathcal{E}}{N_0}\right)\right] I_0\left(\sqrt{\frac{2\mathcal{E}}{N_0}}\, x\right) dx \end{aligned} \tag{7.30}$$

This can be expressed in terms of the *Q function*, which is defined (see Appendix D) as

$$Q(\gamma,\beta) \triangleq \int_{\beta}^{\infty} x \exp\left[-\tfrac{1}{2}(x^2 + \gamma^2)\right] I_0(\gamma x)\, dx$$

With the necessary assignments, the detection probability in terms of the Q-function is

$$P_{11} = Q\left(\sqrt{\frac{2\mathcal{E}}{N_0}}, \sqrt{2 \ln \frac{1}{\alpha}}\right) \tag{7.31}$$

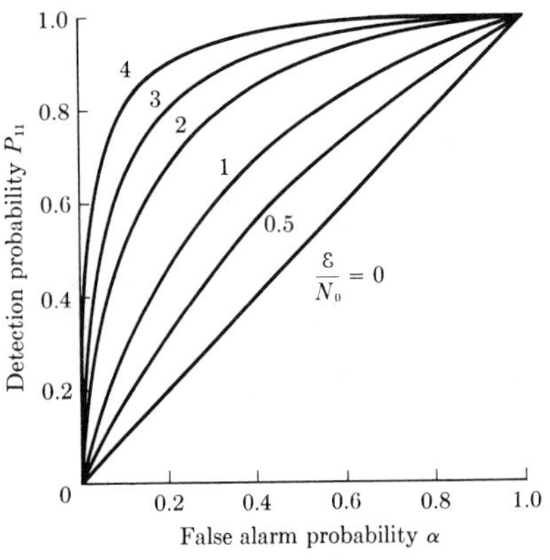

Fig. 7.4 Receiver operating characteristics for the detection system in Example 7.1.

The detection probability P_{11} is plotted against the false-alarm probability α for various signal-to-noise ratios $\mathcal{E}/N_0$ in Fig. 7.4. This is the *receiving operating characteristic* of the detectors.

The effectiveness of a detection system is often conveniently described in terms of the minimum signal-to-noise ratio necessary to attain a specified detection probability for a given false-alarm probability. In the above example (Fig. 7.4), for instance, for a FAP of 0.1, if a detection probability of 0.8 is required, the signal-to-noise ratio $\mathcal{E}/N_0$ of the *minimal detectable signal* is 4. Or what is more, with the transmitted average power and the signal duration time specified, we can determine via the radar range equation the maximum range for which such a signal may be detected with this detection probability.

Example 7.2 Detection with the complex-envelope representation We now consider the optimal detection of the somewhat more general signal

$$s(t,\phi) = A(t) \cos [\omega_c t + \theta(t) + \phi] \qquad 0 \le t \le T \tag{7.32}$$

where the amplitude and phase fluctuations, $A(t)$ and $\theta(t)$, are assumed to be narrowband with respect to the carrier frequency ω_c. The carrier frequency is assumed known and unchanged from that which is transmitted. This corresponds to detecting a stationary target, where the doppler frequency shift is zero. We again assume known arrival time and unknown reference phase ϕ, which is a random variable uniformly distributed over the interval $(-\pi,\pi)$. The convenient representation of the complex envelope will be introduced in this example, and we shall see that the addition of the amplitude and phase variations in (7.32) does not introduce any mathematical complications.

The notion of the complex envelope is easily introduced by noting that

$$\begin{aligned} s(t,\phi) &= A(t) \cos [\omega_c t + \theta(t) + \phi] \\ &= \text{Re}\,\{Z(t) \exp [j(\omega_c t + \phi)]\} \end{aligned} \tag{7.33}$$

where

$$Z(t) \triangleq A(t) \exp j\theta(t)$$

$Z(t)$ is known as the *complex envelope* of the signal, and Re means the *real part of*. The actual envelope or amplitude modulation is

$$A(t) = |Z(t)|$$

Although the representation in (7.33) is valid for any bandwidth of the amplitude and phase modulations, the interpretation has more meaning when they are narrowband with respect to ω_c, for then the representation is a product of the slowly varying part of the overall waveform, namely, the complex envelope and the carrier-frequency terms $\exp j\omega_c t$.

The optimal detector is expressable in terms of the complex envelope of the received waveform $y(t)$ and the complex envelope $Z(t)$. Consider the integral

$$\int_0^T y(t)s(t;\psi)\,dt$$

in the general expression for the optimal detector in (7.9). Substitution yields

$$\int_0^T y(t)\,\mathrm{Re}\,\{Z(t)\exp[j(\omega_c t + \psi)]\}\,dt = e_c \cos\psi - e_s \sin\psi \qquad (7.34)$$

where

$$e_c \triangleq \int_0^T y(t)\,\mathrm{Re}\,[Z(t)\exp j\omega_c t]\,dt$$

and

$$e_s \triangleq \int_0^T y(t)\,\mathrm{Im}\,[Z(t)\exp j\omega_c t]\,dt \qquad (7.35)$$

where *Im* means *the imaginary part of*. As in Example 7.1, the optimal detector can be reduced to forming

$$R \triangleq \sqrt{e_c^2 + e_s^2} \qquad (7.36)$$

and comparing R to the threshold R_T. The test statistic R can also be expressed as

$$\begin{aligned} R &= |e_c + je_s| = \left| \int_0^T y(t)\{\mathrm{Re}\,[Z(t)\exp j\omega_c t] + j\,\mathrm{Im}\,[Z(t)\exp j\omega_c t]\}\,dt \right| \\ &= \left| \int_0^T y(t)Z(t)\exp j\omega_c t\,dt \right| \end{aligned} \qquad (7.37)$$

Thus the optimal decision function has been expressed in terms of the complex envelope of the transmitted waveform. We can also express $y(t)$ in terms of an amplitude modulation $V(t)$ and phase modulation $\Delta(t)$ with center frequency ω_c:

$$\begin{aligned} y(t) &= V(t)\cos[\omega_c t + \Delta(t)] \\ &= \mathrm{Re}\,[E(t)\exp j\omega_c t] \end{aligned} \qquad (7.38)$$

where

$$E(t) \triangleq V(t)\exp[j\,\Delta(t)]$$

is defined as the complex envelope of the received waveform. If we substitute into (7.37), the detector can be expressed as[1]

$$R = \left| \int_0^T \operatorname{Re}\,[E(t) \exp j\omega_c t] Z(t) \exp (j\omega_c t)\, dt \right|$$
$$= \frac{1}{2} \left| \int_0^T \overline{E(t)} Z(t)\, dt \right| \tag{7.39}$$

where the overbar denotes the *complex conjugate of*. Therefore the optimal detector is the absolute value of the inner product of the complex envelope of the transmitted signal with the conjugated complex envelope of the received waveform. The significance of complex envelope is that it takes into account the slow variations of the phase modulation as well as the amplitude modulation. Hence the optimal detector is expressible in terms of only the slowly varying portions of the transmitted signal and the received waveform. It can be readily shown that R also has the representation

$$R = \frac{1}{2} \left(\left\{ \int_0^T V(t) A(t) \cos\,[\Delta(t) - \theta(t)]\, dt \right\}^2 + \left\{ \int_0^T V(t) A(t) \sin\,[\theta(t) - \Delta(t)]\, dt \right\}^2 \right)^{\frac{1}{2}} \tag{7.40}$$

The system performance characteristics are as in Example 7.1.

Example 7.3 Binary noncoherent communication system As a final example let us consider a communication environment in which both hypotheses are composite. More specifically, let us consider the binary communication problem in which the observed waveform is of the form

$$y(t) = A_i(t) \cos\,[\omega_c t + \theta_i(t) + \phi] + n(t) \qquad 0 \leq t \leq T;\ i = 0, 1 \tag{7.41}$$

where $n(t)$ is white gaussian noise with one-sided spectral density N_0. It is assumed that no knowledge of the phase of the transmitted waveforms is available, so that its a priori probability density function is

$$p(\phi) = \frac{1}{2\pi} \qquad -\pi \leq \phi < \pi$$

By "noncoherent" we mean that this phase information is unavailable.

The transmittable amplitude modulations $\{A_i(t);\ i = 0, 1\}$ and phase modulations $\{\theta_i(t);\ i = 0, 1\}$ are assumed to be narrowband with respect to the rf carrier ω_c, and their a priori probabilities are π_0 and π_1, respectively.

In view of the development at the beginning of the chapter, it is

[1] The $2\omega_c$ terms have been assumed negligible. This follows if the noise is assumed to be of sufficient bandwidth that the white-noise assumption is applicable and yet sufficiently narrowband that the noise spectral density is negligible at and beyond $2\omega_c$. The usual insertion of predetection filters makes this assumption realistic.

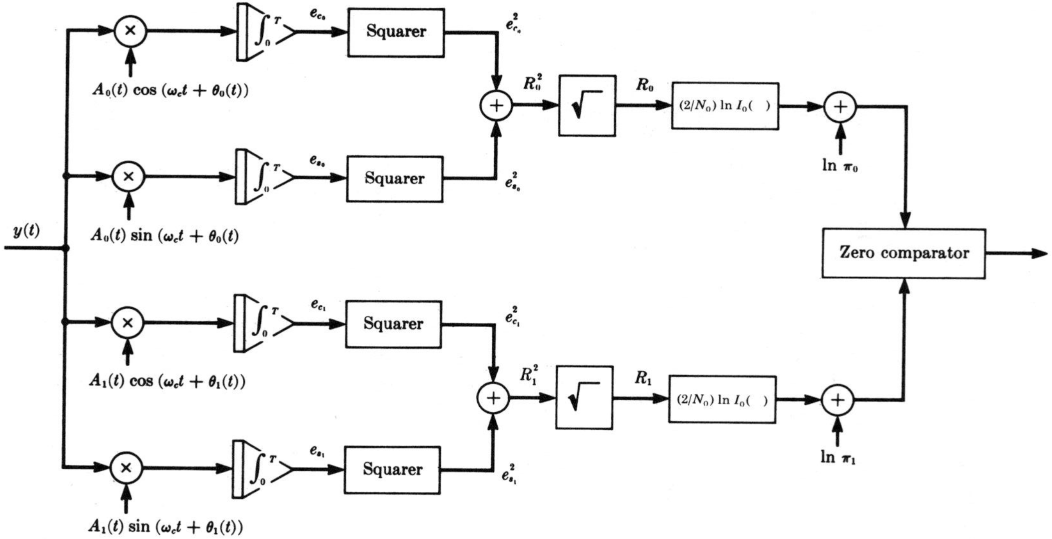

Fig. 7.5 Optimal noncoherent binary receiver.

readily apparent that with probability of error as the performance criterion the optimal decision function forms the generalized likelihood ratio

$$L(y(t); 0 \leq t \leq T) = \frac{(1/2\pi) \int_{-\pi}^{\pi} d\psi \exp\left((-1/N_0) \int_0^T \{y(t) - A_1(t) \cos [\omega_c t + \theta_1(t) + \psi]\}^2 dt\right)}{(1/2\pi) \int_{-\pi}^{\pi} d\psi \exp\left((-1/N_0) \int_0^T \{y(t) - A_0(t) \cos [\omega_c t + \theta_0(t) + \psi]\}^2 dt\right)}$$

and decides a_1, or a 1, if $L(y)$ is greater than the threshold π_0/π_1 and a_0, or a 0, if $L(y)$ is less than the threshold. We simplify as in Example 7.2 and express the optimal receiver equivalently as

$$\frac{(1/2\pi) \int_{-\pi}^{\pi} d\psi \exp\left\{(2/N_0) \int_0^T y(t)A_1(t) \cos [\omega_c t + \theta_1(t) + \psi]\, dt\right\}}{(1/2\pi) \int_{-\pi}^{\pi} d\psi \exp\left\{(2/N_0) \int_0^T y(t)A_0(t) \cos [\omega_c t + \theta_0(t) + \psi]\, dt\right\}} \gtrless \frac{\pi_0}{\pi_1}$$

where we have assumed that both transmittable waveforms have the same average power.

In terms of modified Bessel functions, and finally taking logarithms, the optimal receiver becomes

$$\ln I_0\left(\frac{2}{N_0} R_1\right) - \ln I_0\left(\frac{2}{N_0} R_0\right) \gtrless \ln \frac{\pi_0}{\pi_1} \tag{7.42}$$

where

$$\begin{aligned} R_0 &\triangleq \sqrt{e_{c_0}{}^2 + e_{s_0}{}^2} \\ R_1 &\triangleq \sqrt{e_{c_1}{}^2 + e_{s_1}{}^2} \\ e_{c_i} &\triangleq \int_0^T y(t)A_i(t) \cos [\omega_c t + \theta_i(t)]\, dt \qquad i = 0, 1 \\ e_{s_i} &\triangleq \int_0^T y(t)A_i(t) \sin [\omega_c t + \theta_i(t)]\, dt \qquad i = 0, 1 \end{aligned} \tag{7.43}$$

A block diagram of the optimal receiver is given in Fig. 7.5.

If the a priori probabilities are equally likely, the threshold level becomes zero. Then, since ln and I_0 are both strictly monotonically increasing functions, their inverses exist and the simplified optimal receiver results:

$$\begin{aligned} &\text{If } R_1 > R_0 \qquad \text{decide } a_1 \\ &\text{If } R_1 < R_0 \qquad \text{decide } a_0 \end{aligned} \tag{7.44}$$

We now evaluate the probability of detection for the optimal receiver given by (7.44), where the a priori probabilities are equally likely and the two transmittable waveforms have the same energy $\mathcal{E}$, given by

$$\mathcal{E} \triangleq E\left(\int_0^T \{A_i(t) \cos [\omega_c t + \theta_i(t) + \phi]\}^2 \, dt\right) = \frac{1}{2}\int_0^T A_i^2(t) \, dt \qquad i = 0, 1 \quad (7.45)$$

We shall also assume that the transmittable waveforms are orthogonal; that is,

$$\begin{aligned}
\int_0^T A_0(t) \cos [\omega_c t + \theta_0(t)] A_1(t) \cos [\omega_c t + \theta_1(t)] \, dt &= 0 \\
\int_0^T A_0(t) \cos [\omega_c t + \theta_0(t)] A_1(t) \sin [\omega_c t + \theta_1(t)] \, dt &= 0 \\
\int_0^T A_0(t) \sin [\omega_c t + \theta_0(t)] A_1(t) \cos [\omega_c t + \theta_1(t)] \, dt &= 0 \\
\int_0^T A_0(t) \sin [\omega_c t + \theta_0(t)] A_1(t) \sin [\omega_c t + \theta_1(t)] \, dt &= 0
\end{aligned} \qquad (7.46)$$

In addition, because of symmetry, P_{ii} = Pr [deciding $a_i | s_i(t,\phi)$ was transmitted] is independent of i. Hence we need evaluate only P_{00} or P_{11}, since

$$P_d = P_{00} = P_{11} \qquad (7.47)$$

Let us choose P_{00}.

The method usually employed in evaluating the system performance is to determine the probability density function of the random vector $(e_{c_0}, e_{s_0}, e_{c_1}, e_{s_1})^T$ conditional on the random phase and then average over ϕ. With this density the probability of detection can be calculated. Since the noise is additive, gaussian, and independent of the signals, the density of

$$\begin{bmatrix} e_{c_0} \\ e_{s_0} \\ e_{c_1} \\ e_{s_1} \end{bmatrix}$$

conditioned on $s_0(t,\phi)$ is gaussian.

From (7.43), the conditional mean[1] vector is

$$E\left[\begin{array}{c|c} e_{c_0} & \\ e_{s_0} & s_0(t,\phi) \\ e_{c_1} & \\ e_{s_1} & \end{array}\right] = \begin{bmatrix} \mathcal{E} \cos \phi \\ -\mathcal{E} \sin \phi \\ 0 \\ 0 \end{bmatrix}$$

and the conditional covariance matrix is

$$\frac{\mathcal{E}N_0}{2} \mathbf{I}$$

[1] The $2\omega_c$ terms are negligible and are therefore neglected throughout.

where **I** is the 4-by-4 identity matrix.

The probability of detection for a given ϕ can thus be expressed as

$$P_d(\phi) \triangleq \Pr\,[R_0 > R_1 | y(t) = s_0(t,\phi) + n(t)]$$
$$= \Pr\,[\sqrt{e_{c_0}{}^2 + e_{s_0}{}^2} > \sqrt{e_{c_1}{}^2 + e_{s_1}{}^2}\,|y(t) = s_0(t,\phi) + n(t)]$$
$$= \iint_{-\infty}^{\infty} de_{c_0}\,de_{s_0} \iint_{C} de_{c_1}\,de_{s_1}$$
$$\frac{\exp\{(-1/\mathcal{E}N_0)[(e_{c_0} - \mathcal{E}\cos\phi)^2 + (e_{s_0} + \mathcal{E}\sin\phi)^2 + e_{c_1}{}^2 + e_{s_1}{}^2]\}}{(2\pi)^2\left(\dfrac{\mathcal{E}N_0}{2}\right)^2} \tag{7.48}$$

where C is a circle centered at the origin with radius $\sqrt{e_{c_0}{}^2 + e_{s_0}{}^2}$. We perform the integration with respect to e_{c_1} and e_{s_1} and obtain

$$P_d(\phi) = 1 - \iint_{-\infty}^{\infty} de_{c_0}\,de_{s_0}$$
$$\frac{\exp\{(-1/\mathcal{E}N_0)[(e_{c_0} - \mathcal{E}\cos\phi)^2 + (e_{s_0} + \mathcal{E}\sin\phi)^2]\}}{2\pi(\mathcal{E}N_0/2)}$$
$$\exp\frac{-(e_{c_0}{}^2 + e_{s_0}{}^2)}{\mathcal{E}N_0} \tag{7.49}$$

The integral in (7.49) can be written in terms of the probability of error as

$$P_e(\phi) = E\left(\exp\frac{-e_{c_0}{}^2}{\mathcal{E}N_0}\right)E\left(\exp\frac{-e_{s_0}{}^2}{\mathcal{E}N_0}\right)$$

These integrals can be carried out by appropriately completing the square,[1] with the result that

$$P_e(\phi) = \frac{\exp\dfrac{-\mathcal{E}\cos^2\phi}{2N_0}}{\sqrt{2}}\,\frac{\exp\dfrac{-\mathcal{E}\sin^2\phi}{2N_0}}{\sqrt{2}} = \frac{1}{2}\exp\frac{-\mathcal{E}}{2N_0}$$

Thus $P_e(\phi)$ is independent of ϕ. Hence the probability of error is

$$P_e = \frac{1}{2\pi}\int_0^{2\pi} P_e(\phi)\,d\phi = P_e(\phi)$$

or finally,

$$P_e = \frac{1}{2}\exp\frac{-\mathcal{E}}{2N_0} \tag{7.50}$$

when the signal waveforms are equally likely and have the same energy.

[1] If x is a gaussian random variable with mean m_x and variance $\sigma_x{}^2$, then, if $a < 1/2\sigma_x{}^2$,

$$E(\exp ax^2) = \frac{1}{\sqrt{1 - 2a\sigma_x{}^2}}\exp\frac{am_x{}^2}{1 - 2a\sigma_x{}^2}$$

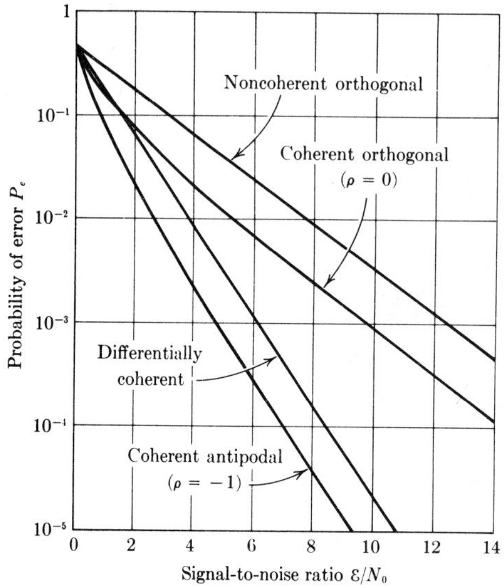

Fig. 7.6 Performance comparison of several binary systems.

Figure 7.6 presents a comparison of certain of the binary communication systems that have been discussed. The probability of error for the binary coherent systems with $\rho = -1$ (antipodal) and $\rho = 0$ (orthogonal) were derived in Chap. 4. The noncoherent binary system with orthogonal signals has system performance given by (7.50). The differentially coherent system is described in Prob. 7.14.

PROBLEMS

7.1 What is the optimal decision function which results from the minimization of Eq. (7.3) when an arbitrary cost matrix is used? Is the result still a likelihood ratio? What is the threshold?

7.2 Verify Eqs. (7.20), (7.21), and (7.24).

7.3 Verify Eqs. (7.27) to (7.30).

7.4 Verify the optimal detector representation in (7.39).

7.5 Verify that the optimal detector representation as given by (7.40) can be obtained from the representation in (7.39), and that it can also be directly obtained from the general representation in (7.9).

7.6 What is the optimal receiver representation of the noncoherent binary communication system in Example 7.3 in terms of complex envelopes?

7.7 (*a*) Determine the alteration that is necessary in the comparator in Example 7.3 when the two transmitted waveforms do not have the same energy.

(*b*) When the a priori probabilities are equally likely, is knowledge of the received signal energy necessary in order to implement the optimal receiver in Example 7.3? What if the a priori probabilities are not equally likely?

7.8 (*a*) Show that the optimal radar detector, as developed in the Example 7.2, is unchanged if the observation interval is assumed known but the signal amplitude is a random variable; that is, let

$$s(t;\phi;V) = VA(t)\cos[\omega_c t + \theta(t) + \phi] \qquad 0 \le t \le T$$

where the normalized amplitude modulation $A(t)$ and phase modulation $\theta(t)$ are assumed known, and V is a nonnegative random variable with a priori probability density function $p(V)$.

(*b*) Determine the detection probability for this system.

Hint: The integral

$$\int_0^\infty p(V) I_0\left(\frac{2V}{N_0} R\right) \exp(-b^2V^2)\, dV$$

is a strictly monotonically increasing function of R.

7.9 *Signal detection based on K received pulses* Consider a detection problem in which the presence or absence of a signal in the presence of additive noise is to be based on observing K echoes or returns, each of T sec duration. Assume that the observed waveform is

$$y_k(t) = A_k \cos(\omega_c t + \phi_k) + n(t) \qquad T_k \le t \le T_k + T,\ k = 1, \ldots, K$$

where the time intervals $[T_k, T_k + T]$ are mutually nonoverlapping, the $\{\phi_k;\ k = 1, \ldots, K\}$ are mutually independent, uniformly distributed random phases over $(-\pi,\pi)$, and the $\{A_k;\ k = 1, \ldots, K\}$ and $\{T_k;\ k = 1, \ldots, K\}$ are assumed known. $n(t)$ is white gaussian noise with two-sided spectral density $N_0/2$.

(*a*) Using the Neyman-Pearson criterion, show that the optimal detector forms the test statistic

$$\sum_{k=1}^{K} \ln I_0\left(\frac{2A_kR_k}{N_0}\right)$$

which is compared to a threshold where

$$R_k \triangleq \sqrt{e_{c_k}^2 + e_{s_k}^2}$$

in which

$$e_{c_k} \triangleq \int_{T_k}^{T_k+T} y_k(t)\cos\omega_c t\, dt$$

and

$$e_{s_k} \triangleq \int_{T_k}^{T_k+T} y_k(t)\sin\omega_c t\, dt$$

(*b*) If the $\{A_k\}$ are all equal, show that at small signal-to-noise ratio the optimal detector can be approximated by the square-law detector

$$\sum_{k=1}^{K} R_k^2$$

and at large signal-to-noise ratios by the linear detector

$$\sum_{k=1}^{K} R_k$$

(c) Determine the threshold setting for both the square-law and linear detector in (b) for a given false-alarm probability.

(d) Determine the detection probability of the square-law detectors (see Helstrom [7.1]).

7.10 Consider again Prob. 7.9 and state which of the following two detectors will have the better performance.

(a) In system A the decision *signal present* or *signal absent* is based on one pulse ($K = 1$) whose duration is $2T$ sec.

(b) In system B the decision *signal present* or *signal absent* is based on two pulses ($K = 2$) each of duration T sec. Note that the total received signal energy is the same in both systems.

7.11 Consider the binary communication system where the received waveform is

$$y(t) = s_i(t,\phi) + n(t) \qquad 0 \le t \le T$$

where

$$s_i(t,\phi) = A_i(t) \cos [\omega_c t + \theta_i(t) + \phi] \qquad i = 1, 2, 3, 4$$

The $\{A_i(t)\}$ and $\{\theta_i(t)\}$ are narrowband with respect to ω_c. $s_1(t;\phi)$ or $s_2(t;\phi)$ corresponds to a 0, and $s_3(t;\phi)$ or $s_4(t;\phi)$ corresponds to a 1.

(a) Determine the optimal receiver which minimizes the probability of error for the cases where (1) the system is coherent and (2) the system is noncoherent and ϕ is uniformly distributed over $(-\pi,\pi)$, with $n(t)$ assumed to be white gaussian noise with two-sided spectral density $N_0/2$.

(b) Determine the probability of error when the four transmittable waveforms are equally likely and mutually orthogonal for the coherent and noncoherent cases, and compare the results.

7.12 Consider again the binary communication problem in Ex. 7.3. This time assume that the received carrier phase ϕ has probability density

$$p(\phi) = \frac{\exp(\alpha \cos \phi)}{2\pi I_0(\alpha)} \qquad -\pi \le \phi \le \pi$$

This is the probability density function that results when the receiver tracks the incoming random phase with a phase-locked loop.

(a) Show that the optimal receiver is the one that forms the test statistic

$$\ln I_0\left(\sqrt{\left(\alpha + \frac{2}{N_0} e_{c_1}\right)^2 + \left(\frac{2}{N_0} e_{s_1}\right)^2}\right) - \ln I_0\left(\sqrt{\left(\alpha + \frac{2}{N_0} e_{c_0}\right)^2 + \left(\frac{2}{N_0} e_{s_0}\right)^2}\right)$$

and decides that signal $s_1(t,\phi)$ was transmitted if this is greater than the threshold $\ln(\pi_0/\pi_1)$.

(b) As $\alpha \to 0$, what does $p(\phi)$ become and what is the form of the receiver?

(c) As $\alpha \to \infty$, what does $p(\phi)$ become and what is the form of the receiver?

(d) Knowledge of which parameter is required by the optimal receiver in this problem?

(e) If the two signal waveforms have the same energy $\mathcal{E}$, are equally likely, and are antipodal [that is, $s_0(t,\phi) = -s_1(t,\phi)$], show that the probability of error is

$$P_e = \int_{-\pi}^{\pi} \frac{\exp(\alpha \cos \phi)}{2\pi I_0(\alpha)} \operatorname{erfc}\left(\sqrt{\frac{2\mathcal{E}}{N_0}} \cos \phi\right) d\phi$$

(f) If the two signal waveforms have the same energy $\mathcal{E}$, are equally likely, and are orthogonal, show that the probability of error is

$$P_e = \int_{-\pi}^{\pi} \frac{\exp(\alpha \cos \phi)}{2\pi I_0(\alpha)} P_e(\phi)\, d\phi$$

where

$$P_e(\phi) = Q\left(\frac{\delta_1}{\sqrt{2}}, \frac{\delta_0}{\sqrt{2}}\right) - \frac{1}{2}\exp\left[-\tfrac{1}{4}(\delta_0^2 + \delta_1^2)\right] I_0\left(\frac{\delta_0 \delta_1}{2}\right)$$

$$\delta_1 = \frac{\alpha}{\sqrt{2\mathcal{E}/N_0}}$$

and

$$\delta_0 = \frac{\sqrt{\alpha^2 + (2\mathcal{E}/N_0)^2 + 4\alpha(\mathcal{E}/N_0)\cos\phi}}{\sqrt{2\mathcal{E}/N_0}}$$

Here Q is *Marcum's Q function* (see Appendix D). *Hint:* See Viterbi [7.4].

7.13 Consider the following binary communication system, which consists of a data signal and a phase reference signal. Assume that the received waveform which contains the data signal is

$$y(t) = A_i(t) \cos[\omega_c t + \theta_i(t) + \phi] + n(t) \qquad 0 \le t \le T,\ i = 0, 1$$

where $\{A_i(t); i = 0, 1\}$ and $\{\theta_i(t); i = 0, 1\}$ are narrowband waveforms with respect to ω_c. The signal waveforms are equally likely, have the same energy $\mathcal{E}$, and their inner product is $\rho\mathcal{E}$ for $-1 \le \rho \le 1$. The noise $n(t)$ is gaussian with two-sided spectral density $N_0/2$ watts/Hz.

The observed waveform which contains the phase reference signal is

$$y_r(t) = \sqrt{\frac{2\mathcal{E}_r}{T}} \cos(\omega_c t + \phi) + n(t) \qquad -KT \le t \le 0$$

where ϕ is a uniformly distributed random variable over $(-\pi,\pi)$, $\mathcal{E}_r$ is the received reference signal energy in T sec and K is a positive integer.

(*a*) Find the receiver operation that minimizes the probability of error.

(*b*) Determine the probability of error for the optimal receiver.

(*c*) For a given $\mathcal{E}$, $\mathcal{E}_r$, and N_0, which choice of ρ minimizes the probability of error?

7.14 In a binary communication system let the received waveform be

$$y(t) = A\cos(\omega_c t + \theta_i + \phi) + n(t) \qquad 0 \le t \le T$$

where $\theta_0 = 0$ and $\theta_1 = \pi$ with equal probability, ϕ is a random variable uniformly distributed over $(-\pi,\pi)$, and $n(t)$ is white gaussian noise with spectral density $N_0/2$. Let a 1 be conveyed to the receiver by a change in phase from that transmitted during the previous T-sec interval, and let a 0 be conveyed by no change in phase. Thus a decision is made after every T-sec interval on the basis of the observed waveform over the previous $2T$ sec.

When the optimal receiver is employed, show that the probability of error for this system is given by[1]

$$P_e = \frac{1}{2}\exp\frac{-\mathcal{E}}{N_0}$$

[1] Intersymbol dependence is neglected here (see, for example, Bussgang and Leiter [7.6], [7.7]).

REFERENCES

7.1 Helstrom, C.: "Statistical Theory of Signal Detection," Pergamon Press, New York, 1960.

7.2 Middleton, D.: "Introduction to Statistical Communication Theory," McGraw-Hill, New York, 1960.

7.3 Hancock, J. C., and P. A. Wintz: "Signal Detection Theory," McGraw-Hill, New York, 1966.

7.4 Viterbi, A. J.: "Principles of Coherent Communication," McGraw-Hill, New York, 1966.

7.5 Wozencraft, J. M., and I. M. Jacobs: "Principles of Communication Engineering," Wiley, New York, 1965.

7.6 Bussgang, J. J., and M. Leiter: Error Rate Approximation for Differential Phase-Shift Keying, *IEEE Trans. Commun. Systems*, vol. CS-2, pp. 18–27, March, 1964.

7.7 Bussgang, J. J., and M. Leiter: Phase Shift Keying with a Transmitted Reference, *IEEE Trans. Commun. Technol.*, vol. Com 14, no. 1, pp. 14–22, February, 1966.

7.8 Wainstein, L. A., and Zubakov, V. D.: "Extraction of Signals from Noise," Prentice-Hall, Englewood Cliffs, N.J., 1962.

8

Detection and Communication in Colored Noise

In this chapter we shall consider the binary decision problem further. The detection and communication problems will both be developed, where the additive noise is still gaussian but is no longer assumed to have flat spectral density for all frequencies. Instead, the zero-mean noise is assumed to have covariance function $R_n(t,\tau)$, which is well behaved and strictly positive-definite. Such noises are often called *nonwhite*, or *colored*.

One method of determining the optimal receiver in such cases is by sampling the received waveform m times over the observation interval $[0,T]$, determining the optimal receiver for the resulting vector model, and finally, letting the number of time samples increase and writing the limit in integral form. However, because of the correlation that now exists between the noise samples, and particularly because the time interval $[0,T]$ consists of an uncountably infinite set, mathematical complications arise in taking the limit.

The second method is to use the eigenfunctions of the noise covariance function as a set of basis functions and the first m coefficients of these eigenfunctions as the observed vector. As the number of coefficients is increased, an integral representation results in the limit as $m \to \infty$ which is the same as that obtained by the sampling method. The passage to the limit is somewhat simpler here, since the coefficients of the basis functions are independent random variables when the eigenfunctions are used as the basis functions and the process is gaussian.

First we shall consider the detection of a known signal in additive colored gaussian noise and obtain the optimal detector by both the methods indicated above. Then we shall examine some coherent and noncoherent binary communication problems.

8.1 DETECTION IN COLORED NOISE

Consider a detection problem in which the received real-valued waveform is

$$y(t) = as(t) + n(t) \qquad 0 \le t \le T \tag{8.1}$$

where $s(t)$ is a deterministic signal known to the receiver and $n(t)$ is a zero mean gaussian stochastic process with covariance function[1] $R_n(t,\tau)$. At time $t = T$ the receiver is to decide between $a = a_0$ and $a = 0$, using the Neyman-Pearson criterion. Equivalently, a decision must be made between the hypothesis that $y(t)$ consists of signal plus noise and the alternative that $y(t)$ consists of noise only.

We first determine the optimal receiver by sampling the received waveform m times, $\{t_1,t_2, \ldots ,t_m\}$, with the samples spaced uniformly over the observation interval. Let

$$\mathbf{Y} \triangleq \begin{bmatrix} y(t_1) \\ \cdot \\ \cdot \\ \cdot \\ y(t_m) \end{bmatrix} \qquad \mathbf{S} \triangleq \begin{bmatrix} s(t_1) \\ \cdot \\ \cdot \\ \cdot \\ s(t_m) \end{bmatrix} \qquad \mathbf{N} \triangleq \begin{bmatrix} n(t_1) \\ \cdot \\ \cdot \\ \cdot \\ n(t_m) \end{bmatrix}$$

The noise vector $\mathbf{N}$ has covariance matrix given by the m-by-m matrix

$$\mathbf{R}_n = \{R_n(t_i,t_j)\}$$

Using the fact that the Neyman-Pearson criterion results in an optimal receiver which forms the likelihood ratio to be compared to a

[1] No assumption about the stationarity of the noise process is needed in this section. $R_n(t,\tau)$ is real-valued and symmetric, however, and therefore self-adjoint.

threshold, the optimal receiver can be explicitly represented[1] as

$$L_m(\mathbf{Y}) = \frac{f(\mathbf{Y}|\mathbf{Y} = a_0\mathbf{S} + \mathbf{N})}{f(\mathbf{Y}|\mathbf{Y} = \mathbf{N})} = \frac{f_N(\mathbf{Y} - a_0\mathbf{S})}{f_N(\mathbf{Y})}$$

$$= \frac{[1/(\sqrt{2\pi})^m \sqrt{|\mathbf{R}_n|}] \exp[-\frac{1}{2}(\mathbf{Y} - a_0\mathbf{S})^T\mathbf{R}_n^{-1}(\mathbf{Y} - a_0\mathbf{S})]}{[1/(\sqrt{2\pi})^m \sqrt{|\mathbf{R}_n|}] \exp(-\frac{1}{2}\mathbf{Y}^T\mathbf{R}_n^{-1}\mathbf{Y})} \gtrless T_T \quad (8.2)$$

where the threshold T_T will be determined from the false-alarm probability. Taking logarithms and simplifying, we obtain

$$\ln L_m(\mathbf{Y}) = a_0\mathbf{S}^T\mathbf{R}_n^{-1}\mathbf{Y} - \frac{a_0^2}{2}\mathbf{S}^T\mathbf{R}_n^{-1}\mathbf{S} \gtrless \ln T_T \quad (8.3)$$

Since $\mathbf{S}$ is known, the optimal detector can be simplified to

$$\text{If } \mathbf{S}^T\mathbf{R}_n^{-1}\mathbf{Y} > R_T \qquad \text{decide } \textit{signal is present} \quad (8.4)$$

where the threshold R_T has absorbed all the deterministic terms in (8.3). Since R_T will be determined from the false-alarm probability, we see that the optimal detector does not require knowledge of the signal strength given by a_0.

Let us now define the vector

$$\mathbf{G} \triangleq \begin{bmatrix} g(t_1) \\ \cdot \\ \cdot \\ \cdot \\ g(t_m) \end{bmatrix} \quad (8.5)$$

as the solution of the vector equation

$$\mathbf{R}_n\mathbf{G}\,\Delta t = \mathbf{S} \quad (8.6)$$

where Δt is the distance between the uniformly spaced samples; that is,

$$\Delta t = t_{i+1} - t_i \qquad i = 0, \ldots, m - 1$$

Since $\mathbf{R}_n$ is presently being assumed to be positive-definite, (8.6) can be expressed as

$$\mathbf{G}\,\Delta t = \mathbf{R}_n^{-1}\mathbf{S}$$

[1] The explicit representation requires the assumption that the covariance matrix $\mathbf{R}_n$ be positive-definite. We make this assumption here with the realization that any difficulties encountered concerning the positive-definiteness of the noise covariance matrix are overcome in the second method.

which, when substituted into (8.4), gives the optimal detector in terms of **G** as

$$\text{If } \mathbf{G}^T\mathbf{Y}\,\Delta t > R_T \qquad \text{decide } \textit{signal present} \tag{8.7}$$

where **G** is the solution to (8.6).

The sampling interval Δt can now be permitted to become arbitrarily small, while simultaneously the number of samples m increases, maintaining the condition

$$m\,\Delta t = T$$

so that in the limit as $\Delta t \to 0$ and $m \to \infty$ the test statistic in (8.7) becomes formally the integral

$$\int_0^T g(t)y(t)\,dt \tag{8.8}$$

where $g(t)$ is the solution of the integral equation

$$\int_0^T R_n(t,\tau)g(\tau)\,d\tau = s(t) \qquad 0 \le t \le T \tag{8.9}$$

and the limit has been formally taken and assumed to exist in (8.6) as well as in (8.7).

The second, and more rigorous, method of obtaining this result consists of representing the random processes $n(t)$ and $y(t)$ in terms of the Karhunen-Loeve series (see Appendix B) and initially using the first m coefficients as the observations. For this we have that the gaussian noise process $n(t)$ can be expressed in the mean-square sense as

$$n(t) = \underset{m\to\infty}{\text{l.i.m.}} \sum_1^m n_i\varphi_i(t) \qquad 0 \le t \le T \tag{8.10}$$

where "l.i.m." is the mean square probability limit and the n_i are independent gaussian random variables given by

$$n_i \triangleq \int_0^T n(t)\varphi_i(t)\,dt \triangleq [n,\varphi_i] \qquad i = 1, 2, \ldots \tag{8.11}$$

which have zero mean and variance λ_i. The $\{\lambda_i\}$ are the eigenvalues of the covariance function $R_n(t,\tau)$ whose corresponding eigenfunctions are the $\{\varphi_i(t)\}$.

Since the $\{\varphi_i(t)\}$ constitute a complete orthonormal set of basis functions[1] for $L_2[0,T]$, $s(t)$ may also be represented in terms of the $\{\varphi_i(t)\}$ as

$$s(t) = \sum_{i=1}^{\infty} s_i\varphi_i(t) \tag{8.12}$$

[1] The class of square integrable functions in the interval $[0,T]$.

where[1]

$$s_i \triangleq [s,\varphi_i]$$

Therefore, when

$$y(t) = a_0 s(t) + n(t)$$

$y(t)$ has the representation

$$y(t) = \underset{m\to\infty}{\text{l.i.m.}} \sum_{i=1}^{m} y_i\varphi_i(t) \tag{8.13}$$

where

$$y_i \triangleq a_0 s_i + n_i \tag{8.14}$$

Since $R_n(t,\tau)$ is assumed known and the $\{\varphi_i(t)\}$ are either known or determinable, the information in a particular sample function of $y(t)$ is completely contained in the set of random variables $\{y_i\}$. Therefore the problem of continuously observing the received waveform over the interval $[0,T]$ has been reduced to having observed the countably infinite number[2] of random variables $\{y_i\}$.

We consider initially the first m coordinates $y_1, \ldots, y_m$ and base the decision of whether the signal is present or absent on the vector

$$\mathbf{Y} \triangleq \begin{bmatrix} y_1 \\ \cdot \\ \cdot \\ \cdot \\ y_m \end{bmatrix} \tag{8.15}$$

The likelihood ratio for this is

$$L_m(\mathbf{Y}) = \frac{f(\mathbf{Y}|\mathbf{Y} = a_0\mathbf{S} + \mathbf{N})}{f(\mathbf{Y}|\mathbf{Y} = \mathbf{N})} = \frac{f_N(\mathbf{Y} - a_0\mathbf{S})}{f_N(\mathbf{Y})}$$
$$= \frac{\prod_{i=1}^{m} (1/\sqrt{2\pi\lambda_i}) \exp\left[-(y_i - a_0 s_i)^2/2\lambda_i\right]}{\prod_{j=1}^{m} (1/\sqrt{2\pi\lambda_j}) \exp\left(-y_j^2/2\lambda_j\right)} \tag{8.16}$$

[1] In this chapter all representations such as (8.12) are in the L_2 sense. That is, the convergence is in the sense that

$$\lim_{n\to\infty} \left\| s(t) - \sum_{k=1}^{n} s_k\varphi_k(t) \right\|^2 = 0$$

[2] Formally, at any rate, the sampling approach also accomplished this.

Taking logarithms and simplifying, we obtain

$$\ln L_m(\mathbf{Y}) = a_0 \sum_{i=1}^{m} \frac{y_i s_i}{\lambda_i} - \frac{1}{2} a_0^2 \sum_{i=1}^{m} \frac{s_i^2}{\lambda_i} \tag{8.17}$$

We can immediately allow $m \to \infty$, with the result that the optimal detector can be represented as

$$\text{If } \sum_{i=1}^{\infty} \frac{y_i s_i}{\lambda_i} > R_T \qquad \text{decide } \textit{signal present} \tag{8.18}$$

We now claim that (8.18) has the same integral representation as obtained by the sampling method:

$$\sum_{i=1}^{\infty} \frac{y_i s_i}{\lambda_i} = \int_0^T g(t) y(t)\, dt \tag{8.19}$$

where $g(t)$ satisfies (8.9). For this we note that from *Mercer's theorem*[1]

$$R_n(t,\tau) = \sum_{i=1}^{\infty} \lambda_i \varphi_i(t) \varphi_i(\tau) \qquad 0 \leq (t,\tau) \leq T \tag{8.20}$$

With the assumption that $g(t)$ is in $L_2[0,T]$, so that it has the representation

$$g(t) = \sum_{i=1}^{\infty} g_i \varphi_i(t) \qquad 0 \leq t \leq T \tag{8.21}$$

where

$$g_i = [g,\varphi_i] \qquad 0 \leq t \leq T$$

(8.20) and (8.21) may be substituted into (8.9), from which we obtain

$$\int_0^T \left[\sum_{i=1}^{\infty} \lambda_i \varphi_i(t) \varphi_i(\tau) \right] \left[\sum_{j=1}^{\infty} g_j \varphi_j(\tau) \right] d\tau = \sum_{j=1}^{\infty} s_j \varphi_j(t) \qquad 0 \leq t \leq T$$

[1] **Mercer's Theorem:** Let A be an integral operator which has kernel $R(t,\tau)$ for $0 < (t,\tau) < T$, that is, continuous, symmetric, and nonnegative-definite. Let $\{\varphi_k(t); k = 1, 2, \ldots\}$ be the eigenfunctions of $R(t,\tau)$, corresponding to the nonzero (therefore positive) eigenvalues $\{\lambda_k\}$, respectively, where $\|\varphi_k(t)\| = 1$ for all k. Then $R(t,\tau)$ has the representation

$$R(t,\tau) = \sum_{k=1}^{\infty} \lambda_k \varphi_k(t) \varphi_k(\tau)$$

in the sense of uniform convergence. For the proof of this theorem see, for example, Reisz and Nagy [8.5] or Ash [8.6].

or, equivalently,

$$\sum_{j=1}^{\infty} \lambda_j g_j \varphi_j(t) = \sum_{j=1}^{\infty} s_j \varphi_j(t) \qquad 0 \leq t \leq T$$

Therefore

$$g_j = \frac{s_j}{\lambda_j} \qquad j = 1, 2, \ldots$$

Hence

$$[g(t),y(t)] = \int_0^T \Big[\sum_{j=1}^{\infty} \frac{s_j}{\lambda_j} \varphi_j(t) \Big] \Big[\sum_{k=1}^{\infty} y_k \varphi_k(t) \Big] \, dt = \sum_{j=1}^{\infty} \frac{s_j y_j}{\lambda_j} \tag{8.22}$$

which verifies (8.19).

Therefore the optimal decision function decides *signal present* if

$$[g,y] > R_T$$

where g must be the solution of the *detection integral equation*[1]

$$R_n g = s \tag{8.23}$$

Formally at least, the decision function may be expressed as

$$[y, R_n^{-1}s] = [R_n^{-1}y, s] = [R_n^{-\frac{1}{2}}y, R_n^{-\frac{1}{2}}s] > R_T \tag{8.24}$$

This can be interpreted as sending the received waveform through a prewhitening filter $R_n^{-\frac{1}{2}}$ and then correlating or matching the filter output $R_n^{-\frac{1}{2}}y$ with the equivalent transmitted signal $R_n^{-\frac{1}{2}}s$.

The detection integral equation (8.23) is a Fredholm equation of the first kind. It can be solved for the class of stationary noise processes whose autocorrelation function $R_n(\tau)$ is the Fourier transform of a spectral density which is a rational function in frequency. The solution will then in general involve Dirac delta functions and their derivatives, and exponential terms. Specifically, Zadeh and Ragazzini [8.7, 8.8] have shown that the solution of

$$s(t) = \int_0^T R_n(t - \tau) g(\tau) \, d\tau \qquad 0 \leq t \leq T$$

where

$$S_n(\omega) = \int_{-\infty}^{\infty} R_n(\tau) \exp(-j\omega\tau) \, d\tau = \frac{N(\omega)}{D(\omega)}$$

and $N(\omega)$ and $D(\omega)$ are polynomials in ω^2 of degrees m and n, respectively,

[1] $R_n g = s$, by which we mean

$$\int_0^T R_n(t,\tau) g(\tau) \, d\tau = s(t) \qquad 0 \leq t \leq T$$

is of the form

$$g(t) = g_0(t) + \sum_{k=0}^{n-m-1} (-1)^k[a_k\delta^{(k)}(t) + b_k\delta^{(k)}(T - t)]$$
$$+ \sum_{k=0}^{m} \{c_k \exp[-h_k(t)] + d_k \exp[-h_k(T - t)]\}$$

In this solution, $g_0(t)$ is the solution of the Wiener-Hopf equation

$$s(t) = \int_{-\infty}^{\infty} R_n(t - \tau)g_0(\tau)\, d\tau \qquad -\infty < t < \infty$$

Also, $\delta^{(k)}(t)$ is the kth derivative of the delta function and $h_k(t)$ are the zeros of the noise spectral density $S_n(\omega)$. The $2n$ unknown coefficients, $\{a_k\}$, $\{b_k\}$, $\{c_k\}$, $\{d_k\}$, are determined by various methods, described by Zadeh and Ragazzini [8.8], Middleton [8.9], Laning and Battin [8.10], and Helstrom [8.11]. The methods essentially involve a set of linear simultaneous equations, with Helstrom's recent technique being the simplest.

If the additional constraint

$$\sum_{k=1}^{\infty} \frac{s_k^2}{\lambda_k^2} < \infty$$

is satisfied [8.1, 8.12], then the solution of (8.23) is square integrable.

We have assumed throughout that $s(t)$ is expandable in terms of the set of basis functions $\{\varphi_i(t)\}$. Suppose now that this is not the case; that is, suppose there exists an $s_0(t)$ such that

$$[s_0,\varphi_i] = 0 \qquad i = 1, 2, \ldots \tag{8.25}$$

Then

$$[n,s_0] \equiv 0$$

and if $s_0(t)$ is used as a weighting function, the noise is completely eliminated. Therefore, if

$$y(t) = s_0(t) + n(t)$$

the detector would form

$$[y,s_0] = [s_0,s_0] > 0$$

and if

$$y(t) = n(t)$$

the detector output would be identically zero, which implies that we can *detect perfectly*. Such cases are known as *singular detection* and do not physically exist. Whenever the detection probability is strictly less than 1, the term *nonsingular* is used.

When $s_0(t)$ is such that (8.25) is satisfied, $s_0(t)$ is said to be in the *null space* of the linear integral operator with kernel $R_n(t,\tau)$. Such signals can be found when $R_n(t,\tau)$ is not a positive-definite kernel.[1] This is a sufficient condition for singular detection but not necessary. What has been considered is a special case of the more general condition for singular detection, namely;

$$\sum_{k=1}^{\infty} \frac{s_k^2}{\lambda_k} = \infty \tag{8.26}$$

This can be conveniently demonstrated by considering judgment based on the first m terms in the sum representing the optimal detector in (8.18). With the assumption that signal is present, the random variable

$$\eta \triangleq \sum_{k=1}^{m} \frac{y_k s_k}{\lambda_k}$$

is gaussian with mean and variance given by

$$K_m \triangleq \sum_{k=1}^{m} \frac{s_k^2}{\lambda_k}$$

which, after normalization, leads to the detection probability

$$P_{11} = \operatorname{erfc}\left[-a_0\sqrt{K_m} + \frac{R_T}{\sqrt{K_m}}\right] \tag{8.27}$$

where R_T is fixed by the FAP. Therefore as $m \to \infty$, singular detection will exist if and only if (8.26) is satisfied.

It has been shown (see Kailath [8.13, 8.14]) that the class of signal waveforms which leads to nonsingular detection can be simply and directly obtained in terms of a special space of functions called the *reproducing kernel Hilbert space* of the noise covariance function.

Returning now to the realistic nonsingular-detection case, we have found the best detector for a given signal $s(t)$. The question immediately arises: If there exists a choice of signals, what is the most preferred choice? Heuristically, a signal should be chosen which is orthogonal to the noise components, or at least as close as possible to orthogonal, for then

$$\int_0^T n(t)s(t)\,dt$$

would be small.

[1] $R_n(t,\tau)$ is always nonnegative-definite.

With this in mind, assume that the eigenvalues $\{\lambda_i\}$ of $R_n(t,\tau)$ have been indexed in decreasing order,

$$\lambda_1 \geq \lambda_2 \geq \lambda_3 \geq \cdots$$

and let

$$s(t) = \varphi_l(t)$$

for some fixed l. The optimal detector then forms

$$\int_0^T y(t)\varphi_l(t)\,dt$$

which is compared to a threshold. When the signal is absent, so that $y(t) = n(t)$, the variance of $[n,\varphi_l]$ is

$$E([n,\varphi_l]^2) = \lambda_l$$

Hence, the larger the index l, the smaller the eigenvalue λ_l, and the smaller the variance of the test statistic.

Therefore, if there is a choice of signals, we should choose eigenfunctions which have eigenvalues that are as small as possible. When implementation considerations are taken into account, however, there are limitations on the extent to which this idea can be pursued.

8.2 COHERENT BINARY COMMUNICATION IN COLORED NOISE

The optimal binary receiver for a coherent system in which the channel adds gaussian noise with zero mean and covariance function $R_n(t,\tau)$ can be determined in a manner similar to that for the coherent detector discussed in the previous section. The observed real-valued waveform is

$$y(t) = s_j(t) + n(t) \qquad 0 \leq t \leq T$$

where j is 0 or 1 with a priori probabilities π_0 and π_1, respectively.

With the sampling method the likelihood ratio

$$\frac{f(\mathbf{Y}|\mathbf{S}_1)}{f(\mathbf{Y}|\mathbf{S}_0)}$$

becomes

$$L_m(\mathbf{Y}) = \frac{\exp\left[-\frac{1}{2}(\mathbf{Y} - \mathbf{S}_1)^T\mathbf{R}_n^{-1}(\mathbf{Y} - \mathbf{S}_1)\right]}{\exp\left[-\frac{1}{2}(\mathbf{Y} - \mathbf{S}_0)^T\mathbf{R}_n^{-1}(\mathbf{Y} - \mathbf{S}_0)\right]} \tag{8.28}$$

which is compared with the threshold π_0/π_1. If we take logarithms and simplify, we find that the optimal decision function decides $\mathbf{S}_1$ if

$$\mathbf{S}_1^T\mathbf{R}_n^{-1}\mathbf{Y} - \tfrac{1}{2}\mathbf{S}_1^T\mathbf{R}_n^{-1}\mathbf{S}_1 + \ln \pi_1 > \mathbf{S}_0^T\mathbf{R}_n^{-1}\mathbf{Y} - \tfrac{1}{2}\mathbf{S}_0^T\mathbf{R}_n^{-1}\mathbf{S}_0 + \ln \pi_0 \tag{8.29}$$

If the vectors $\mathbf{G}_0$ and $\mathbf{G}_1$ are defined to be the solutions of the equations

$$\begin{aligned} \mathbf{R}_n\mathbf{G}_0\,\Delta t &= \mathbf{S}_0 \\ \mathbf{R}_n\mathbf{G}_1\,\Delta t &= \mathbf{S}_1 \end{aligned} \tag{8.30}$$

the receiver then becomes

$$\mathbf{G}_1^T\mathbf{Y}\,\Delta t - \tfrac{1}{2}\mathbf{G}_1^T\mathbf{S}_1\,\Delta t + \ln \pi_1 \gtrless \mathbf{G}_0^T\mathbf{Y}\,\Delta t - \tfrac{1}{2}\mathbf{G}_0^T\mathbf{S}_0\,\Delta t + \ln \pi_0 \tag{8.31}$$

where $\mathbf{G}_0$ and $\mathbf{G}_1$ satisfy (8.30). If we now let the sampling interval Δt approach zero, we obtain in the limit

$$\int_0^T g_1(t)y(t)\,dt - \frac{1}{2}\int_0^T g_1(t)s_1(t)\,dt + \ln \pi_1 \gtrless \int_0^T g_0(t)y(t)\,dt - \frac{1}{2}\int_0^T g_0(t)s_0(t)\,dt + \ln \pi_0 \tag{8.32}$$

where $g_0(t)$ and $g_1(t)$ are solutions of the integral equations

$$\int_0^T R_n(t,\tau)g_j(\tau)\,d\tau = s_j(t) \qquad 0 \le t \le T,\ j = 0, 1 \tag{8.33}$$

which are obtained as the formal limits of (8.30).

The performance of this system is now obtained when the a priori probabilities are assumed to be equal. The probability of a correct decision is then

$$P_d = \pi_1 P_{11} + \pi_0 P_{00} = \tfrac{1}{2}(P_{11} + P_{00}) = P_{11} \tag{8.34}$$

where it can be shown that $P_{11} = P_{00}$ when $\pi_0 = \pi_1$. We have that

$$\begin{aligned} P_{11} &= \Pr\{[g_1,y] - \tfrac{1}{2}[g_1,s_1] > [g_0,y] - \tfrac{1}{2}[g_0,s_0] \mid y(t) = s_1(t) + n(t)\} \\ &= \Pr\{[g_1 - g_0, n] > -\tfrac{1}{2}[s_1 - s_0, R_n^{-1}(s_1 - s_0)]\} \end{aligned} \tag{8.35}$$

If we define

$$\xi \triangleq [g_1 - g_0, n] \tag{8.36}$$

then ξ is a gaussian random variable with zero mean and variance

$$\begin{aligned} \sigma_\xi^2 &= [g_1 - g_0, R_n(g_1 - g_0)] \\ &= [s_1 - s_0, R_n^{-1}(s_1 - s_0)] \end{aligned} \tag{8.37}$$

After normalizing ξ, we obtain finally that

$$P_{11} = \operatorname{erfc}\left(-\tfrac{1}{2}\sqrt{[s_1 - s_0, R_n^{-1}(s_1 - s_0)]}\right) \tag{8.38}$$

or, equivalently, that the probability of error is

$$P_e = \operatorname{erfc}\left(\tfrac{1}{2}\sqrt{[s_1 - s_0, R_n^{-1}(s_1 - s_0)]}\right) \tag{8.39}$$

8.3 NONCOHERENT BINARY COMMUNICATION IN COLORED NOISE

In this section we shall determine the optimal binary receiver for a noncoherent system in which the channel adds stationary gaussian noise with zero mean and covariance function $R_n(\tau)$. The observed waveform will be expressed in terms of the complex-envelope representation[1] as

$$y(t) = s_j(t,\phi) + n(t) = \operatorname{Re}[E(t) \exp i\omega_c t] \qquad 0 \le t \le T \tag{8.40}$$

where $j = 0$ or 1 with a priori probabilities π_0 and π_1, respectively, and $E(t)$, the complex envelope of the received waveform, is

$$E(t) = \tilde{s}_j(t,\phi) + \tilde{n}(t) \tag{8.41}$$

in which $\tilde{s}(t,\phi)$ and $\tilde{n}(t)$ are the complex envelopes of the signal and noise, respectively. The signal waveforms are assumed to be narrowband with respect to ω_c, and their complex envelopes $\{\tilde{s}_j(t,\phi)\}$ may be expressed in terms of their real and imaginary parts as

$$\tilde{s}_j(t,\phi) = s_{R_j}(t,\phi) + is_{I_j}(t,\phi) = A_j(t) \exp[i(\theta_j(t) + \phi)] \qquad j = 0, 1 \tag{8.42}$$

where the $\{A_j(t)\}$ and $\{\theta_j(t)\}$ are the signal amplitude and phase modulations. The unknown reference phase ϕ is assumed to be a random variable uniformly distributed over $[0,2\pi]$.

The noise is assumed to be stationary and narrowband in the sense that its spectral density is negligible for $|f| > 2f_c$, where $\omega_c = 2\pi f_c$. We then have the representation

$$n(t) = \operatorname{Re}[\tilde{n}(t) \exp i\omega_c t] \tag{8.43}$$

where

$$\tilde{n}(t) \triangleq n_R(t) + in_I(t)$$

in which $n_R(t)$ and $n_I(t)$ are independent real gaussian stochastic processes[2] with identical covariance function $R_1(\tau)$, where

$$R_n(\tau) = R_1(\tau) \cos \omega_c \tau$$

If the eigenfunctions and eigenvalues of $R_1(\tau)$ are denoted by

[1] For the development of this optimal receiver using time sampling methods see Viterbi [8.3].

[2] This is discussed in detail in Viterbi [8.3].

$\{\varphi_k(t)\}$ and $\{\lambda_k\}$, respectively, then $\tilde{n}(t)$ can be represented in terms of the Karhunen-Loeve series (see Appendix B) as

$$\tilde{n}(t) = \underset{K\to\infty}{\text{l.i.m.}} \sum_{k=1}^{K} (n_{R_k} + in_{I_k})\varphi_k(t) \tag{8.44}$$

where

$$n_{R_k} \triangleq [n_R, \varphi_k] \qquad k = 1, 2, \ldots$$

and

$$n_{I_k} \triangleq [n_I, \varphi_k] \qquad k = 1, 2, \ldots$$

The $\{n_{R_k}\}$ and $\{n_{I_k}\}$ are mutually independent zero-mean gaussian random variables with variances $\{\lambda_k\}$, respectively.

The signal waveforms can also be expanded in terms of this complete orthonormal set of eigenfunctions, with the representation

$$\tilde{s}_j(t,\phi) = \sum_{k=1}^{\infty} (s_{R_k}{}^j(\phi) + is_{I_k}{}^j(\phi))\varphi_k(t) \qquad 0 \le t \le T; j = 0, 1 \tag{8.45}$$

where

$$s_{R_k}{}^j(\phi) \triangleq [s_{R_j}(t,\phi), \varphi_k(t)]$$

and

$$s_{I_k}{}^j(\phi) \triangleq [s_{I_j}(t,\phi), \varphi_k(t)]$$

inally, if we define

$$u_k \triangleq s_{R_k}{}^j(\phi) + n_{R_k} \qquad k = 1, 2, \ldots$$

and (8.46)

$$v_k \triangleq s_{I_k}{}^j(\phi) + n_{I_k} \qquad k = 1, 2, \ldots$$

then the complex envelope of the received waveform has the series representation

$$E(t) = \underset{K\to\infty}{\text{l.i.m.}} \sum_{k=1}^{K} (u_k + iv_k)\varphi_k(t) \tag{8.47}$$

With these representations we now obtain the optimal receiver operation. As in the second method discussed in Sec. 8.1, the receiver will be obtained by using the first m coordinates of the series representation of the received waveform. We shall use the first m coordinates of both the $\{u_k\}$ and the $\{v_k\}$, for which the optimal receiver, the generalized likelihood ratio, is

$$L_m(\mathbf{Y}) = \frac{\dfrac{1}{2\pi}\displaystyle\int_{-\pi}^{\pi} d\psi \frac{1}{(\sqrt{2\pi})^m \sqrt{\prod_{l=1}^{m} \lambda_l}} \exp\left\{-\frac{1}{2}\sum_{k=1}^{m} \frac{[u_k - s_{R_k}{}^1(\psi)]^2}{\lambda_k}\right\} \dfrac{1}{(\sqrt{2\pi})^m \sqrt{\prod_{l=1}^{m} \lambda_l}} \exp\left\{-\frac{1}{2}\sum_{k=1}^{m} \frac{[v_k - s_{I_k}{}^1(\psi)]^2}{\lambda_k}\right\}}{\dfrac{1}{2\pi}\displaystyle\int_{-\pi}^{\pi} d\psi \frac{1}{(\sqrt{2\pi})^m \sqrt{\prod_{l=1}^{m} \lambda_l}} \exp\left\{-\frac{1}{2}\sum_{k=1}^{m} \frac{[u_k - s_{R_k}{}^0(\psi)]^2}{\lambda_k}\right\} \dfrac{1}{(\sqrt{2\pi})^m \sqrt{\prod_{l=1}^{m} \lambda_l}} \exp\left\{-\frac{1}{2}\sum_{k=1}^{m} \frac{[v_k - s_{I_k}{}^0(\psi)]^2}{\lambda_k}\right\}} \tag{8.48}$$

which, with the criterion of probability of error, is compared with the threshold π_0/π_1.

Simplifying (8.48), we obtain

$$L_m(\mathbf{Y}) = \frac{\dfrac{1}{2\pi}\displaystyle\int_{-\pi}^{\pi} d\psi \exp\left\{\sum_{k=1}^{m} \frac{u_k s_{R_k}{}^1(\psi) + v_k s_{I_k}{}^1(\psi)}{\lambda_k} - \frac{1}{2}\sum_{k=1}^{m} \frac{[s_{R_k}{}^1(\psi)]^2 + [s_{I_k}{}^1(\psi)]^2}{\lambda_k}\right\}}{\dfrac{1}{2\pi}\displaystyle\int_{-\pi}^{\pi} d\psi \exp\left\{\sum_{k=1}^{m} \frac{u_k s_{R_k}{}^0(\psi) + v_k s_{I_k}{}^0(\psi)}{\lambda_k} - \frac{1}{2}\sum_{k=1}^{m} \frac{[s_{R_k}{}^0(\psi)]^2 + [s_{I_k}{}^0(\psi)]^2}{\lambda_k}\right\}} \tag{8.49}$$

As the number of coordinates is increased and we finally allow m to approach infinity, we have that

$$\lim_{m\to\infty} \sum_{k=1}^{m} \frac{u_k s_{R_k}{}^j(\psi) + v_k s_{I_k}{}^j(\psi)}{\lambda_k} = \operatorname{Re}\left[\int_0^T E(t)\overline{g_j(t,\psi)}\, dt\right] \qquad j = 0, 1 \tag{8.50}$$

where $g(t,\psi)$ is defined to be the solution of the integral equation

$$\int_0^T R_1(t,\tau)g_j(\tau,\psi)\, d\tau = \mathfrak{s}(t,\psi) \qquad 0 \le t \le T,\ -\pi \le \psi \le \pi,\ j = 0, 1 \tag{8.51}$$

To show this we note initially that

$$\sum_{k=1}^{m} \frac{u_k s_{R_k}{}^j(\psi) + v_k s_{I_k}{}^j(\psi)}{\lambda_k} = \operatorname{Re}\left[\sum_{k=1}^{m} \frac{s_{R_k}{}^j(\psi) - i s_{I_k}{}^j(\psi)}{\lambda_k}(u_k + iv_k)\right] \qquad j = 0, 1 \quad (8.52)$$

From (8.51), we may write

$$\int_0^T \left[\sum_k \lambda_k \varphi_k(t)\varphi_k(\tau)\right]\left[\sum_l g_l{}^j(\psi)\varphi_l(\tau)\right] d\tau = \sum_k [s_{R_k}{}^j(\psi) + i s_{I_k}{}^j(\psi)]\,\varphi_k(t)$$

which implies formally that

$$g_j(t,\psi) \triangleq \sum_k g_k{}^j(\psi)\varphi_k(t) = \sum_k \frac{s_{R_k}{}^j(\psi) + i s_{I_k}{}^j(\psi)}{\lambda_k}\varphi_k(t)$$

Upon substitution we have that

$$\int_0^T E(t)\overline{g_j(t,\psi)}\, dt = \int_0^T \sum_k \frac{s_{R_k}{}^j(\psi) - i s_{I_k}{}^j(\psi)}{\lambda_k}\varphi_k(t) \sum_l (u_l + iv_l)\varphi_l(t)\, dt$$
$$= \sum_k \frac{s_{R_k}{}^j(\psi) - i s_{I_k}{}^j(\psi)}{\lambda_k}(u_k + iv_k) \qquad j = 0, 1$$

which, with (8.52), demonstrates (8.50).

Similarly, it can be shown that

$$\sum_k \frac{[s_{R_k}{}^j(\psi)]^2}{\lambda_k} = [s_{R_j}(t,\psi),\, R_1{}^{-1} s_{R_j}(t,\psi)] \qquad j = 0, 1 \qquad (8.53a)$$

and

$$\sum_k \frac{[s_{I_k}{}^j(\psi)]^2}{\lambda_k} = [s_{I_j}(t,\psi),\, R_1{}^{-1} s_{I_j}(t,\psi)] \qquad j = 0, 1 \qquad (8.53b)$$

Therefore, from (8.53),

$$\sum_k \frac{[s_{R_k}{}^j(\psi)]^2 + [s_{I_k}{}^j(\psi)]^2}{\lambda_k} = \operatorname{Re}\,[\tilde{s}_j, \overline{R_1{}^{-1}\tilde{s}_j}] \qquad j = 0, 1 \qquad (8.54)$$

If we substitute the integral representations for the infinite sums into (8.49) as $m \to \infty$, the optimal receiver becomes

$$L(y(t); 0 \le t \le T) = \frac{(1/2\pi)\int_{-\pi}^{\pi} d\psi\, \exp\{\operatorname{Re}\,[E,\bar{g}_1] - \frac{1}{2}\operatorname{Re}\,[\tilde{s}_1,\bar{g}_1]\}}{(1/2\pi)\int_{-\pi}^{\pi} d\psi\, \exp\{\operatorname{Re}\,[E,\bar{g}_0] - \frac{1}{2}\operatorname{Re}\,[\tilde{s}_0,\bar{g}_0]\}} \qquad (8.55)$$

Since $\tilde{s}_j(t,\phi)$ is of the form

$$s_j(t,\psi) = S_j(t)\exp i\psi$$

where

$$S_j(t) \triangleq A_j(t) \exp i\theta_j(t) \qquad j = 0, 1$$

then

$$g_j(t,\psi) = R_1^{-1}S_j(t) \exp i\psi$$

and

$$\text{Re}\,[\tilde{s}_j,\overline{g_j}] = [S_j(t),R_1^{-1}\overline{S_j(t)}] \qquad j = 0, 1 \tag{8.56}$$

which is real and independent of ψ.

Similarly, if we express

$$\text{Re}\,[E,\overline{g_j}] = \text{Re}\,[E,R_1^{-1}\bar{S}_j] \cos\psi + \text{Im}\,[E,R_1^{-1}\bar{S}_j] \sin\psi \tag{8.57}$$

and substitute (8.56) and (8.57) into (8.55), the optimal receiver reduces to

$$L(y(t);\, 0 \le t \le T) = \frac{\exp\{-\frac{1}{2}[S_1,R_1^{-1}\bar{S}_1]\}\, I_0(|[E,R_1^{-1}\bar{S}_1]|)}{\exp\{-\frac{1}{2}[S_0,R_1^{-1}\bar{S}_0]\}\, I_0(|[E,R_1^{-1}\bar{S}_0]|)} \gtrless \frac{\pi_0}{\pi_1} \tag{8.58}$$

We take logarithms, and the optimal receiver decides a_1 if

$$\begin{aligned} \ln I_0(|[E,R_1^{-1}\bar{S}_1]|) &- \tfrac{1}{2}[S_1,R_1^{-1}\bar{S}_1] + \ln \pi_1 \\ &> \ln I_0(|[E,R_1^{-1}\bar{S}_0]|) - \tfrac{1}{2}[S_0,R_1^{-1}\bar{S}_0] + \ln \pi_0 \end{aligned} \tag{8.59}$$

A significantly simpler suboptimal system is the one which compares the outputs of the two *envelope correlation receivers* and decides a_1 if

$$|[E,R_1^{-1}\bar{S}_1]| > |[E,R_1^{-1}\bar{S}_0]| \tag{8.60}$$

The evaluation of the performance of this system can be carried out in the same way as in the white-noise case.

Envelope correlation detectors can be interpreted as the envelope

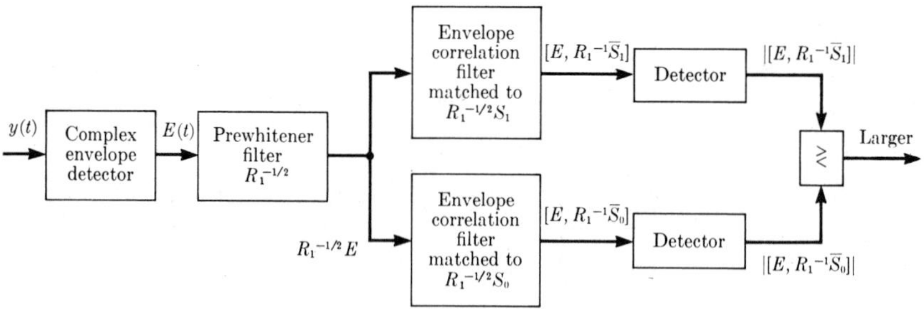

Fig. 8.1 Envelope correlation receiver for the noncoherent-binary-communication-system receiver with colored gaussian noise.

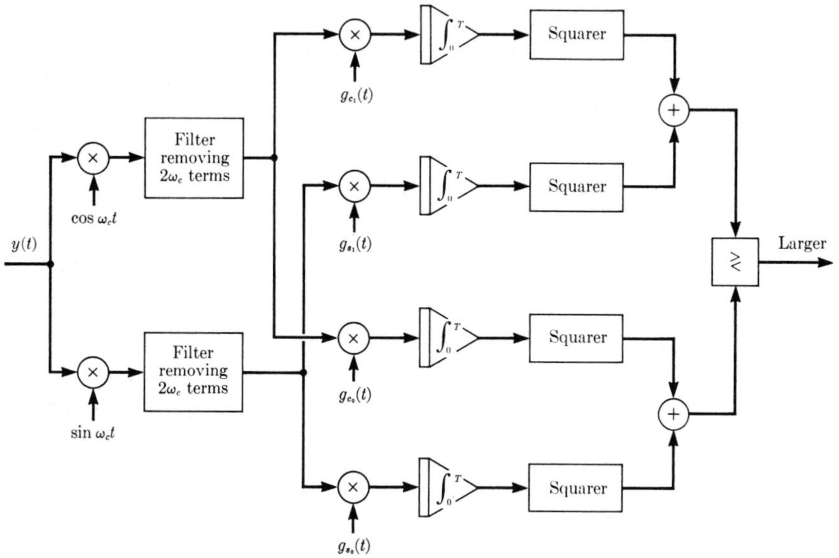

Fig. 8.2 An equivalent diagram of the noncoherent-binary-communication-system receiver with colored gaussian noise.

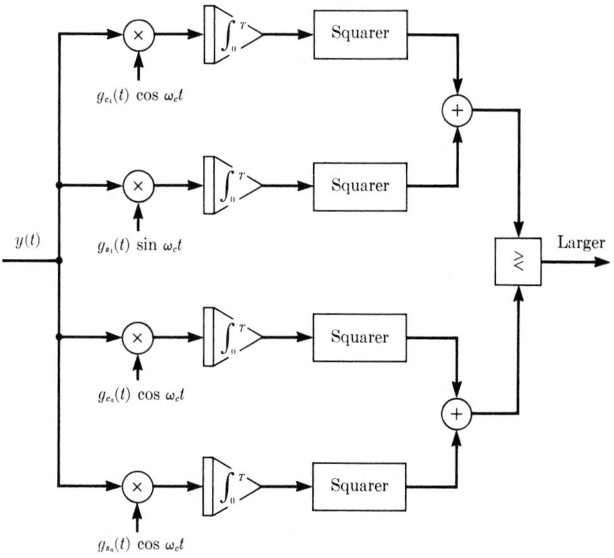

Fig. 8.3 Simplified version of the noncoherent receiver.

$E(t)$ being sent through a prewhitener filter, whose operation is given by the kernel $R_1^{-\frac{1}{2}}(\tau)$. The output is then sent through two envelope correlation filters or envelope matched filters. Their outputs are then compared, and the largest is chosen. A block diagram of this interpretation is given in Fig. 8.1.

The system is equivalently shown in Fig. 8.2, where the functions $g_{c_j}(t)$ and $g_{s_j}(t)$ are defined to be the solutions of the integral equations

$$\int_0^T R_1(t-\tau)g_{c_j}(\tau)\,d\tau = A_j(t)\cos\theta_j(t) \qquad 0 \le t \le T; j = 0, 1$$

and

$$\int_0^T R_1(t-\tau)g_{s_j}(\tau)\,d\tau = A_j(t)\sin\theta_j(t) \qquad 0 \le t \le T; j = 0, 1 \tag{8.61}$$

A simplified version of the diagram in Fig. 8.2 is given in Fig. 8.3.

PROBLEMS

8.1 In (8.17), as $m \to \infty$, show that

$$\sum_{k=1}^{\infty} \frac{s_k^2}{\lambda_k} = [g(t),s(t)]$$

where

$$\int_0^T R(t,\tau)g(\tau)\,d\tau = s(t) \qquad 0 \le t \le T$$

8.2 (*a*) Show that the threshold level R_T in the detector in (8.24) is given by

$$\operatorname{erfc}\frac{R_T}{\sqrt{[s,R_n^{-1}s]}} = \alpha$$

where α is the false-alarm probability.

(*b*) Using the threshold given in (*a*), show that the resulting detection probability is given by

$$P_{11} = \operatorname{erfc}\{\operatorname{erfc}^{-1}\alpha - a_0\sqrt{[s, R_n^{-1}s]}\}$$

8.3 Show that in a colored-noise-detection problem, if

$$[s_0,\varphi_k] = 0 \qquad k = 1, 2, \ldots$$

then $s_0(t)$ is necessarily in the null space of the noise covariance function $R_n(t,\tau)$.

8.4 Show that in a colored-noise-detection problem, if the noise is stationary and its spectrum is rational, then the detection is nonsingular.

8.5 (*a*) Show that in a colored-noise-detection problem, when the received waveform is $y(t) = a_0 s(t) + n(t)$ and the optimal detector is being used, the output signal-to-noise ratio is given by

$$\frac{a_0^2[s,g]^2}{E([n,g]^2)} = a_0 \sqrt{[s,R_n^{-1}s]}$$

(b) If $s(t) = \varphi_l(t)$, show that the output signal-to-noise ratio is a_0^2/λ_l.

(c) If $s(t) = \sum_{k=1}^{m} a_k\varphi_k(t)$, what is the output signal-to-noise ratio?

8.6 Verify Eq. (8.35).

8.7 Verify Eq. (8.37).

8.8 Verify Eqs. (8.38) and (8.39).

8.9 Verify Eqs. (8.53) and (8.54).

8.10 Verify Eq. (8.56).

8.11 Determine the probability of error for the system operation described by (8.60).

8.12 Show that the representation of the binary communication system in Fig. 8.2 is the same as that in Fig. 8.1.

8.13 Consider detecting the presence or absence of the signal

$$s(t) = \exp(-\alpha t) \qquad 0 \le t \le T$$

when the observed waveform is

$$y(t) = s(t) + n(t) \qquad 0 \le t \le T$$

where $n(t)$ is a sample waveform from a zero-mean gaussian stochastic process with covariance function

$$R_n(t,\tau) = \exp(-\alpha|t-\tau|) \qquad 0 \le (t, \tau) \le T$$

(a) Show that the test statistic has weighting function $g(t) = \delta(t)$ (which is not square integrable).

(b) Show that the performance of the resulting system is determined by $s(0)$.

(c) Show that

$$\sum_{k=1}^{\infty} \frac{s_k^2}{\lambda_k} < \infty$$

implying that the detection is nonsingular even though

$$\sum_{k=1}^{\infty} \frac{s_k^2}{\lambda_k^2} = \infty$$

REFERENCES

8.1 Grenander, U.: Stochastic Processes and Statistical Interference, *Arkiv Mathematik*, vol. 17, no. 1, pp. 195–277, 1950.

8.2 Helstrom, C.: "Statistical Theory of Signal Detection," Pergamon Press, New York, 1960.

8.3 Viterbi, A. J.: "Principles of Coherent Communication," McGraw-Hill, New York, 1966.

8.4 Selin, I.: "Detection Theory," Princeton University Press, Princeton, N.J., 1965.

8.5 Riesz, F., and B. S. Nagy: "Functional Analysis," Ungar, New York, 1955.

8.6 Ash, R. B.: "Information Theory," Interscience, New York, 1966.

8.7 Zadeh, L. A., and J. R. Ragazzini: An Extension of Wiener's Theory of Prediction, *J. Appl. Phys.*, vol. 21, pp. 645–655, July, 1950.

8.8 Zadeh, L. A., and J. R. Ragazzini: Optimum Filters for the Detection of Signals in Noise, *Proc. IRE*, vol. 40, pp. 1223–1231, October, 1952.

8.9 Middleton, D.: "An Introduction to Statistical Communication Theory," appendix 2, McGraw-Hill, New York, 1960.

8.10 Laning, J. H., and R. H. Battin: "Random Processes in Automatic Control," Chapter 8, McGraw-Hill, New York, 1956.

8.11 Helstrom, C. W.: Solution of the Detection Integral Equation for Stationary Filtered White Noise, *IEEE Trans. Inform. Theory*, vol. IT-11, no. 3, pp. 335–339, July 1965.

8.12 Kelly, E. J., I. S. Reed, and W. L. Root: The Detection of Radar Echoes in Noise -I, *J. Soc. Ind. Appl. Math*, vol. 8, no. 2, pp. 309–341, June, 1960.

8.13 Kailath, T.: A Projection Method for Signal Detection in Colored Gaussian Noise, *IEEE Trans. Inform. Theory*, vol. IT-13, no. 3, pp. 441–447, July 1967.

8.14 Kailath, T.: Some Results on Singular Detection, *Information and Control*, vol. 9, no. 2, pp. 130–152, 1966.

8.15 Root, W. L.: Singular Gaussian Measures in Detection Theory, in M. Rosenblatt (ed.), "Proceedings of the Symposium on Time Series Analysis," pp. 327–340, Wiley, New York, 1962.

8.16 Root, W. L.: The Detection of Signals in Gaussian Noise, in A. V. Balakrishnan (ed.), "Communication Theory," pp. 160–191, McGraw-Hill, New York, 1968

9

Detecting a Stochastic Signal in Noise

In the detection models considered thus far we have assumed that the receiver either had complete knowledge of the transmittable waveforms or knew them except for parameters such as phase, amplitude, and/or doppler rf carrier shift. However, in communication systems which, for example, employ the reflecting characteristics of the ionospheric layers of the earth's atmosphere, the transmitted signals become distorted to such an extent that their received version can be described only statistically, in which cases they have the appearance of a narrowband stochastic process. Such systems are called *scatter* or *multipath communication systems* and are often used for conveying information beyond the horizon.

In a *passive sonar system*, for example, the receiver is required to determine the presence or absence of a stochastic process. The problem is often made more difficult by the presence of background noise or receiver front-end noise, such as was assumed in the models discussed in previous chapters.

We shall consider first the detection of a random vector in the presence of an additive noise vector and then extend these results to the continuous cases. We restrict ourselves here to the cases of detection and binary communication, and in particular to determining the necessary likelihood ratios, with the understanding that the threshold can be determined by selecting an appropriate criterion of performance.

9.1 DETECTION OF A RANDOM VECTOR

In considering the detection of a random vector in the presence of additive noise, let us assume that the observed m-dimensional vector $\mathbf{Y}$ is of the form

$$\mathbf{Y} = a\mathbf{S} + \mathbf{N} \tag{9.1}$$

where $\mathbf{N}$ is a gaussian random vector with zero-mean vector and covariance matrix given by $\mathbf{\Lambda}_n$, $\mathbf{S}$ is the signal vector, independent of the noise vector, which is also assumed to be a zero-mean gaussian random vector with covariance matrix $\mathbf{\Lambda}_s$, and the scaler a can take on one of two values, either $a_1 \neq 0$, representing *signal present*, or $a_1 = 0$, representing *signal absent*. Initially we shall assume that a_1 is a known constant. The case of a_1 unknown corresponds to attempting to determine the presence or absence of a signal vector with known statistics, with the exception of average power.

As in several of our previous models, $\mathbf{Y}$ may be interpreted as a sequence of m samples from an observed stochastic process with continuous time parameter.

The optimal decision function for this binary detection problem is the likelihood ratio

$$\frac{p(\mathbf{Y}|\mathbf{Y} = a_1\mathbf{S} + \mathbf{N})}{p(\mathbf{Y}|\mathbf{Y} = \mathbf{N})}$$

which is compared with a threshold. For the present let us assume that the Neyman-Pearson approach is to be employed, so that the threshold will be determined from the false-alarm probability.

Under the hypothesis *signal present* the covariance matrix of $\mathbf{Y}$ is $a_1{}^2\mathbf{\Lambda}_s + \mathbf{\Lambda}_n$; it is $\mathbf{\Lambda}_n$ for *signal absent*, so that the likelihood ratio may be expressed explicitly as

$$\begin{aligned} L(\mathbf{Y}) &= \frac{p(\mathbf{Y}|\mathbf{Y} = a_1\mathbf{S} + \mathbf{N})}{p(\mathbf{Y}|\mathbf{Y} = \mathbf{N})} \\ &= \left(\frac{|\mathbf{\Lambda}_n|}{|a_1{}^2\mathbf{\Lambda}_s + \mathbf{\Lambda}_n|}\right)^{\frac{1}{2}} \frac{\exp\left[-\frac{1}{2}\mathbf{Y}^T(a_1{}^2\mathbf{\Lambda}_s + \mathbf{\Lambda}_n)^{-1}\mathbf{Y}\right]}{\exp\left(-\frac{1}{2}\mathbf{Y}^T\mathbf{\Lambda}_n{}^{-1}\mathbf{Y}\right)} \end{aligned} \tag{9.2}$$

where it is assumed throughout that both $\mathbf{\Lambda}_s$ and $\mathbf{\Lambda}_n$ are strictly positive-definite, so that the appropriate inverse operations may be performed. The ratio of determinants in (9.2) is then nonzero and can be assumed to be absorbed in the threshold. We take logarithms, and the decision function immediately reduces to comparing the difference of two quadratic forms to a threshold; that is,

$$\text{If } \mathbf{Y}^T\mathbf{\Lambda}_n^{-1}\mathbf{Y} - \mathbf{Y}^T(a_1^2\mathbf{\Lambda}_s + \mathbf{\Lambda}_n)^{-1}\mathbf{Y} > T_T \qquad \text{decide } \textit{signal present} \tag{9.3}$$

where the coefficient $\frac{1}{2}$ has also been absorbed by the threshold T_T. Therefore the optimal detection of a gaussian signal vector in the presence of an additive gaussian noise vector involves the formation of the quadratic form

$$\mathbf{Y}^T[\mathbf{\Lambda}_n^{-1} - (a_1^2\mathbf{\Lambda}_s + \mathbf{\Lambda}_n)^{-1}]\mathbf{Y} \tag{9.4}$$

called a *quadratic detector*, and the comparison of this functional of $\mathbf{Y}$ to a threshold.

There are some important simplifications of this detector. With the identity

$$(a_1^2\mathbf{\Lambda}_s + \mathbf{\Lambda}_n)^{-1} = \mathbf{\Lambda}_n^{-1}(\mathbf{I} + a_1^2\mathbf{\Lambda}_s\mathbf{\Lambda}_n^{-1})^{-1} \tag{9.5}$$

where $\mathbf{I}$ is the m-by-m identity matrix, the quadratic detector (9.4) can be expressed as

$$\mathbf{Y}^T\mathbf{\Lambda}_n^{-1}[\mathbf{I} - (\mathbf{I} + a_1^2\mathbf{\Lambda}_s\mathbf{\Lambda}_n^{-1})^{-1}]\mathbf{Y} \tag{9.6}$$

In many cases of practical interest (particularly in sonar) we are concerned with the detection of a signal vector whose average power is much less than that of the accompanying noise. In terms of the covariance matrices of the signal and noise, this is the same as saying that the magnitude of the covariance matrix of the signals, including its coefficient a_1^2, is much less than the magnitude of the covariance matrix of the noise, where the magnitude of a matrix $\mathbf{\Lambda}$ is defined as

$$\begin{aligned} \|\mathbf{\Lambda}\| &= \sup_{\mathbf{x}} \|\mathbf{\Lambda x}\| \qquad \|\mathbf{x}\| = 1 \\ &= \sup_{\mathbf{x}} \frac{\|\mathbf{\Lambda x}\|}{\|\mathbf{x}\|} \qquad \|\mathbf{x}\| \neq 0 \end{aligned} \tag{9.7}$$

Therefore the assumption that the signal power is much less than the noise power implies[1]

$$a_1^2\|\mathbf{\Lambda}_s\| \ll \|\mathbf{\Lambda}_n\| \tag{9.8}$$

from which we can conclude that

$$a_1^2\|\mathbf{\Lambda}_s\mathbf{\Lambda}_n^{-1}\| \ll 1 \tag{9.9}$$

[1] This is equivalent to saying that the largest eigenvalue of $\mathbf{\Lambda}_n$ is much larger than a_1^2 times the largest eigenvalue of $\mathbf{\Lambda}_s$.

The advantage of this frequently realistic assumption is that we can now utilize the following theorem from the theory of linear vector spaces.

THEOREM 9.1 *Given any square matrix* $\mathbf{\Lambda}$, *if for all vectors* $\mathbf{x}$ *there exists a constant* γ, *independent of* $\mathbf{x}$, *such that* $0 < \gamma < 1$ *and such that*

$$\|\mathbf{\Lambda x}\| < \gamma \|\mathbf{x}\|$$

then

$$(\mathbf{I} + \mathbf{\Lambda})^{-1} = \mathbf{I} - \mathbf{\Lambda} + \mathbf{\Lambda}^2 - \mathbf{\Lambda}^3 + \cdots \tag{9.10}$$

For the proof of this theorem see, for example, Friedman [9.1]. For small γ the first two terms provide the approximation

$$(\mathbf{I} + \mathbf{\Lambda})^{-1} \approx \mathbf{I} - \mathbf{\Lambda} \tag{9.11}$$

If we substitute this approximation in (9.6), the quadratic detector can be approximated by

$$\text{If } \mathbf{Y}^T\mathbf{\Lambda}_n^{-1}\mathbf{\Lambda}_s\mathbf{\Lambda}_n^{-1}\mathbf{Y} > T_T \qquad \text{decide } \textit{signal present} \tag{9.12}$$

where the factor a_1^2 has been absorbed in the threshold. Therefore at small signal-to-noise ratios the quadratic detector has the approximation given by the quadratic form in (9.12), which has the distinct advantage of being independent of the signal power when the threshold is being determined by the false-alarm probability. It is clear that the exact representation in (9.6) requires knowledge of the signal power.

In the special case when the noise components are uncorrelated, and hence statistically independent because of the gaussian assumption, and when the variance of each noise component is the same, the appropriate solution reduces to

$$\mathbf{Y}^T\mathbf{\Lambda}_s\mathbf{Y} > T_T' \tag{9.13}$$

This is, of course, not optimal on account of the above approximation. In fact, it can easily be shown (Prob. 9.2) that when

$$\mathbf{\Lambda}_n = \sigma^2\mathbf{I}$$

the optimal detector forms

$$\mathbf{Y}^T\mathbf{A}\mathbf{Y}$$

where the matrix $\mathbf{A}$ is the solution of the matrix equation

$$\sigma^4\mathbf{A} + a_1^2\sigma^2\mathbf{\Lambda}_s\mathbf{A} = a_1^2\mathbf{\Lambda}_s \tag{9.14}$$

When both the signal and noise vectors consist of uncorrelated components, the threshold can easily be chosen and the resulting probability of error easily evaluated (see Prob. 9.3).

This approximate solution (9.12) can be shown (see Baker [9.5]) to be the optimal solution when the model is as indicated here, and we look for the detector of the form

$$D(\mathbf{Y}) \triangleq \mathbf{Y}^T\mathbf{W}\mathbf{Y}$$

where the criterion of optimality is to choose that matrix $\mathbf{W}$ which maximizes the *deflection* H, defined as

$$H \triangleq \frac{[E(D|\mathbf{Y} = \mathbf{S} + \mathbf{N}) - E(D|\mathbf{Y} = \mathbf{N})]^2}{E(D^2|\mathbf{Y} = \mathbf{N}) - [E(D|\mathbf{Y} = \mathbf{N})]^2}$$

9.2 DETECTION OF A STOCHASTIC PROCESS IN NOISE

Let us now extend these concepts to cases of continuous sampling of the respective stochastic processes. Specifically, we shall assume that the observed waveform is

$$y(t) = as(t) + n(t) \qquad 0 \leq t \leq T \tag{9.15}$$

where $s(t)$ and $n(t)$ are independent zero-mean gaussian stochastic processes and have covariance functions $R_s(t,\tau)$ and $R_n(t,\tau)$, respectively. We are asked to optimally detect the presence ($a = a_1 > 0$) or absence ($a = 0$) of the signal $s(t)$. We shall assume throughout that the covariance functions, as well as the interval $[0,T]$, are known by the receiver with the realization that in specific applications there will be additional more difficult considerations required.[1]

In the passive sonar example the sea noise is represented by $n(t)$, and vibrations from a submarine in the surrounding area are represented by $s(t)$. Clearly, the detector is to make an optimal distinction between the presence or absence of the submarine. This model also applies in coherent low-data-rate telemetry systems when the length of the bit time causes extensive carrier-phase jitter, resulting in so much distortion of the transmitted signal that the received waveforms may be represented as narrowband stochastic processes. Communication via the ionosphere produces the same effect, as previously indicated.

The following formal derivation of the optimal detector follows the development of Kadota [9.12], to whom the interested reader is referred for a rigorous presentation.

Under the assumption *signal absent* the received waveform is noise only, in which case the covariance function is $R_n(t,\tau)$ for $0 \leq (t,\tau) \leq T$, whose eigenvalues we designate as $\{\lambda_k\}$ with corresponding orthonormal

[1] The problem of knowledge of necessary statistical information is one which is prevalent throughout the development of the theory.

eigenfunctions $\{\varphi_k(t)\}$. Throughout we assume that the indexing of the eigenvalues is such that they decrease as a function of k, that is,

$$\lambda_1 \geq \lambda_2 \geq \lambda_3 \geq \cdots$$

In the case *signal present* the covariance function of the received waveform is

$$R_p(t,\tau) = a_1{}^2 R_s(t,\tau) + R_n(t,\tau) \tag{9.16}$$

with eigenvalues and orthonormal eigenfunctions designated by $\{\eta_k\}$ and $\{\psi_k(t)\}$, respectively.

When the received waveform is

$$y(t) = a_1 s(t) + n(t)$$

$y(t)$ has the Karhunen-Loeve expansion (see Appendix B)

$$y(t) = \underset{K\to\infty}{\text{l.i.m.}} \sum_{k=1}^{K} y_k \psi_k(t) \qquad 0 \leq t \leq T \tag{9.17}$$

where the

$$y_k \triangleq [y(t),\psi_k(t)]\Big|_{y(t)=a_1 s(t)+n(t)} \qquad k = 1, 2, \ldots \tag{9.18}$$

are mutually independent gaussian random variables with zero-mean values and variances

$$\sigma_{y_k}{}^2 = \eta_k \qquad k = 1, 2, \ldots \tag{9.19}$$

Under the hypothesis *signal absent* the received waveform has the representation

$$y(t) = n(t) = \underset{K\to\infty}{\text{l.i.m.}} \sum_{k=1}^{K} y'_k \varphi_k(t) \qquad 0 \leq t \leq T \tag{9.20}$$

where

$$y'_k \triangleq [y(t),\varphi_k(t)]\Big|_{y(t)=n(t)} \qquad k = 1, 2, \ldots$$

are mutually independent zero-mean gaussian random variables with variances

$$\sigma^2_{y'_k} = \lambda_k \qquad k = 1, 2, \ldots \tag{9.21}$$

respectively.

With the *continuous method* of arriving at the optimal detector, the likelihood ratio is formed with the first m coefficients of the Karhunen-Loeve expansion. For it to be a legitimate likelihood ratio, the same m random variables must be employed under both hypotheses.

Let us first consider the likelihood ratio under the hypothesis *signal present*, which we shall form using the random variables[1]

$$\{y_k;\ k = 1, \ldots, m\}$$

given by

$$y_k = [y(t),\psi_k(t)] \tag{9.22}$$

Therefore, since we are using the eigenfunctions of $R_p(t,\tau)$, when

$$y(t) = a_1 s(t) + n(t)$$

the $\{y_k\}$ will then be statistically independent zero-mean gaussian random variables with variances

$$\sigma_{y_k}{}^2 = \eta_k \qquad k = 1, \ldots, m \tag{9.23}$$

However, when $y(t) = n(t)$ the $\{y_k\}$ given by (9.22) are zero-mean gaussian random variables with covariances given by

$$r_{jk} \triangleq E(y_j y_k | y(t) = n(t)) = \iint_0^T R_n(t,\tau)\psi_j(t)\psi_k(\tau)\,dt\,d\tau = \sum_{l=1}^{\infty} \lambda_l \nu_{lj} \nu_{lk} \tag{9.24}$$

where

$$\nu_{lj} \triangleq [\varphi_l(t),\psi_j(t)] \tag{9.25}$$

If we define the covariance matrices

$$\mathbf{R}_p{}^{(m)} \triangleq \begin{bmatrix} \eta_1 & & 0 \\ & \eta_2 & \\ & & \ddots \\ 0 & & & \eta_m \end{bmatrix}$$

and

$$\mathbf{R}_a{}^{(m)} \triangleq [r_{ij}]$$

of

$$\mathbf{Y} \triangleq \begin{bmatrix} y_1 \\ \cdot \\ \cdot \\ \cdot \\ y_m \end{bmatrix}$$

under the *signal present* and *signal absent* hypotheses, respectively, we can

[1] The $\{y_k\}$ are not intended to represent the received process; they are used as a scheme by which the likelihood ratio may be formed and the necessary limiting operations carried out.

express the likelihood ratio as

$$L_m(\mathbf{Y}) = \left(\frac{|\mathbf{R}_a{}^{(m)}|}{|\mathbf{R}_p{}^{(m)}|}\right)^{\frac{1}{2}} \frac{\exp -\frac{1}{2}\mathbf{Y}^T(\mathbf{R}_p{}^{(m)})^{-1}\mathbf{Y}}{\exp -\frac{1}{2}\mathbf{Y}^T(\mathbf{R}_a{}^{(m)})^{-1}\mathbf{Y}}$$

which is compared with a threshold. After we have taken logarithms, the optimal decision function reduces to

$$\text{If } \frac{1}{2} \ln \frac{|\mathbf{R}_a{}^{(m)}|}{|\mathbf{R}_p{}^{(m)}|} + \frac{1}{2} Q_m > T_m \qquad \text{decide } \textit{signal present} \tag{9.26}$$

where the threshold T_m is dependent upon the use made of the receiver, and

$$Q_m \triangleq \mathbf{Y}^T[(\mathbf{R}_a{}^{(m)})^{-1} - (\mathbf{R}_p{}^{(m)})^{-1}]\mathbf{Y} \tag{9.27}$$

The goal now is to allow the dimensionality m in the quadratic detector to increase and obtain an integral representation for the receiver operation in the limit. Note first that if the inequality

$$0 < \lim_{m\to\infty} \frac{|\mathbf{R}_a{}^{(m)}|}{|\mathbf{R}_p{}^{(m)}|} < \infty \tag{9.28}$$

is not satisfied, we have a singular detection case. In fact, perfect detection can be attained if and only if (9.28) is not satisfied (see Kadota [9.12]). On the assumption that (9.28) is satisfied, we define

$$w^{(m)}(t,\tau) \triangleq \sum_{j=1}^{m} \sum_{k=1}^{m} w_{jk}{}^{(m)}\psi_j(t)\psi_k(\tau) \tag{9.29}$$

where the matrix of coefficients $[w_{jk}{}^{(m)}]$ is

$$\mathbf{W}^{(m)} \triangleq [w_{jk}{}^{(m)}] \triangleq (\mathbf{R}_a{}^{(m)})^{-1} - (\mathbf{R}_p{}^{(m)})^{-1} \tag{9.30}$$

Then the optimal quadratic detector in (9.27) can be written as

$$Q_m = \iint_0^T y(t)y(\tau)w^{(m)}(t,\tau)\, dt\, d\tau \tag{9.31}$$

The optimal detector, when all the data available in the received waveform is used, is given by

$$Q_\infty \triangleq \lim_{m\to\infty} Q_m$$

where

$$Q_\infty \triangleq \iint_0^T y(t)y(\tau)w(t,\tau)\, dt\, d\tau \tag{9.32}$$

with

$$w(t,\tau) \triangleq \lim_{m\to\infty} w^{(m)}(t,\tau) \tag{9.33}$$

To determine the integral equation whose solution is the weighting function in the optimal quadratic detector in (9.30) we note that $\mathbf{W}^{(m)}$ is the solution of the matrix equation

$$\mathbf{R}_a{}^{(m)}\mathbf{W}^{(m)}\mathbf{R}_p{}^{(m)} = \mathbf{R}_p{}^{(m)} - \mathbf{R}_a{}^{(m)} \tag{9.34}$$

which can be expressed equivalently (see Prob. 9.9) by the integral equation

$$\iint_0^T R_a{}^{(m)}(t,u)w^{(m)}(u,v)R_p{}^{(m)}(v,\tau)\,du\,dv = R_p{}^{(m)}(t,\tau) - R_a{}^{(m)}(t,\tau) \qquad 0 \leq (t,\tau) \leq T \tag{9.35}$$

where

$$R_a{}^{(m)}(t,\tau) \triangleq \sum_{k=1}^{m}\sum_{l=1}^{m} r_{kl}\psi_k(t)\psi_l(\tau) \tag{9.36}$$

and

$$R_p{}^{(m)}(t,\tau) \triangleq \sum_{k=1}^{m} \eta_k\psi_k(t)\psi_k(\tau) \tag{9.37}$$

In (9.35) we are now solving directly for the desired weighting function, rather than via the coefficients in its series representation (9.29). Formally taking limits in (9.35) yields the integral equation whose solution is the optimal weighting function.

The optimal detector thus decides *signal present* if

$$\iint_0^T y(t)w(t,\tau)y(\tau)\,dt\,d\tau > T_T \tag{9.38}$$

where $w(t,\tau)$ is the solution of the integral equation

$$\iint_0^T R_a(t,u)w(u,v)R_p(v,\tau)\,du\,dv = R_p(t,\tau) - R_a(t,\tau) \qquad 0 \leq (t,\tau) \leq T \tag{9.39}$$

and in the detection model the threshold T_T is determined from the false-alarm probability.

The conditions under which the above limiting operation is valid have been provided by Kadota [9.12]. The other formulation of the problem is that of forming the likelihood ratio with the random variables obtained from the eigenfunctions of the noise covariance function:

$$y_k' = [y(t),\varphi_k(t)] \qquad k = 1, 2, \ldots$$

This provides the same results and follows immediately from the symmetry of the model. This noise-in-noise model could equally well have been formulated as a binary communication system (see Prob. 9.11).

Examples of the optimal detector for specific covariance functions are given by Helstrom [9.6], Slepian [9.9], and Kadota [9.12]; some are given in the problems at the end of the chapter. Kadota [9.12] has determined the conditions under which the integral equation (9.39) has a square integrable solution, $w(t,\tau)$,

$$\iint_0^T w^2(t,\tau)\,dt\,d\tau < \infty$$

which is such that

$$w(t,\tau) = \sum_{j=1}^{\infty} \sum_{k=1}^{\infty} w_{jk}\psi_j(t)\psi_k(\tau)$$

Balakrishnan [9.11] has shown that if the signal and noise covariance kernels commute, the form of the optimal detector simplifies to

$$w(t,\tau) = \sum_{k=1}^{\infty} w_k f_k(t) f_k(\tau)$$

where the $\{f_k(t)\}$ are the eigenfunctions of the linear operator $R_n^{-1}R_s$; that is, the $f_k(t)$ satisfy the characteristic equation

$$\int_0^T R_s(t,\tau)f_k(\tau)\,d\tau = \gamma_k \int_0^T R_n(t,\tau)f_k(\tau)\,d\tau \qquad k = 1, 2, \ldots; \quad 0 \le t \le T$$

where γ_k is the eigenvalue with eigenfunction $f_k(t)$. Finally, necessary and sufficient conditions under which this simplification can be obtained have been found by Baker [9.13], who determined necessary and sufficient conditions under which the operators R_n and R_s commute.

In conclusion, the proper choice of an acceptable model for the noise-in-noise problem is a difficult one, inasmuch as seemingly realistic models in certain cases provide arbitrarily small probabilities of error (see, for example, Prob. 9.11 and Slepian [9.9]). For a model of the detection of noise in noise which is realistic for practical applications, it may be necessary to take into account the incomplete knowledge of the statistics of the processes involved as well as the receiver's inability to distinguish the received waveform with arbitrary precision. Slepian [9.9] has pointed this out, indicating that results which do not take these physical limitations into account must be applied carefully.

PROBLEMS

9.1 Given (9.8), verify the inequality (9.9).

9.2 Verify equation (9.14) by means of (9.6).

9.3 In a detection environment let the observed vector be

$$\mathbf{Y} = a\mathbf{S} + \mathbf{N}$$

where $\mathbf{S}$ and $\mathbf{N}$ are independent zero-mean gaussian random vectors with covariance matrices $\mathbf{\Lambda}_s = \sigma_s^2\mathbf{I}$ and $\mathbf{\Lambda}_n = \sigma_n^2\mathbf{I}$, respectively. The scaler a is either $a_1 > 0$ or $a_1 = 0$.

(*a*) Show that the optimal detector forms

$$\|\mathbf{Y}\|^2$$

which is compared with a threshold.

(*b*) Evaluate the threshold based upon a false-alarm probability equal to α.

(*c*) Determine the corresponding detection probability in terms of tabulated functions. Hint: See the chi-squared distribution (Appendix F).

9.4 Consider Prob. 9.3 again. This time assume that the signal covariance matrix is

$$\mathbf{\Lambda}_s = \sigma_s^2 \begin{bmatrix} 1 & \rho & 0 & \cdots & 0 \\ \rho & 1 & \ddots & \ddots & \vdots \\ 0 & \ddots & \ddots & \ddots & 0 \\ \vdots & \ddots & \ddots & \ddots & \rho \\ 0 & \cdots & 0 & \rho & 1 \end{bmatrix} \qquad \rho < 1$$

Find representations for the threshold and resulting detection probability.

9.5 *Binary communication via random vectors* Specify the optimal receiver using probability of error as the criterion when the received vector is

$$\mathbf{Y} = \mathbf{S} + \mathbf{N}$$

where the signal vector $\mathbf{S}$ is a gaussian random vector with zero mean and one of two possible covariance matrices, $\mathbf{\Lambda}_1$ or $\mathbf{\Lambda}_2$, which have a priori probabilities π_1 and π_2, respectively. First assume the zero-mean gaussian noise vector has arbitrary covariance matrix, say, $\mathbf{\Lambda}_n$, then determine the simplifications which arise when this covariance is assumed to be of the form

$$\mathbf{\Lambda}_n = \sigma_n^2\mathbf{I}$$

where $\mathbf{I}$ is the identity matrix.

9.6 *Perfect detectability of noise in noise* If the observed waveform is

$$y(t) = s(t) + n(t)$$

over the finite interval $0 \le t \le T$, where $s(t)$ and $n(t)$ are independent white gaussian noise processes, show that the energy detector

$$\int_0^T y^2(t)\,dt$$

is optimum, and that with this detector the presence or absence of $s(t)$ can be determined without error.

9.7 In determining the presence of the signal $s(t)$ consider the case when the observed waveform is

$$y(t) = as(t)n_1(t) + n_2(t) \qquad 0 \le t \le T$$

where $n_2(t)$ is white gaussian noise with one-sided spectral density N_0, $n_1(t)$ is a stationary gaussian stochastic process with zero mean and spectral density

$$S_{n_1}(\omega) = \frac{2}{1 + \omega^2}$$

$s(t) = t$, and a is either $a_1 > 0$ or $a_1 = 0$.

(*a*) Given that a_1 is a small positive constant, what is the optimal detector for determining the presence or absence of $s(t)$ for a given false-alarm probability? Describe one technique for system implementation.

(*b*) Determine a simple approximate solution for large values of a_1. Determine whether perfect detection is possible, and if so, how. Hint: The eigenfunctions of $\exp -b|\tau|$ are sine and cosine functions.

9.8 Consider the diagram in Fig. P9.8. The signal $s(t)$ is a zero-mean stationary gaussian process with covariance function $R_s(\tau)$. The additive noise $n(t)$ is white and gaussian, with one-sided spectral density N_0, Δ is a fixed known time delay and $a < 1$ is a fixed known attenuation. We wish to determine the presence or absence of the delayed and attenuated path (*switch open* or *switch closed*) after observing

$$y(t) \qquad 0 \leq t \leq T$$

(*a*) Find the structure of the optimal detector explicitly in terms of integral operations on the observed waveform $y(t)$ by means of the Neyman-Pearson criterion. Assume a false-alarm probability α and assume

$$N_0 \gg R_s(0)$$

(*b*) Give more specific details when

$$R_s(\tau) = P \cos \frac{2\pi\tau}{T}$$

(*c*) Find the optimal detector when $N_0 = 0$.

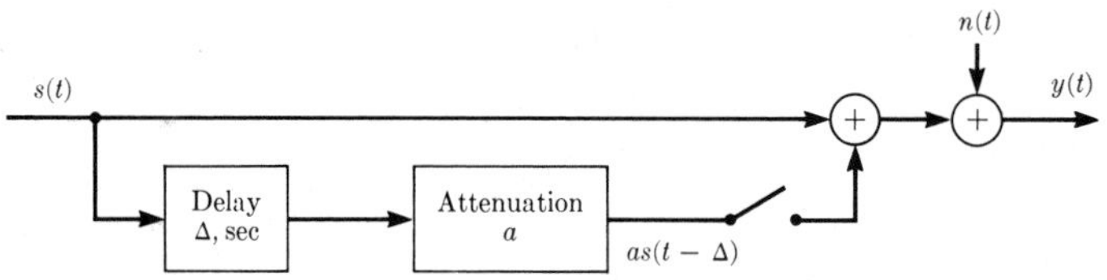

Fig. P9.8

9.9 Derive the integral equation (9.35) from the matrix equation (9.34). This can be done by premultiplying and postmultiplying (9.34) by the vectors

$$[\psi_1(t), \ldots, \psi_m(t)] \qquad \text{and} \qquad [\psi_1(\tau), \ldots, \psi_m(\tau)]^T$$

respectively, making the appropriate substitutions for the matrices in (9.34), and simplifying.

9.10 What is the integral equation whose solution is the optimal weighting function for the detection of the presence or absence of a gaussian stochastic process whose mean is $m_s(t)$, when the mean of the additive gaussian noise process is $m_n(t)$? (See Kadota [9.12].)

9.11 *Binary communication via stochastic signals* Assume that the channel of a binary communication system distorts the transmittable signals so extensively that they can appropriately be considered as members of one of two possible stochastic processes upon arrival at the receiver. In particular, let the received waveform be

$$y(t) = s(t) \qquad 0 \leq t \leq T$$

where the zero-mean process has covariance function $R_1(t,\tau)$ or $R_0(t,\tau)$ with a priori probabilities π_1 and π_0, respectively.

(*a*) Determine the optimal receiver structure in this case, where the additive noise has been assumed negligible.

(*b*) Show that perfect communication is possible when

$$R_1(t,\tau) = \sigma^2 R_0(t,\tau) \qquad \sigma^2 \neq 1$$

9.12 Verify Eq. (9.24).

9.13 Show that the covariance matrices $\mathbf{R}_p^{(m)}$ and $\mathbf{R}_a^{(m)}$ as defined in (9.25) and (9.26), respectively, are positive-definite when $R_p(t,\tau)$ and $R_a(t,\tau)$ are positive-definite covariance functions.

9.14 Verify Eq. (9.31).

REFERENCES

9.1 Friedman, B.: "Principles and Techniques of Applied Mathematics," Wiley, New York, 1956.

9.2 Grenander, U.: Stochastic Processes and Statistical Inference, *Arkro Matemat.*, vol. 17, no. 1, 1950, pp. 195–277.

9.3 Price, R.: Optimum Detection of Stochastic Signals in Noise with Applications to Scatter-multipath Communications, *IRE Trans. Inform. Theory*, vol. IT-2, pp. 125–135, 1956.

9.4 Davis, R. C.: The Detectability of Random Signals in the Presence of Noise, *IRE Trans. Inform. Theory*, vol. PGIT-3, pp. 52–62, 1957.

9.5 Baker, C. R.: Optimum Quadratic Detection of a Random Vector in a Gaussian Noise, *IEEE Trans. Commun. Technol.*, vol. Com-14, no. 6, December, 1966.

9.6 Helstrom, C. W.: "Statistical Theory of Signal Detection," Pergamon Press, New York, 1960.

9.7 Middleton, D.: On the Detection of Stochastic Signals in Additive Normal Noise, part I, *IRE Trans. Inform. Theory*, vol. IT-3, June, 1957; part II, vol. IT-6, June, 1960.

9.8 Middleton, D.: On Singular and Nonsingular Optimum (Bayes) Tests for the Detection of Normal Stochastic Signals in Normal Noise, *IRE Trans. Inform. Theory*, vol. IT-7, April, 1961.

9.9 Slepian, D.: Some Comments on the Detection of Gaussian Signals in Gaussian Noise, *IRE Trans. Inform. Theory*, vol. IT-4, 1958.

9.10 Bello, P.: Some Results on the Problem of Discriminating between Two Gaussian Processes, *IRE Trans. Inform. Theory*, vol. IT-7, 1961.

9.11 Balakrishnan, A. V.: On a Class of Nonlinear Estimation Problems, *IEEE Trans. Inform. Theory*, vol. IT-10, no. 4, October, 1964.

9.12 Kadota, T. T.: Optimum Reception of Binary Gaussian Signals, *Bell Systems Tech. J.*, vol. 43, 1964, pp. 2767–2810; vol. 44, 1965, pp. 1621–1658.

9.13 Baker, C. R.: "Simultaneous Reduction of Covariance Operators and the Noise-in-Noise Problems of Communication Theory," doctoral dissertation, University of California, Los Angeles, Department of Engineering, June, 1967.

9.14 Root, W. L.: The Detection of Signals in Gaussian Noise, in A. V. Balakrishnan (ed.), "Communication Theory," pp. 160–191, McGraw-Hill, New York, 1968.

10
M-ary Digital Communication Systems

Thus far we have considered only binary decision problems, that is, models where the decision space consists of two elements. In such problems the signal space either consisted of two elements or was partitioned into two disjoint subspaces. Let us now develop the more general situation in which the signal set consists of M possible signals or M classes of signals with a priori probabilities $\{\pi_i; i = 1, \ldots, M\}$. Such a model arises in certain communication systems and in pattern-recognition problems. For example, in communications, instead of recognizing or detecting a binary sequence bit by bit, as we have done previously, we might delay the decision until a sequence of length n has been observed, at which point the receiver attempts to decide which signal has been transmitted from a class of $M = 2^n$ possibilities, that is the class of possible sequences of binary bits of length n. This is known as *block encoding.*

We shall restrict ourselves in this development to the Bayes decision rules. The possible decisions are $\{a_1, \ldots, a_M\}$, corresponding to $\{s_1, \ldots, s_M\}$, respectively. The question posed is: For a given set of

$\{s_i\}$, what is the optimal receiver in the Bayes sense? For this we recall the general representation of the average risk developed in Chap. 3,

$$R(\pi,\delta) = \int_{\mathbf{\Omega}} \pi(s)\, ds \int_{\mathbf{\Gamma}} dy \int_{\mathbf{A}} da\, C(s,a)\delta(a|y)f(y|s) \tag{10.1}$$

As the problem has been presented, $\mathbf{\Omega}$ and $\mathbf{A}$ contain a finite number of elements. Setting $\pi(s_i) = \pi_i$, we can write

$$R(\pi,\delta) = \sum_{i=1}^{M} \pi_i \int_{\mathbf{\Gamma}} \sum_{j=1}^{M} C(s_i,a_j)\delta(a_j|y)f(y|s_i)\, dy \tag{10.2}$$

where we must maintain the constraint

$$\sum_{j=1}^{M} \delta(a_j|y) = 1$$

for every $y \in \mathbf{\Gamma}$. Equivalently, $R(\pi,\delta)$ can be expressed as

$$R(\pi,\delta) = \sum_{j=1}^{M} \int_{\mathbf{\Gamma}} A_j(y)\delta(a_j|y)\, dy \tag{10.3}$$

where

$$A_j(y) \triangleq \sum_{i=1}^{M} \pi_i C(s_i,a_j)f(y|s_i) \tag{10.4}$$

For a given y and j, $A_j(y)$ is specified by the conditions of the problem and is unchangeable.

We define

$$m(y) \triangleq \min_k A_k(y) \tag{10.5}$$

and let

$$R_m \triangleq \int_{\mathbf{\Gamma}} \sum_{j=1}^{M} m(y)\delta(a_j|y)\, dy = \int_{\mathbf{\Gamma}} m(y)\, dy \tag{10.6}$$

a number that cannot be changed, since it is determined only by unadjustable portions of the model. Then, by adding and subtracting R_m in (10.3), we may express $R(\pi,\delta)$ as

$$R(\pi,\delta) = \sum_{j=1}^{M} \int_{\mathbf{\Gamma}} [A_j(y) - m(y)]\delta(a_j|y)\, dy + R_m \tag{10.7}$$

Since $A_j(y)$ will never be less than $m(y)$, the optimal choice of $\delta(a_j|y)$ is

$$\delta_B(a_j|y) = \begin{cases} 0 & \text{for all } y \text{ such that } A_j(y) > m(y) \\ 1 & \text{for all } y \text{ such that } A_j(y) = m(y) \end{cases} \tag{10.8}$$

Whenever the set of $y \in \mathbf{\Gamma}$ for which $A_k(y) = A_l(y)$ for some $k \neq l$ is a set of measure 0, which is generally the case, we can conclude that there will be a unique minimum among the $A_j(y)$ with probability 1. Hence the optimal decision function is nonrandom, and δ_B uniquely minimizes $R(\pi,\delta)$. That is,

$$\min_{\delta} R(\pi,\delta) = R(\pi,\delta_B) = R_m \tag{10.9}$$

Consider the specialization[1]

$$C(s_i,a_j) = 1 - \delta_{ij}$$

for which the average risk becomes the probability of error. Then

$$\begin{aligned} A_j(y) &= \sum_{i \neq j} \pi_i f(y|s_i) \\ &= f(y) - \pi_j f(y|s_j) \end{aligned}$$

and minimizing

$$A_j(y) = \sum_{i \neq j} \pi_i f(y|s_i) \qquad j = 1, \ldots, M$$

is equivalent to maximizing

$$\pi_k f(y|s_k) \qquad k = 1, \ldots, M$$

Therefore, if the cost matrix

$$\mathbf{C} = [C(s_i,a_j)]$$

is that for the probability of error,

$$\mathbf{C} = \begin{bmatrix} 0 & & 1 \\ & \ddots & \\ 1 & & 0 \end{bmatrix} \tag{10.10}$$

we can equivalently write the optimal decision function as

$$\delta_B(a_j|y) = \begin{cases} 1 & \text{if } \pi_j f(y|s_j) = \max_k \pi_k f(y|s_k) \\ 0 & \text{otherwise} \end{cases} \tag{10.11}$$

which is the same as

$$\delta_B(a_j|y) = \begin{cases} 1 & \text{if } f(s_j|y) = \max_k f(s_k|y) \\ 0 & \text{otherwise} \end{cases} \tag{10.12}$$

The $\{f(s_k|y)\}$ are the *a posteriori probabilities*. Hence the decision function

[1] δ_{ij} is the *Kronecker delta*.

which minimizes the probability of error consists of a choice of that signal which maximizes the a posteriori probabilities.

10.1 COHERENT M-ARY COMMUNICATION

Consider the special case where the observable is the m-dimensional random vector of the form

$$\mathbf{Y} = \mathbf{S}_j + \mathbf{N}$$

where $\mathbf{S}_j$ is one of M possible signal vectors, with a priori probabilities $\{\pi_i; i = 1, \ldots, M\}$. The additive noise is assumed independent of the signal and gaussian, with zero mean and covariance matrix given by

$$E(\mathbf{N}\mathbf{N}^T) = \sigma^2[\mathbf{I}]$$

where $\mathbf{I}$ is the m-by-m identity matrix.

From (10.11), for the cost matrix in (10.10) the optimal decision function is, after observing $\mathbf{Y}$,

$$\text{If } \pi_k f(\mathbf{Y}|\mathbf{S}_k) = \max_i \pi_i f(\mathbf{Y}|\mathbf{S}_i) \qquad \text{decide } a_k \tag{10.13}$$

Since the noise is additive and independent of the signal, the decision function can equivalently be expressed as

$$\text{If } \pi_k f_N(\mathbf{Y} - \mathbf{S}_k) = \max_i [\pi_i f_N(\mathbf{Y} - \mathbf{S}_i)] \qquad \text{decide } a_k$$

or, more specifically, decide

$$\pi_k \exp\left(\frac{-1}{2\sigma^2}\|\mathbf{Y} - \mathbf{S}_k\|^2\right) = \max_i \left[\pi_i \exp\left(\frac{-1}{2\sigma^2}\|\mathbf{Y} - \mathbf{S}_i\|^2\right)\right] \tag{10.14}$$

which corresponds to deciding a_k if

$$\tfrac{1}{2}\|\mathbf{Y} - \mathbf{S}_k\|^2 - \sigma^2 \ln \pi_k = \min_i \left(\tfrac{1}{2}\|\mathbf{Y} - \mathbf{S}_i\|^2 - \sigma^2 \ln \pi_i\right) \tag{10.15}$$

The $\|\mathbf{Y}\|^2$ terms in (10.15) can be discarded, since they do not affect the decision.

Defining $\|\mathbf{S}_i\|^2 \triangleq \mathcal{E}_i$, the optimal decision becomes

$$\begin{aligned}\text{If } &\mathbf{S}_k^T\mathbf{Y} - \tfrac{1}{2}\mathcal{E}_k + \sigma^2 \ln \pi_k \\ &= \max_i \left(\mathbf{S}_i^T\mathbf{Y} - \tfrac{1}{2}\mathcal{E}_i + \sigma^2 \ln \pi_i\right) \qquad \text{decide } a_k\end{aligned} \tag{10.16}$$

This is the *vector form of the matched filter*. The observed vector is *correlated*, or *matched*, with each of the possible signal vectors. These values are adjusted by the signal energies and a priori probabilities as indicated in (10.16), and that a_k is chosen which corresponds to the largest of these adjusted values. It is clear that if the a priori probabilities are equally likely and the signal vectors each have the same energy, the receiver becomes

$$\text{If } \mathbf{Y}^T\mathbf{S}_k = \max_i \mathbf{Y}^T\mathbf{S}_i \qquad \text{decide } a_k \tag{10.17}$$

The probability of correct decision for the system in (10.16) can be expressed as

$$\begin{aligned} P_c &= \sum_{k=1}^{M} \pi_k \Pr(a_k \text{ was decided} \,|\mathbf{S}_k \text{ was transmitted}) \\ &= \sum_{k=1}^{M} \pi_k \Pr[\mathbf{S}_k{}^T\mathbf{Y} - \tfrac{1}{2}\mathcal{E}_k + \sigma^2 \ln \pi_k \\ &\qquad\qquad = \max_i (\mathbf{S}_i{}^T\mathbf{Y} - \tfrac{1}{2}\mathcal{E}_i + \sigma^2 \ln \pi_i)|\mathbf{Y} = \mathbf{S}_k + N] \end{aligned} \tag{10.18}$$

This result can be extended in the same way as in the binary case to continuously observed waveforms over the known interval of time $[0,T]$, with the following result. For a coherent communication system in which the observed waveform is

$$y(t) = s_j(t) + n(t) \qquad 0 \le t \le T \tag{10.19}$$

where $s_j(t)$ is one of M transmittable signal waveforms with a priori probabilities $\{\pi_i;\ i = 1, \ldots, M\}$ and total energies

$$\mathcal{E}_i = \int_0^T s_i^2(t)\,dt \qquad i = 1, \ldots, M$$

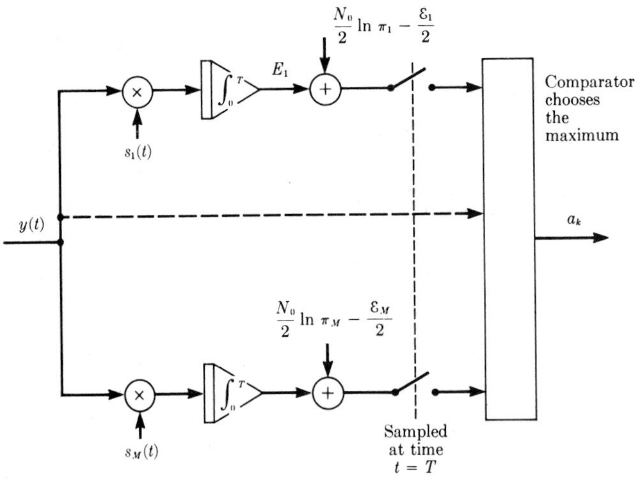

Fig. 10.1 The optimal M-ary receiver for a coherent communication system in white gaussian noise.

the receiver which minimizes the probability of error is the one that forms

$$E_i \triangleq \int_0^T y(t)s_i(t)\,dt \qquad i = 1, \ldots, M \tag{10.20}$$

and decides a_k [or that $s_k(t)$ was transmitted] if

$$E_k - \frac{\mathcal{E}_k}{2} + \frac{N_0}{2}\ln \pi_k = \max_i \left(E_i - \frac{\mathcal{E}_i}{2} + \frac{N_0}{2}\ln \pi_i\right) \tag{10.21}$$

The receiver which forms the inner products or correlations in (10.20) is called the *matched-filter receiver*. A block diagram of this system is given in Fig. 10.1. If the signal energies are all equal and they are a priori equally likely, the receiver operation reduces to deciding a_k if

$$E_k = \max_i E_i \tag{10.22}$$

The decision function in (10.22) is independent of the signal energy and noise spectral density. If the a priori probabilities are not equally likely, then implementation of the optimal receiver requires knowledge of the noise spectral density.

In Part Two general expressions for the probability of detection will be considered in depth for these systems, as well as for certain specific examples of M-ary coherent communication systems.

10.2 NONCOHERENT M-ARY COMMUNICATION

The extension from binary to M-ary noncoherent communication systems follows in much the same way as in coherent systems. The received waveform is expressed as

$$y(t) = s_j(t;\phi) + n(t) \qquad 0 \le t \le T \tag{10.23}$$

where

$$s_j(t;\phi) = A_j(t)\cos\left[\omega_0 t + \theta_j(t) + \phi\right] \qquad 0 \le t \le T;\ j = 1, \ldots, M$$

is one of M transmittable signal waveforms with a priori probabilities $\{\pi_i;\ i = 1, \ldots, M\}$ and energies given by

$$\mathcal{E}_i \triangleq \int_0^T s_i^2(t;\phi)\,dt = \frac{1}{2}\int_0^T A_i^2(t)\,dt \qquad i = 1, \ldots, M$$

where the assumption has been made that the amplitude and phase modulations are narrowband with respect to the carrier frequency. The unknown random phase is uniformly distributed over $(-\pi,\pi)$, and the additive noise is white and gaussian with two-sided spectral density $N_0/2$.

Proceeding in the same manner as in the binary noncoherent example, with the exception of maximizing over M a posteriori probabilities instead of two, we can express the receiver which minimizes the probability of error as

$$\text{If } \pi_k \frac{1}{2\pi} \int_{-\pi}^{\pi} d\psi \exp\left(\frac{-1}{N_0} \int_0^T \{y(t) - A_k(t) \cos [\omega_c t + \theta_k(t) + \psi]\}^2 \, dt\right)$$
$$= \max_i \pi_i \frac{1}{2\pi} \int_{-\pi}^{\pi} d\psi \exp\left(\frac{-1}{N_0} \int_0^T \{y(t)$$
$$- A_i(t) \cos [\omega_c t + \theta_i(t) + \psi]\}^2 \, dt\right) \qquad \text{decide } a_k \quad (10.24)$$

Taking logarithms and simplifying results in the decision function

$$\text{If } \ln \pi_k + \ln\left(\frac{1}{2\pi} \int_{-\pi}^{\pi} d\psi \exp\left\{-\frac{\mathcal{E}_k}{N_0} + \frac{2}{N_0} \int_0^T y(t) A_k(t) \cos [\omega_c t\right.\right.$$
$$\left.\left. + \theta_k(t) + \psi] \, dt\right\}\right) = \max_i \left[\ln \pi_i + \ln\left(\frac{1}{2\pi} \int_{-\pi}^{\pi} d\psi\right.\right.$$
$$\left.\left.\exp\left\{-\frac{\mathcal{E}_i}{N_0} + \frac{2}{N_0} \int_0^T y(t) A_i(t) \cos [\omega_c t + \theta_i(t) + \psi] \, dt\right\}\right)\right]$$
$$\text{decide } a_k \quad (10.25)$$

In terms of the modified Bessel function, we obtain the result

$$\text{If } \ln I_0\left(\frac{2R_k}{N_0}\right) - \frac{\mathcal{E}_k}{N_0} + \ln \pi_k = \max_i \left[\ln I_0\left(\frac{2R_i}{N_0}\right) - \frac{\mathcal{E}_i}{N_0} + \ln \pi_i\right]$$
$$\text{decide } a_k \quad (10.26)$$

where

$$R_i \triangleq \sqrt{e_{c_i}{}^2 + e_{s_i}{}^2} \qquad i = 1, \ldots, M$$
$$e_{c_i} \triangleq \int_0^T y(t) A_i(t) \cos [\omega_c t + \theta_i(t)] \, dt \qquad i = 1, \ldots, M$$

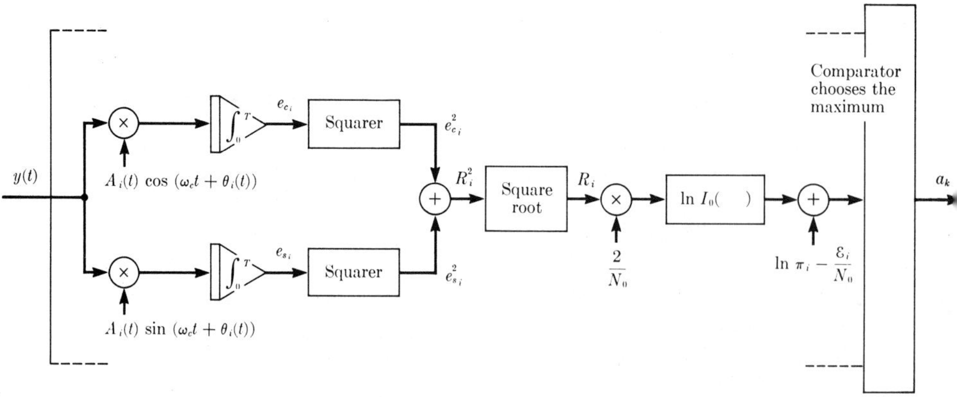

Fig. 10.2 The ith branch of the optimal M-ary receiver for a noncoherent communication system in white gaussian noise.

and

$$e_{s_i} \triangleq \int_0^T y(t)A_i(t) \sin [\omega_c t + \theta_i(t)]\, dt \qquad i = 1, \ldots, M \qquad (10.27)$$

A block diagram of this system is given in Fig. 10.2. The probability of detection for the system is considered in Chap. 19.

PROBLEMS

10.1 In the M-ary detection problem the maximum-likelihood decision function is defined as the one which decides $\mathbf{S}_j$ was transmitted if

$$f(\mathbf{Y}|\mathbf{S}_j) = \max_i f(\mathbf{Y}|\mathbf{S}_i)$$

where $\mathbf{Y}$ is the observation. Show that the maximum-likelihood decision function corresponds to the Bayes decision function when the costs for an error are all equal and when the a priori probabilities are all equal.

10.2 Express the optimal noncoherent M-ary receiver in terms of complex envelopes.

10.3 Consider the following 3-ary communication system, using the m-dimensional sampling model

$$\mathbf{Y} = \mathbf{S}_j + \mathbf{N}$$

where

$$\|\mathbf{S}_1\|^2 = \|\mathbf{S}_2\|^2 = \mathcal{E}$$
$$\mathbf{S}_1{}^T\mathbf{S}_2 = \rho\mathcal{E} \qquad -1 \le \rho \le 1$$
$$\mathbf{S}_3 = \mathbf{0}$$

The additive noise is gaussian with mean zero and covariance matrix given by $\sigma^2\mathbf{I}$, where $\mathbf{I}$ is the m-by-m unit matrix.

(*a*) Determine the optimal Bayes receiver for a generalized cost matrix and a priori probabilities $\{\pi_i;\ i = 1, 2, 3\}$.

(*b*) Determine the probability of error if all signals are equally likely.

(*c*) What is the optimal choice for ρ when the criterion is probability of error?

10.4 Let a received waveform be of the form

$$y(t) = s_j(t) + n(t) \qquad 0 \le t \le 2\pi;\ j = 1, 2, 3$$

where

$$s_1(t) = \cos t$$
$$s_2(t) = \cos 2t$$
$$s_3(t) = \cos 3t$$

and $n(t) = x_1 \cos 4t + x_2 \sin 4t$, with x_1 and x_2 independent gaussian random variables with zero mean and variance σ^2. Find a function $h(t)$ and the necessary thresholds so that with probability 1 the transmitted signal can be correctly detected when the receiver forms

$$\int_0^{2\pi} h(t)y(t)\, dt$$

and uses this test statistic for its decision.

10.5 Consider the M-ary coherent communication system with a received waveform

$$y(t) = vs_j(t) + n(t)$$

where the transmittable waveforms have energies $\{\mathcal{E}_i;\ i = 1, \ldots, M\}$ and a priori probabilities $\{\pi_i;\ i = 1, \ldots, M\}$. The received signal amplitude is a nonnegative random variable with probability density function $p(v)$. The additive noise is white gaussian with two-sided spectral density $N_0/2$.

(*a*) Show that the receiver which minimizes the probability of error forms

$$C_i \triangleq \pi_i \int_0^\infty p(v) \exp \frac{-\mathcal{E}_i v^2}{N_0} \exp \left[\frac{2v}{N_0} \int_0^T y(t)s_i(t)\, dt \right] dv \qquad i = 1, \ldots, M$$

and decides s_k if

$$C_k = \max_i C_i$$

(*b*) If $p(v)$ is Rayleigh distributed, that is,

$$p(v) = \frac{v}{\sigma_v^2} \exp \frac{-v^2}{2\sigma_v^2}$$

show that

$$C_i = \pi_i \int_0^\infty \frac{v}{\sigma_v^2} \exp \left[-\left(\frac{1}{2\sigma_v^2} + \frac{\mathcal{E}_i}{N_0} \right) v^2 \right] \exp \left[\frac{2v}{N_0} \int_0^T y(t)s_i(t)\, dt \right] \qquad i = 1, \ldots, M$$

and that this is approximately the same as choosing the largest[1] of

$$\pi_j d_j E_j \exp (d_j E_j^2)$$

for large signal-to-noise ratios, where

$$d_j \triangleq \frac{1}{N_0^2[(1/2\sigma_v^2) + (\mathcal{E}_j/N_0)]}$$

and

$$E_i \triangleq \int_0^T y(t)s_i(t)\, dt$$

10.6 *Noncoherent Rayleigh fading channel* Consider the M-ary noncoherent communication system with a received waveform

$$y(t) = vs_j(t,\phi) + n(t)$$

where the transmittable waveforms

$$s_j(t,\phi) = A_j(t) \cos [\omega_c t + \theta_j(t) + \phi]$$

have narrowband amplitude and phase modulation, with energies

$$\mathcal{E}_i = \frac{1}{2} \int_0^T A_i^2(t)\, dt \qquad i = 1, \ldots, M$$

[1] *Hint:*

$$\int_0^\infty x \exp [-(ax^2 + 2bx)]\, dx = \frac{1}{2a} - \frac{b}{a^{\frac{3}{2}}} \exp \left(\frac{b^2}{a} \right) \int_{b/\sqrt{a}}^\infty \exp (-y^2)\, dy$$

and a priori probabilities $\{\pi_i; i = 1, \ldots, M\}$. The received signal amplitude is a nonnegative random variable with probability density function $p(v)$. The phase ϕ is uniformly distributed over $(-\pi,\pi)$. The additive noise is white and gaussian with one-sided spectral density N_0.

(a) Show that the receiver which minimizes the probability of error forms

$$F_i \triangleq \pi_i \int_0^\infty p(v) \exp\left(\frac{-\mathcal{E}_i v^2}{N_0}\right) I_0\left(\frac{2v}{N_0} R_i\right) dv \qquad i = 1, \ldots, M$$

and decides that s_k was sent if

$$F_k = \max_i F_i$$

where

$$R_i \triangleq \sqrt{e_{c_i}{}^2 + e_{s_i}{}^2}$$

$$e_{c_i} \triangleq \int_0^T y(t)A_i(t) \cos\left[\omega_c t + \theta_i(t)\right] dt$$

$$e_{s_i} \triangleq \int_0^T y(t)A_i(t) \sin\left[\omega_c t + \theta_i(t)\right] dt \qquad i = 1, \ldots, M$$

(b) If $p(v)$ is Rayleigh distributed (as in the previous problem), show that $\{F_i\}$ reduces[1] to

$$F_i = \frac{\pi_i \exp\{2\sigma_v{}^2 R_i{}^2/[N_0{}^2(1 + 2\sigma_v{}^2\mathcal{E}_i/N_0)]\}}{1 + 2\sigma_v{}^2\mathcal{E}_i/N_0} \qquad i = 1, \ldots, M$$

(c) What further simplification is attainable when the transmittable signals have the same energy and are equally likely?

10.7 *M-ary partially coherent communication* Consider the M-ary communication system with a received waveform

$$y(t) = s_j(t,\phi) + n(t) \qquad 0 \leq t \leq T$$

where the transmitted signal waveforms are given by

$$s_j(t,\phi) = A_j(t) \cos\left(\omega_c t + \theta_j(t) + \phi\right) \qquad j = 1, \ldots, M$$

The $\{s_j(t,\phi)\}$ are equally likely narrowband signals with the same energy

$$\mathcal{E} = \frac{1}{2}\int_0^T A_j{}^2(t)\, dt \qquad j = 1, \ldots, M$$

and $n(t)$ is white gaussian noise with one-sided spectral density N_0. The received carried phase ϕ has probability density function

$$p(\phi) = \frac{\exp(\alpha \cos \phi)}{2\pi I_0(\alpha)} \qquad -\pi \leq \phi \leq \pi$$

(a) Find the optimal receiver for this system.

(b) Find the resulting probability of error when the transmitted signals are orthogonal.

[1] *Hint:*

$$\int_0^\infty x \exp(-a^2x^2)\, I_0(bx)\, dx = \frac{1}{2a^2} \exp\frac{b^2}{4a^2}$$

Hint: See Prob. 7.12 and Ref. [10.2].

10.8 *M-ary phase-shift keying with transmitted reference tone* Consider the M-ary digital communication system in which the receiver observes

$$y(t) = A \cos \left(\omega_0 t + \frac{2\pi l}{M} + \phi \right) + n(t)$$

over the interval $[0,T]$, where $l = 1, 2, \ldots, M$ with equally likely a priori probabilities. A phase reference tone is also received,

$$y_r(t) = A_r \cos (\omega_0 t + \phi) + n_r(t)$$

over the interval $(-qT,0)$, where q is some positive integer, and ϕ is the same in both received waveforms and is uniformly distributed over $(-\pi,\pi)$. The two noise processes are white and gaussian, each with spectral density $N_0/2$ and independent of one another.

(*a*) Determine the best a posteriori receiver for this M-ary communication system.

(*b*) Evaluate the resulting probability of error.

(*c*) For a fixed amount of total energy, determine the optimal percentages to be placed in the information-bearing signal and in the reference tone, respectively (see Bussgang and Leiter [10.11, 10.12].)

10.9 *M-ary coherent systems with nonwhite gaussian noise* Consider the M-ary communication system with a received waveform

$$y(t) = s_j(t) + n(t) \qquad 0 \le t \le T$$

where the transmitted waveforms $\{s_j(t)\}$ have a priori probabilities $\{\pi_j; j = 1, \ldots, M\}$ and energies $\{\mathcal{E}_j; j = 1, \ldots, M\}$, respectively. The additive noise $n(t)$ is gaussian with zero mean and covariance function $R_n(t,\tau)$ for $0 \le (t,\tau) \le T$. Which receiver operation minimizes the probability of error?

10.10 *M-ary noncoherent systems with nonwhite gaussian noise* Consider the M-ary communication system where the received waveform is

$$y(t) = s_j(t,\varphi) + n(t) \qquad 0 \le t \le T$$

and the transmitted waveforms are given by

$$s_j(t,\varphi) = A_j(t) \cos [\omega_c t + \theta_j(t) + \varphi] \qquad j = 1, \ldots, M$$

The $\{s_j(t,\varphi)\}$ have a priori probabilities $\{\pi_j; j = 1, \ldots, M\}$ and are narrowband signals with energies

$$\mathcal{E}_j = \frac{1}{2} \int_0^T A_j^2(t)\, dt \qquad j = 1, \ldots, M$$

respectively. The noise $n(t)$ is gaussian with zero mean and covariance function $R_n(t,\tau)$ for $0 \le (t,\tau) \le T$. What is the optimal receiver operation with probability of error as the performance criterion?

10.11 *Estimation* Consider observing a sequence of 1s and 0s of length m, where the probability of a 1 is q and the probability of a 0 is $1 - q$. Assume that successive elements of the sequence are independent of one another and that prior to the sequence q is chosen at random according to the a priori distribution $\pi(q)$. Estimate q after observing the sequence and choose the estimate so as to minimize the average risk when the cost function is $(q - a)^2$, where $a = \hat{q}$ is the best estimate.

Answer:

$$\hat{q} = \frac{\int_0^1 q^{n+1}(1-q)^{m-n}\pi(q)\,dq}{\int_0^1 q^n(1-q)^{m-n}\pi(q)\,dq}$$

where n is the number of 1s in the sequence.

Now let $\mathbf{\Omega}$ be finite, i.e., $\mathbf{\Omega} = \{s_1, \ldots, s_M\}$ with a priori probabilities $\pi_1, \ldots, \pi_M$, where $\pi_k > 0$, $k = 1, \ldots, M$. What is the best estimate?

REFERENCES

10.1 Middleton, D.: "Introduction to Statistical Communication Theory," McGraw-Hill, New York, 1960.

10.2 Viterbi, A. J.: "Principles of Coherent Communications," McGraw-Hill, New York, 1966.

10.3 Wozencraft, J. M., and I. M. Jacobs: "Principles of Communication Engineering," Wiley, New York, 1965.

10.4 Hancock, J. C., and P. A. Wintz: "Signal Detection Theory," McGraw-Hill, New York, 1966.

10.5 Golomb, S., et al.: "Digital Communications," Prentice-Hall, Englewood Cliffs, N.J., 1964.

10.6 Arthurs, E., and H. Dym: On the Optimum Detection of Digital Signals in the Presence of White Gaussian Noise: A Geometric Interpretation and a Study of Three Basic Data Transmission Systems, *IRE Trans. Commun. Systems*, vol. CS-10, December, 1962.

10.7 Nuttall, A.: Error Probabilities for Equi Correlated *M*-ary Signals under Phase Coherent and Phase Incoherent Reception, *IRE Trans. Inform. Theory*, vol. IT-8, July, 1962.

10.8 Turin, G. L.: An Introduction to Matched Filters, *IRE Trans. Inform. Theory*, vol. IT-6, June, 1960.

10.9 Price, R.: Optimum Detection of Random Signals in Noise, with Applications to Scatter-multipath Communication, *IRE Trans. Inform. Theory*, vol. IT-2, 1956.

10.10 Helstrom, C. W.: The Resolution of Signals in White Gaussian Noise, *Proc. IRE*, vol. 43, September, 1955.

10.11 Bussgang, J. J., and M. Leiter: Error Rate Approximation for Differential Phase Shift Keying, *IEEE Trans. Commun. Systems*, vol. CS-12, pp. 18–27, March, 1964.

10.12 Bussgang, J. J., and M. Leiter: Phase Shift Keying with a Transmitted Reference, *IEEE Trans. Commun. Technol.*, vol. Com-14, no. 1, pp. 14–22, February, 1966.

part two

Signal Design

11
Introduction

The fundamental purpose of communication and telemetry systems is the transmission of reliable information or data through an unreliable channel. In space communications a major constraint is the limitation of transmitter power available on the spacecraft. Or, equivalently, for a given transmitter size, and hence power, we are interested in maximizing the possible range of the spacecraft while still maintaining telemetry transmission to the earth within a specified error rate. Also, the waveforms of the transmitted signals are limited by the complexity of the coding and transmitting equipment, but more significantly by their allowed bandwidth. The bandwidth normally must be small enough so that transmission over adjacent channels is not affected by the operation in the given channel.

As early as 1949 Shannon [11.1] demonstrated that it should be possible to exchange bandwidth for power at a fixed source rate so that by increasing bandwidth we could, at least in principle, make up for the lack of transmitter power. This can be made slightly more precise by introducing channel capacity, as derived by Shannon, for a given bandwidth B,

$$C = B \log_2 \left(1 + \frac{P}{N}\right) \tag{11.1}$$

where P is the signal power and N is the noise power. When the communication is that of transmission and reception of electromagnetic signals, either microwave or optical, between a satellite or a deep-space vehicle and the earth, the channel disturbances can be represented by additive white gaussian noise. This is because the principle source of disturbances in space communications is galactic noise, which has a very wide bandwidth and near-constant spectral density, representable as white noise. The advent of the *maser* has reduced internal thermal noise in the microwave region to negligibility in comparison with galactic noise. Similarly, in optical systems, with lasers, for instance, the limiting noise due to the "zero-point field" is again white and gaussian. The noise spectral density in either case is given by

$$S(\omega) = \frac{h\omega}{2\left(\exp \frac{h\omega}{kT_s} - 1\right)} \tag{11.2}$$

where

T_s = source temperature
ω = carrier frequency
h = Planck's constant
k = Boltzman's constant

Thus with the advent of space communications we have "real-life" channels which can be statistically described by additive gaussian noise.

Hence, if the one-sided noise spectral density is N_0, then the noise power is

$$N = N_0 B \tag{11.3}$$

and

$$C_B = B \log_2 \left(1 + \frac{P}{N_0 B}\right) \tag{11.4}$$

which is a maximum when B increases without bound. The maximum value is

$$C_\infty = \frac{P}{N_0} \log_2 e \tag{11.5}$$

By Shannon's coding theorems, for a given information rate H we can transmit at rate H with zero error if

$$H < C \tag{11.6}$$

Hence, if bandwidth is no limitation, we should be able to attain transmission with no error if

$$H < \frac{P}{N_0} \log_2 e \tag{11.7}$$

or the power required need not be greater than

$$\frac{HN_0}{\log_2 e} \tag{11.8}$$

However, there are two major drawbacks to this theory. For one thing, the coding theorems do not tell us how to attain zero error transmission. For another, even if we were fortunate enough to find the optimum code, we know that it would require transmission signals of infinite time duration to attain zero error. In a practical system, however, the time duration T of the signal waveforms is fixed and finite. Moreover, regardless of which concept of bandwidth we use, infinite bandwidth is never allowed in a practical system.

The approach we take here is to design optimal signals when they are constrained to a specified average power P and a finite time duration T. We shall state precisely what we mean by finite bandwidth and how this affects optimal signal design. We shall adopt the design criterion of probability of error because of its inherent physical significance based on the law of large numbers. The optimization can be viewed either as minimizing the average power for a given probability of error or as minimizing the probability of error for a given allowable signal power; the results will be the same. We employ the latter optimization here. This places the optimal-signal-design problem in the framework of detection theory, differing from most approaches, which place the problem in the context of information theory or error-correcting codes. Although the information-theoretic notions are not necessary in what follows, we shall often find them useful, if only to indicate the connections.

REFERENCE

11.1 Shannon, C. E.: "The Mathematical Theory of Communication," University of Illinois Press, Urbana, Ill., 1949.

12
Problem Statement for Coherent Channels

12.1 DESCRIPTION IN THE TIME DOMAIN

Let us begin with a description of the kind of communication link with which we are concerned. A basic block diagram of the system is shown in Fig. 12.1. The information or data usually comes from several analog sources which are sampled, digitalized, and arranged in the form of sequences of binary digits, although in general the digitalized symbols could be elements from a K-ary alphabet. The encoder maps sequences of digits of length n one to one onto a set of M time-varying waveforms each of duration T sec. This is known as block encoding. The resulting sequence of time-varying signals is then used to modulate a high-frequency carrier (AM, FM, PM, etc., or combinations of these). Since each member of each sequence can be any element of the alphabet, we necessarily have $M = K^n$, or $M = 2^n$ in the binary case. Thus the primary source may be taken as discrete, with a nominal rate of H bits per second. If the source is binary, as is usually the case, then the maximum possible rate is, of course, the number of bits per second, and this is what is normally taken as H, since

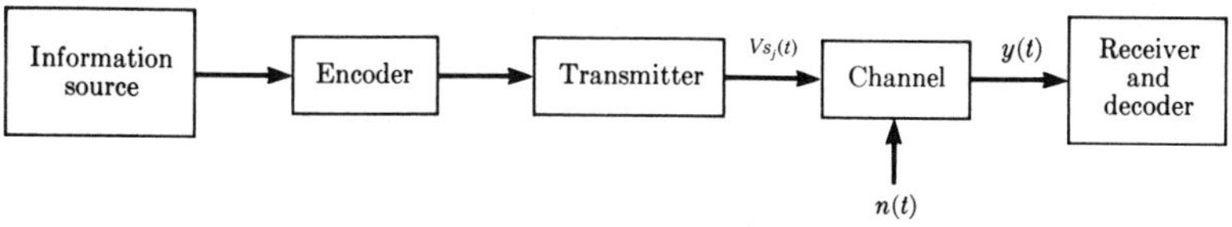

Fig. 12.1 Basic diagram of a digital communication system.

the system must certainly be prepared to handle this rate. Because it is normally impossible to specify quantitatively the precise probabilistic structure of the source, the maximum value of H is assumed, corresponding to equally likely signals:

$$H = \frac{1}{T} \log_2 M$$

Thus in one T-sec interval the total number of possible signals is

$$M = 2^{HT}$$

It is clear that M and T are more basic here than H. For our purposes, we could say that the M transmittable waveforms are equally likely and that each successive waveform is independent of all previous ones (neglecting the form of the data source and the encoder).

The transmitted signal for a particular T-sec interval is assumed to be of the form

$$Vs_j(t)$$

where V is an amplitude scale factor and $s(t)$ is one of M equally likely signals of the form

$$s_j(t) \triangleq A_j(t) \cos [\omega_c t + \theta_j(t)] \qquad 0 \leq t \leq T; j = 1, \ldots, M \tag{12.1}$$

$A_j(t)$ and $\theta_j(t)$ are narrowband signals with respect to the radian carrier frequency ω_c, and the $A_j(t)$ are normalized; that is,

$$\frac{1}{T} \int_0^T A_j^2(t)\, dt = 1 \qquad j = 1, \ldots, M \tag{12.2}$$

so that $V^2/2$ is the average power level of the transmitted waveform. Note that a completely equivalent way to write the transmitted signals would be to introduce complex waveforms, as discussed in Part One.

The received signal for a particular period T will be denoted by

$$y(t) \triangleq Vs_j(t) + n(t) \qquad 0 \leq t \leq T \tag{12.3}$$

where $n(t)$ is white gaussian noise with two-sided spectral density $N_0/2$.

The receiver is assumed to have the following characteristics:

1. It knows the form (except for the amplitude V) of all the transmitted signals, that is, the exact form of each $s_j(t)$ for $j = 1, \ldots, M$.
2. It is synchronized in time (knows the time interval $[0,T]$ during which the signal will arrive) and is phase coherent (knows the carrier phase angle). The recent development of synchronous codes (see Refs. [12.1–12.3]) and phase-locked loops (Refs. [12.4, 12.5]) makes these two assumptions physically attainable.
3 Its sole purpose is to decide at time T which of the M signals $s_j(t)$ has been transmitted, only on the basis of the waveform received during the interval $[0,T]$.

As shown in Chap. 10, for a fixed M-ary signaling alphabet $\{s_j(t)\}$, and with minimum probability of error as the optimization criterion, the optimal receiver is the one that forms the scalar products (see Fig. 12.2)

$$E_i \triangleq [y(t) \cdot s_i(t)] \triangleq \int_0^T y(t)s_i(t)\,dt \qquad i = 1, \ldots, M \tag{12.4}$$

and decides that the kth signal has been transmitted if[1]

$$E_k = \max (E_1, E_2, \ldots, E_M) \tag{12.5}$$

The notation used in (12.4) for the scalar product (or correlation of time functions) will be used throughout.

For every set of M signals of the form specified by (12.1), there exists an optimal receiver and a corresponding probability of detection. In this large class of sets of M signals, all restricted to an energy of $TV^2/2$ in T

[1] Note that the decision rule will not be altered if all the E_i are multiplied by a common positive scale factor.

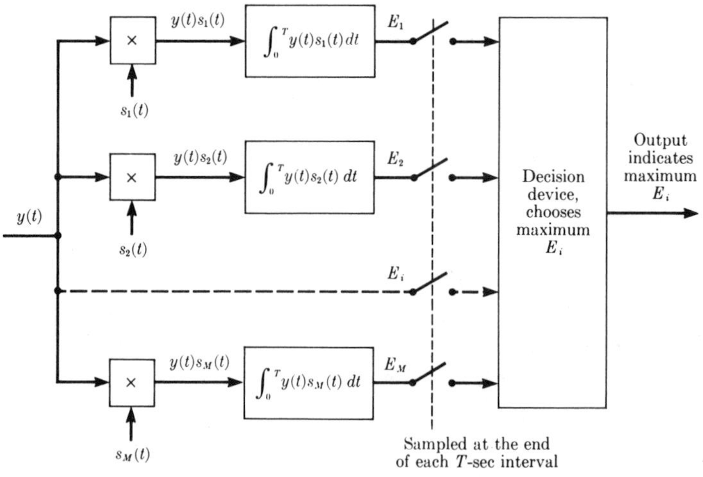

Fig. 12.2 Optimal M-ary coherent receiver.

sec, we wish to find that subclass which has the largest probability of detection. We also want to determine whether this optimal class of sets is independent of the average power level $V^2/2$. The following section will indicate that determining this subclass can be reduced to finding certain classes of D-dimensional vectors on the D-dimensional unit sphere satisfying certain optimization requirements.

12.2 REDUCTION TO FINITE-DIMENSIONAL EUCLIDEAN SPACE

From Eq. (12.2), the average received signal power is given by

$$P_{\text{av}} \triangleq \frac{1}{T}\int_0^T [Vs_j(t)]^2\,dt = \frac{V^2}{2T}\int_0^T A_j^2(t)\,dt + \frac{V^2}{2T}\int_0^T A_j^2(t)\cos\,[2\omega_c t + 2\theta_j(t)]\,dt = \frac{V^2}{2} \tag{12.6}$$

on the assumption that the carrier frequency is high enough that the second integral can be disregarded. Omission throughout of integrals of this form is justified by the narrowband assumption of $A_j(t)$ and $\theta_j(t)$. The signal energy $\mathcal{E}$ received during the interval $[0,T]$ is

$$\mathcal{E} = TP_{\text{av}} = \frac{V^2T}{2} \tag{12.7}$$

From (12.3) and (12.4), the ith correlator output E_i, provided the jth signal has been transmitted, is given by

$$\begin{aligned} E_i &= \int_0^T [Vs_j(t) + n(t)]s_i(t)\,dt \\ &= V\int_0^T A_j(t)A_i(t)\cos\,[\omega_c t + \theta_i(t)]\cos\,[\omega_c t + \theta_j(t)]\,dt \\ &\quad + \int_0^T n(t)A_i(t)\cos\,[\omega_c t + \theta_i(t)]\,dt \qquad i = 1, \ldots, M \end{aligned} \tag{12.8}$$

If we expand and apply the narrowband assumption, we can write E_i as

$$\begin{aligned} E_i &= \int_0^T \frac{V}{2}A_j(t)\cos\theta_j(t)\,A_i(t)\cos\theta_i(t)\,dt \\ &\quad + \int_0^T \frac{V}{2}A_j(t)\sin\theta_j(t)\,A_i(t)\sin\theta_i(t)\,dt \\ &\quad + \int_0^T n(t)\cos\omega_c t\,A_i(t)\cos\theta_i(t)\,dt \\ &\quad - \int_0^T n(t)\sin\omega_c t\,A_i(t)\sin\theta_i(t)\,dt \qquad i = 1, \ldots, M \end{aligned} \tag{12.9}$$

If we let $\mathbf{S}_j(t)$ be the two-dimensional time vector

$$\mathbf{S}_j(t) \triangleq \begin{bmatrix} s_{j1}(t) \\ s_{j2}(t) \end{bmatrix} \triangleq \begin{bmatrix} A_j(t)\cos\theta_j(t) \\ A_j(t)\sin\theta_j(t) \end{bmatrix} \qquad 0 \le t \le T;\, j = 1, \ldots, M \tag{12.10}$$

and specify $\mathbf{N}(t)$ and $\mathbf{Y}(t)$ by

$$\mathbf{N}(t) \triangleq \begin{bmatrix} n_1(t) \\ n_2(t) \end{bmatrix} \triangleq \begin{bmatrix} n(t) \cos \omega_c t \\ -n(t) \sin \omega_c t \end{bmatrix} \tag{12.11}$$

and

$$\mathbf{Y}(t) \triangleq \begin{bmatrix} y_1(t) \\ y_2(t) \end{bmatrix} \triangleq \frac{V}{2} \mathbf{S}_j(t) + \mathbf{N}(t) \tag{12.12}$$

the E_i and corresponding probability of detection can be computed by considering the following equivalent system. Let $\mathbf{S}_j(t)$ be the transmitted signal vector, $\mathbf{N}(t)$ be the additive noise vector, and $\mathbf{Y}(t)$ the received signal vector. The optimal receiver in this case performs the operations

$$E_i = [\mathbf{Y}(t) \cdot \mathbf{S}_i(t)] = [y_1(t) \cdot s_{i1}(t)] + [y_2(t) \cdot s_{i2}(t)] \qquad i = 1, \ldots, M \tag{12.13}$$

which agrees with (12.9). Thus an observer at the receiver output would not be able to distinguish between the two systems, since the set of E_i is identical for each.

Furthermore, notice that the carrier frequency has been eliminated from the signal set in the second model. Because there is a one-to-one correspondence between the signal sets of the two models, an optimal set for one model corresponds to an optimal set for the other. The signal set in the second model does not involve the carrier frequency, so it is eliminated from the optimization, provided it is sufficiently large to allow the narrowband assumptions.

A further reduction is possible to finite-dimensional vectors, since the signal set is finite. That is, we can write

$$\mathbf{S}_j(t) = \sum_{k=1}^{D} s_k{}^j \boldsymbol{\psi}_k(t) \qquad j = 1, \ldots, M \tag{12.14}$$

where

$$\boldsymbol{\psi}_k(t) \triangleq \begin{bmatrix} \psi_{k1}(t) \\ \psi_{k2}(t) \end{bmatrix} \qquad k = 1, \ldots, D \tag{12.15}$$

is an orthogonal set of basis functions such that

$$[\boldsymbol{\psi}_k(t) \cdot \boldsymbol{\psi}_l(t)] = [\psi_{k1}(t) \cdot \psi_{l1}(t)] + [\psi_{k2}(t) \cdot \psi_{l2}(t)] = T\delta_{kl} \tag{12.16}$$

and where

$$s_k{}^i \triangleq \frac{1}{T} [\mathbf{S}_i(t) \cdot \boldsymbol{\psi}_k(t)] \tag{12.17}$$

The dimensionality D of the signal set is at most M and is a measure of the bandwidth of the signal set. This will be discussed in detail in the next section.

From (12.14), each signal can now be characterized by the D-dimensional vector

$$\mathbf{S}_i \triangleq \begin{bmatrix} s_1^i \\ s_2^i \\ \cdot \\ \cdot \\ \cdot \\ s_D^i \end{bmatrix} \qquad i = 1, \ldots, M \tag{12.18}$$

The noise vector $\mathbf{N}(t)$ can also be expanded by use of the orthogonal basis functions; that is, we can express $\mathbf{N}(t)$ as

$$\mathbf{N}(t) = \sum_{j=1}^{D} \zeta_j \boldsymbol{\psi}_j(t) + \boldsymbol{\Delta}(t) \tag{12.19}$$

where the ζ_j are gaussian random variables with

$$E(\zeta_j) = \frac{E([\mathbf{N}(t) \cdot \boldsymbol{\psi}_j(t)])}{T} = 0 \tag{12.20}$$

and

$$E(\zeta_i \zeta_j) = \frac{1}{T^2} E([\mathbf{N}(\alpha) \cdot \boldsymbol{\psi}_i(\alpha)][\mathbf{N}(\beta) \cdot \boldsymbol{\psi}_j(\beta)]) = \frac{N_0}{4T} \delta_{ij} \tag{12.21}$$

and where $\boldsymbol{\Delta}(t)$ is that part of the noise which is orthogonal to the basis functions $\boldsymbol{\psi}_j(t)$ for $j = 1, \ldots, D$. In fact, with probability 1 we have that

$$[\boldsymbol{\Delta}(t) \cdot \boldsymbol{\psi}_j(t)] = 0 \qquad j = 1, \ldots, D \tag{12.22}$$

Since $\boldsymbol{\Delta}(t)$ is orthogonal to each $\boldsymbol{\psi}_j(t)$, $\boldsymbol{\Delta}(t)$ does not affect the value of any of the correlator outputs E_i for $i = 1, \ldots, M$. Therefore for our optimization purposes $\boldsymbol{\Delta}(t)$ can be eliminated from consideration, and if we define

$$\mathbf{N}_0(t) \triangleq \sum_{i=1}^{D} \zeta_i \boldsymbol{\psi}_i(t) \tag{12.23}$$

we can allow the received signal $\mathbf{Y}(t)$ to be

$$\mathbf{Y}(t) = \frac{V}{2} \mathbf{S}_j(t) + \mathbf{N}_0(t) \tag{12.24}$$

$\mathbf{N}_0(t)$ can be characterized by the D-dimensional noise vector

$$\mathbf{Z} \triangleq \begin{bmatrix} \zeta_1 \\ \cdot \\ \cdot \\ \cdot \\ \zeta_D \end{bmatrix} \tag{12.25}$$

If we substitute each of these expansions into (12.13), we can express E_i as

$$E_i = \left[\sum_{k=1}^{D} \left(\frac{V}{2} s_k{}^j + \zeta_k \right) \psi_k(t) \cdot \sum_{l=1}^{D} s_l{}^i \psi_l(t) \right]$$

$$= T \left(\frac{V}{2} \mathbf{S}_j + \mathbf{Z} \right)^T \mathbf{S}_i \qquad i = 1, \ldots, M \tag{12.26}$$

where the superscript T means *the transpose of*. Also,

$$E(\mathbf{Z}) = \mathbf{0}$$

and

$$E(\mathbf{Z}\mathbf{Z}^T) = \frac{N_0}{4T} \mathbf{I} \tag{12.27}$$

where $\mathbf{I}$ is the D-by-D identity matrix.

We shall designate the signal-vector inner products by

$$\mathbf{S}_i{}^T \mathbf{S}_j = \sum_{k=1}^{D} s_k{}^i s_k{}^j = \begin{cases} 1 & \text{if } i = j \\ \lambda_{ij} & \text{if } i \neq j \end{cases}$$

The $\mathbf{S}_j$ are vectors on the D-dimensional unit sphere.

From (12.26) we note that the E_i can be formed by assuming that the transmitted signal is the D-dimensional vector $(V/2)\mathbf{S}_j$, the channel adds the noise vector $\mathbf{Z}$, and the received signal is the vector

$$\mathbf{Y} = \frac{V}{2} \mathbf{S}_j + \mathbf{Z}$$

The optimal receiver forms the scalar products indicated by (12.26) and again decides that $\mathbf{S}_k$ was transmitted if

$$E_k = \max_i E_i$$

The decision rule, and therefore the probability of detection, will not be affected if the E_i are all multiplied by a scale factor. This allows us to multiply the received vector by the factor $\sqrt{4T/N_0}$ and to neglect the T in (12.26). Then we can define a normalized received vector

$$\mathbf{Y}_0 = \frac{V\sqrt{T}}{\sqrt{N_0}} \mathbf{S}_j + \mathbf{Z}_0 \tag{12.28}$$

where

$$\mathbf{Z}_0 = \sqrt{\frac{4T}{N_0}}\, \mathbf{Z}$$

Here $\mathbf{Z}_0$ is a D-dimensional gaussian noise vector with zero mean and covariance matrix equal to the D-by-D identity matrix.

Let

$$\lambda \triangleq \frac{V\sqrt{T}}{\sqrt{N_0}} = \sqrt{\frac{2\mathcal{E}}{N_0}} \tag{12.29}$$

where $\mathcal{E}$ is defined as the *total received signal energy* during the interval $[0,T]$.

Then

$$\mathbf{Y}_0 = \lambda \mathbf{S}_j + \mathbf{Z}_0 \tag{12.30}$$

λ^2 is the ratio of signal energy to noise spectral density, which we shall henceforth call the *signal-to-noise ratio.*

This last formulation of the E_i cannot be distinguished (except for a scale factor) from the basic model formulation by an observer stationed at the receiver output. From (12.30) and (12.5), the probability of detection P_d, or equivalently, the *probability of correct decision*, is given by

$$P_d = \sum_{j=1}^{M} \Pr \mathbf{S}_j \Pr (\mathbf{Y}_0{}^T\mathbf{S}_j = \max_i \mathbf{Y}_0{}^T\mathbf{S}_i | \mathbf{S}_j \text{ was transmitted})$$

or equivalently,

$$P_d = \frac{1}{M} \sum_{j=1}^{M} \Pr (E_j = \max_i E_i | \mathbf{S}_j \text{ was transmitted}) \tag{12.31}$$

The problem of determining the set of $\{\mathbf{S}_j\}$ that maximizes P_d for various M and D and of determining the dependence or independence of these optimal sets on the signal-to-noise ratio is known as the *sphere-packing problem* of communication theory (see Ref. [12.6]). The solution of this problem, also known as the *signal-selection problem,* is our primary goal.

We have shown that the signal-design problem can be reduced to that of finding M unit vectors on a D-dimensional unit sphere which maximize a given functional, the probability of detection. We conclude by indicating that once the optimum set of $\{\mathbf{S}_j\}$ has been determined, the corresponding set of $\{A_i(t)\}$ and $\{\phi_i(t)\}$ can easily be determined. With the optimum set of $\{\mathbf{S}_j\}$,

$$\begin{bmatrix} A_j(t) \cos \phi_j(t) \\ A_j(t) \sin \phi_j(t) \end{bmatrix} = \sum_{i=1}^{D} s_i{}^j \psi_i(t) = \begin{bmatrix} a_j(t) \\ b_j(t) \end{bmatrix}$$

where

$$a_j(t) \triangleq \sum_{i=1}^{D} s_i{}^j \psi_{i1}(t)$$

and

$$b_j(t) \triangleq \sum_{i=1}^{D} s_i{}^j \psi_{i2}(t)$$

Then

$$A_j(t) = \sqrt{a_j{}^2(t) + b_j{}^2(t)}$$

and $\theta_j(t)$ is the principal value of $\tan^{-1} [b_j(t)/a_j(t)]$.

12.3 BANDWIDTH CONSIDERATIONS

In Chap. 14 we shall see that the probability of detection can be written as a function of only the signal-to-noise ratio and the set of inner products $\{\lambda_{ij}\}$ of the signal vectors. That is, the only characteristic of the basis functions used in determining the probability of detection is that they are orthogonal. The actual waveshapes are not a consideration. This means that if we have a second communication system with a different transmittable signal set $\{S_i'(t); i = 1, \ldots, M\}$ which can be expressed in terms of a different orthogonal basis, say $\{\mathbf{\Gamma}_i(t); i = 1, \ldots, D\}$, but with the same linear combinations as those in the original system, that is,

$$\mathbf{S}_j'(t) = \sum_{i=1}^{D} s_i{}^j \mathbf{\Gamma}_j(t) \qquad j = 1, \ldots, M \tag{12.32}$$

where

$$[\mathbf{\Gamma}_i(t) \cdot \mathbf{\Gamma}_j(t)] = T\delta_{ij}$$

and with M, D, V, T, $N_0/2$, and the set of $\{s_i{}^j\}$ the same in both systems, then the probability of detection for both systems is equal. Or, more generally speaking, the probability of detection is a function of λ, M, D, and the set of signal-vector inner products $\{\lambda_{ij}\}$, and depends not at all on which orthogonal basis is used to form the signal waveforms. For this reason the D basis functions can be designed or chosen according to some other criterion, and can be selected independently of the signal-selection problem. The usual criterion for the basis functions is conformity to some specified bandwidth restrictions. The only parameter which affects both the selection of optimal signal vectors and the selection of the orthogonal basis functions is the total number of basis functions D.

A realistic way to arrive at a value for D is the following. For a given time interval $[0,T]$ outside which the waveforms must be identically zero, how many orthogonal waveforms can we design which have a given percentage of energy within a specified bandwidth $[-B,B]$, with the realization, as will be shown later, that the maximum value D need be is $M - 1$? If D has already been determined because of its effect on probability of error (which will also be discussed later in detail), then the criterion for determining basis functions can be rephrased in either of the following ways:

1. For given T and D, what is the minimum bandwidth B required for D orthogonal functions which vanish in time outside $[0,T]$ and have a given percentage of energy within the bandwidth $[-B,B]$?
2. For given T and D, what is the largest percentage of total energy that D orthogonal functions can have within the given bandwidth $[-B,B]$ when the functions vanish in time outside $[0,T]$?

It should be noted that strict time-and-band-limitedness is not possible; however, in some applications (such as telephone channels) strict band-limitedness is required. The problem of determining these optimal waveforms when the criterion is any of the above is discussed by Slepian and Pollak [12.7] and Landau and Pollak [12.8].

The main point is that, except for the dimensionality D, the selection of signal vectors to minimize probability of error is disjoint from the problem of choosing waveforms that conform to certain bandwidth restrictions. For a given T and given percentage of energy restriction, the bandwidth required for D orthogonal waveforms that satisfy the energy requirement will increase as D increases. For this reason D is said to be a measure of the required bandwidth of the signal waveforms.

REFERENCES

12.1 Stiffler, J. J.: Synchronization of Telemetry Codes, *IRE Trans. Space Electron. Telemetry*, vol. SET-8, June, 1962, pp. 112–116.

12.2 Selin, I., and F. Tuteur: Synchronization of Coherent Detectors, *IEEE Trans. Commun. Systems*, vol. CS-11, March, 1963, pp. 100–109.

12.3 Gumacos, C.: Analysis of an Optimum Sync. Search Procedure, *IEEE Trans. Commun. Systems*, vol. CS-11, March, 1963, pp. 88–99.

12.4 Viterbi, A. J.: Phase-locked Loop Dynamics by Fokker-Planck Techniques, *Proc. IEEE*, vol. 51, no. 12, December, 1963, pp. 1737–1753.

12.5 Charles, F. J., and W. C. Lindsey: Some Analytical and Experimental Phase-locked Loop Results for Low Signal-to-noise Ratios, *Proc. IEEE*, vol. 54, no. 9, September, 1966.

12.6 Balakrishnan, A. V.: A Contribution to the Sphere-packing Problems of Communication Theory, *J. Math. Anal. Appl.*, vol. 3, no. 3, December, 1961.

12.7 Slepian, D., and H. O. Pollak: Prolate Spheroidal Wave Functions, Fourier Analysis and Uncertainty, I, *Bell Systems Tech. J.*, January, 1961, pp. 43–64.

12.8 Landau, H. J., and H. O. Pollak: Prolate Spheroidal Wave Functions, Fourier Analysis and Uncertainty, II, *Bell Systems Tech. J.*, January, 1961, pp. 65–84.

13
Signal Design When the Dimensionality of the Signal Set is Restricted to 2

We now study the optimal-signal-design problem for the case when the dimensionality of the signal set is restricted to 2. We shall show that the signal set consisting of equal spacing of the M vectors around the unit circle is the optimum, and that this optimal choice is independent of the signal-to-noise ratio. We shall then look into the question of whether it is possible to have two suboptimal signal sets, say, $\{\mathbf{S}_i;\ i = 1, \ldots, M\}$ and $\{\mathbf{S}'_i;\ i = 1, \ldots, M\}$, such that the probability of detection for $\{\mathbf{S}_i\}$, namely, $P_d(\lambda;\ \{\mathbf{S}_i\})$, is larger than that for $\{\mathbf{S}'_i\}$ for some signal-to-noise ratio, while the reverse is true for other signal-to-noise ratios. We shall see that a subclass of the two-dimensional signal sets does not possess this property, but that there are signal sets for which the preference does indeed depend on the signal-to-noise ratio.

When the signal waveforms are restricted to two dimensions, they can be expressed in the time domain as

$$\mathbf{S}_j(t) = \begin{bmatrix} A_j(t) \cos \theta_j(t) \\ A_j(t) \sin \theta_j(t) \end{bmatrix} = s_1{}^j \psi_1(t) + s_2{}^j \psi_2(t) \qquad j = 1, \ldots, M \tag{13.1}$$

One choice of basis functions are those which are non-time-varying,

$$\psi_1(t) = \begin{bmatrix} 1 \\ 0 \end{bmatrix} \quad \text{and} \quad \psi_2(t) = \begin{bmatrix} 0 \\ 1 \end{bmatrix} \quad 0 \leq t \leq T \tag{13.2}$$

Any choice of basis functions is possible, of course, but in the two-dimensional case these are the most logical. With this choice $\mathbf{S}_j(t)$ is then non-time-varying. Hence $\mathbf{S}_j(t)^T\mathbf{S}_j(t)$ is non-time-varying, which indicates that $A_j(t)$, and therefore $\theta_j(t)$, are constants for each j. Since

$$\frac{1}{T}\int_0^T A_j^2(t)\,dt = 1 \qquad j = 1, \ldots, M$$

and since $A_j(t) = A_j$, we get $A_j = 1$ for $j = 1, \ldots, M$. Therefore

$$\mathbf{S}_j(t) = \begin{bmatrix} \cos\theta_j \\ \sin\theta_j \end{bmatrix} = \begin{bmatrix} s_1^j \\ s_2^j \end{bmatrix} \qquad j = 1, \ldots, M \tag{13.3}$$

The $\mathbf{S}_j$ are unit vectors on the two-dimensional unit circle, as they should be to conform to the development in Chap. 12. The corresponding transmittable signal set is

$$V\cos(\omega_c t + \theta_j) \qquad 0 \leq t \leq T; j = 1, \ldots, M$$

and with this choice of basis functions the phase angle θ_j corresponds to the angle that the signal vector $\mathbf{S}_j$ makes with the positive y_1 axis, where

$$\mathbf{Y} = \begin{bmatrix} y_1 \\ y_2 \end{bmatrix} = \lambda\mathbf{S}_j + \mathbf{N} \tag{13.4}$$

is the normalized received vector (see Fig. 13.1). So, when D is restricted to 2, the only allowed variation between the different transmittable signals is the reference phase angle.

13.1 OPTIMAL SIGNAL SELECTION IN TWO DIMENSIONS

For a fixed set of $\{\mathbf{S}_j\}$ on the two-dimensional unit circle, and with the optimum receiver for this particular signal set (selection of that signal

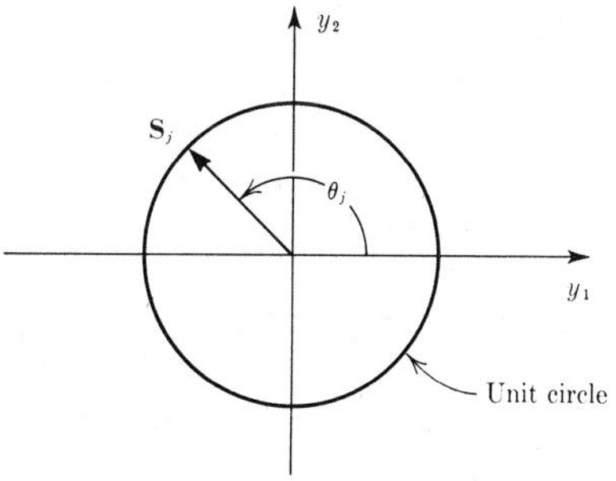

Fig. 13.1 Relating the phase angle in the transmittable signal to the corresponding signal vector in two dimensions.

which corresponds to the maximum correlator output), we have, from (12.31), the probability of detection

$$P_d(\lambda;\{\mathbf{S}_j\}) = \frac{1}{M}\sum_{j=1}^{M} \Pr\,(\mathbf{Y}^T\mathbf{S}_j = \max_i \mathbf{Y}^T\mathbf{S}_i|\mathbf{S}_j \text{ was transmitted})$$

This receiver is optimum for equally likely signals, and the corresponding probability of detection for this receiver does not depend on the a priori probabilities when the optimal signal set is employed.

When P_d is written in this manner there is no restriction on dimensionality. However, in two dimensions we can express P_d in the following way.

THEOREM 13.1 *For the received vector* $\mathbf{Y}$, *given by* (13.4), *where* $\mathbf{N}$ *has a gaussian probability density function with covariance matrix*

$$\begin{bmatrix} 1 & 0 \\ 0 & 1 \end{bmatrix}$$

the probability of detection can be written as

$$P_d(\lambda;\{\mathbf{S}_j\}) = \frac{1}{\pi}\exp\left(-\frac{1}{2}\lambda^2\right)\int_0^\infty dr\, r \exp\left(-\frac{1}{2}r^2\right)\frac{1}{M}\sum_{i=1}^{M}\int_0^{\varphi_i/2}\exp\,(\lambda r\cos\alpha)\,d\alpha \qquad (13.5)$$

where the angle φ_i *is the angle between signal vectors* $\mathbf{S}_i$ *and* $\mathbf{S}_{i+1}$ (*with the signals numbered consecutively around the unit circle and* φ_M *defined as the angle between* $\mathbf{S}_M$ *and* $\mathbf{S}_1$).

Hence we clearly have

$$\sum_{i=1}^{M}\varphi_i = 2\pi \qquad \varphi_i \geq 0;\ i = 1, \ldots, M \qquad (13.6)$$

Proof: Since the additive noise is independent of the a priori signal distribution, we can write

$$p(\mathbf{Y}|\mathbf{S}_i) = p_N(\mathbf{Y} - \lambda\mathbf{S}_i) = \frac{1}{2\pi}\exp\left[-\frac{1}{2}(\mathbf{Y} - \lambda\mathbf{S}_i)^T(\mathbf{Y} - \lambda\mathbf{S}_i)\right] \qquad (13.7)$$

where p_N is the noise probability density function. P_d can then be expressed as

$$P_d(\lambda;\{\mathbf{S}_i\}) = \frac{1}{M}\sum_{i=1}^{M}\int_{R_i} p(\mathbf{Y}|\mathbf{S}_i)\,d\mathbf{Y} \qquad (13.8)$$

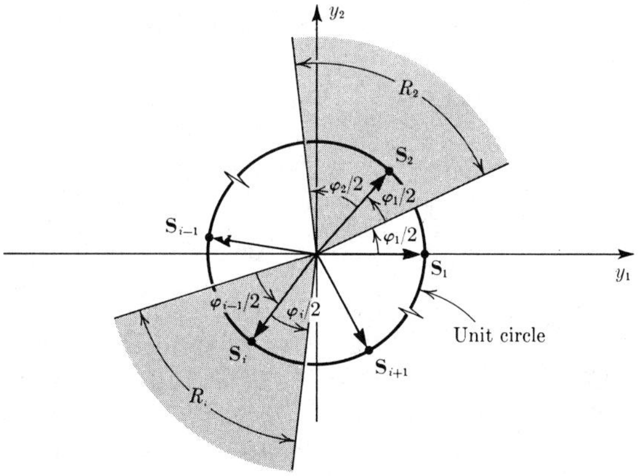

Fig. 13.2 Optimal decision regions for M unit vectors in two dimensions.

where R_i is the region where $\mathbf{Y}^T\mathbf{S}_i \geq \mathbf{Y}^T\mathbf{S}_j$ for $j = 1, \ldots, M$ (see Fig. 13.2). We let

$$\mathbf{Y} = r\begin{bmatrix}\cos\varphi \\ \sin\varphi\end{bmatrix} \tag{13.9}$$

and define

$$\begin{aligned}
\Delta_0 &\triangleq \frac{\varphi_M}{2} \\
\Delta_1 &\triangleq \frac{\varphi_1}{2} \\
\Delta_2 &\triangleq \varphi_1 + \frac{\varphi_2}{2} \\
\Delta_i &\triangleq \sum_{j=1}^{i-1} \varphi_j + \frac{\varphi_i}{2} \qquad i = 2, \ldots, M-1
\end{aligned} \tag{13.10}$$

Also,

$$\mathbf{S}_i = \begin{bmatrix} s_1{}^i \\ \\ s_2{}^i \end{bmatrix} = \begin{bmatrix} \cos \sum_{j=1}^{i-1} \varphi_j \\ \sin \sum_{j=1}^{i-1} \varphi_j \end{bmatrix} \tag{13.11}$$

Substitution yields

$$P_d(\lambda;\{\mathbf{S}_i\}) = \frac{1}{M\pi} \exp\left(-\frac{1}{2}\lambda^2\right) \int_0^\infty dr\, r \exp\left(-\frac{1}{2}r^2\right) \frac{1}{2}\left\{\sum_{i=1}^{M} \int_{\Delta_{i-1}}^{\Delta_i} d\varphi \right.$$
$$\left. \exp\left[\lambda r(s_1{}^i \cos\varphi + s_2{}^i \sin\varphi)\right]\right\} \tag{13.12}$$

which can be further reduced to

$$P_d(\lambda;\{\varphi_i\}) = \frac{1}{M\pi}\exp\left(-\frac{1}{2}\lambda^2\right)\int_0^\infty dr\, r\exp\left(-\frac{1}{2}r^2\right)L(\lambda r)$$

where

$$L(\lambda r) = \sum_{i=1}^{M}\int_0^{\varphi_i/2}\exp(\lambda r\cos\alpha)\,d\alpha \tag{13.13}$$

Hence the probability of detection depends only on the signal-to-noise ratio λ and the set of angles $\{\varphi_i\}$ between the adjacent signal vectors. This indicates that P_d is independent of rotations of the signal vectors about the origin, which is equivalent to saying that P_d depends only on signal-vector inner products and is independent of orthogonal transformations made on them. We shall prove this fact for all dimensions in the next chapter. P_d is also independent of the order in which the signals are placed on the unit circle as long as the angle between $\mathbf{S}_i$ and the signal vector adjacent to it in the counterclockwise direction remains φ_i. This characteristic is true only when $D = 2$, since for $D > 2$ the concept of numbering the signals requires more criteria for ordering.

THEOREM 13.2 *In two dimensions, the optimal signal set (optimal in the sense of maximizing the probability of detection) consists of the M signal vectors equally spaced around the unit circle; moreover, this is the optimum for all signal-to-noise ratios.*

Proof: It is sufficient to prove that $L(\lambda r)$ is maximized at every $\lambda r > 0$ by choosing $\varphi_i = 2\pi/M$ for $i = 1, \ldots, M$ (equal spacing). This is accomplished by showing that any other choice of $\{\varphi_i\}$ will be less.

Let $L_0(\lambda r)$ correspond to the choice

$$\varphi_i = \frac{2\pi}{M} \qquad i = 1, \ldots, M \tag{13.14}$$

and let

$$f(\beta) = \int_0^\beta \exp(k\cos\alpha)\,d\alpha \tag{13.15}$$

$f(\beta)$ is a convex downward function of β for $0 \le \beta \le \pi$ and k positive. This is true because

$$\frac{\partial^2 f(\beta)}{\partial\beta^2} = -k\sin\beta\exp(k\cos\beta)$$

is either always greater than or equal to 0 or always less than or equal to 0, depending on the given value of k. In our case $k = \lambda r$ is always

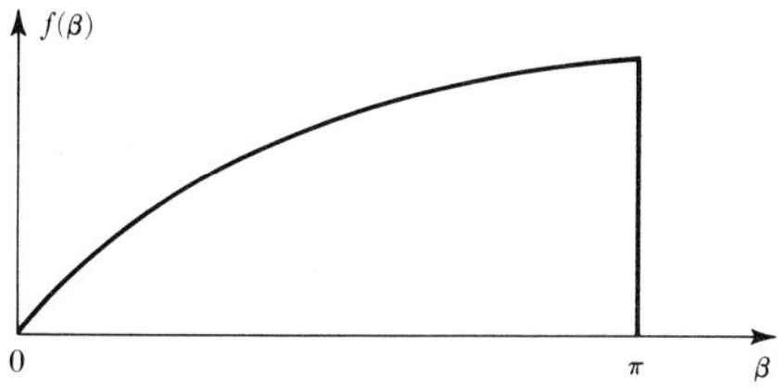

Fig. 13.3 The convex downward function $f(\beta)$ versus β.

greater than or equal to 0. Because of the convexity of $f(\beta)$ (see Fig. 13.3),

$$f\left(\sum_{i=1}^{M} \gamma_i\beta_i\right) \geq \sum_{i=1}^{M} \gamma_i f(\beta_i) \tag{13.16}$$

where

$$\sum_{i=1}^{M} \gamma_i = 1 \qquad \gamma_i \geq 0,\, i = 1, \ldots, M$$

and

$$\beta_i \in [0,\pi] \qquad i = 1, \ldots, M$$

Let

$$\gamma_i = \frac{1}{M} \qquad \text{and} \qquad \beta_i = \frac{\varphi_i}{2}$$

then

$$\sum_{i=1}^{M} \beta_i = \pi$$

Substitution yields

$$Mf\left(\frac{\pi}{M}\right) \geq \sum_{i=1}^{M} f\left(\frac{\varphi_i}{2}\right) \tag{13.17}$$

Thus

$$\begin{aligned} L_0(\lambda r) &= \sum_{i=1}^{M} \int_0^{\pi/M} \exp(\lambda r \cos\alpha)\, d\alpha \\ &\geq \sum_{i=1}^{M} \int_0^{\varphi_i/2} \exp(\lambda r \cos\alpha)\, d\alpha = L(\lambda r) \end{aligned} \tag{13.18}$$

The probability of detection for this optimal signal set is

$$P_d\left(\lambda; \left\{\varphi_i = \frac{2\pi}{M}\right\}\right) = P_{d_{\max}}(\lambda) = \frac{1}{\pi} \exp\left(-\tfrac{1}{2}\lambda^2\right) \int_0^\infty dr\, r \exp\left(-\tfrac{1}{2}r^2\right) \int_0^{\pi/M} \exp\left(\lambda r \cos \alpha\right) d\alpha \quad (13.19)$$

which can be rewritten as

$$P_{d_{\max}}(\lambda) = 2 \int_0^\infty G(y)\, dy \int_{y \cot(\pi/M) - \lambda}^\infty G(x)\, dx \quad (13.20)$$

where the zero mean unit variance Gaussian density is

$$G(x) = \frac{1}{\sqrt{2\pi}} \exp -\tfrac{1}{2}x^2$$

The minimum probability of error $P_{e_{\min}}(\lambda) = 1 - P_{d_{\max}}(\lambda)$ is plotted in Fig. 13.4.

Note that the optimal receiver and optimal signal set were found under the assumption of an equilikely a priori signal distribution. However, the probability of detection of the resulting optimal system is independent of the a priori distribution; this is an excellent property, since in practical situations the a priori distribution is not really known in advance. This does not mean, however, that this system is optimum for a known nonequilikely a priori distribution.

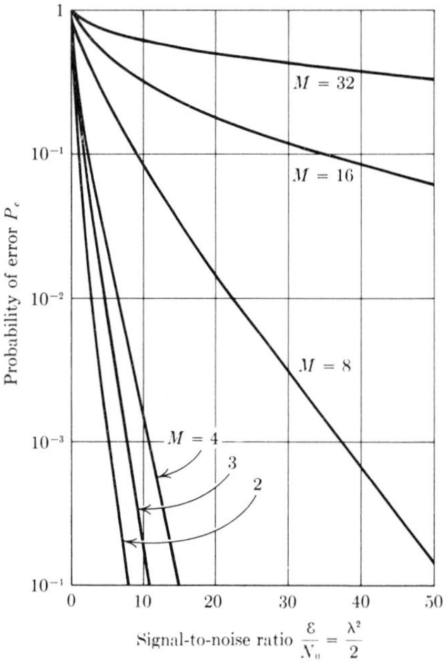

Fig. 13.4 Probability of error versus signal-to-noise ratio for the optimal signal set when $D = 2$.

13.2 COMMUNICATION EFFICIENCY AND CHANNEL CAPACITY FOR TWO-DIMENSIONAL SIGNAL SETS

An excellent figure of merit for comparing communication systems is to compare their communication efficiency [13.9] β, the minimum received signal energy per information bit transmitted that is necessary to maintain a specified probability of error in the presence of white gaussian noise with one-sided spectral density N_0. *Communication efficiency* β is thus defined as

$$\beta \triangleq \frac{\textit{Average received signal power}}{(N_0/2)(\textit{information rate of source in bits/sec})} \tag{13.21}$$

In the basic model we have assumed that successive messages are statistically independent and that they are equally likely. Hence

$$\begin{aligned} H &= \textit{information rate of source} \\ &= \log_2 M \textit{ bits/message} \\ &= \frac{1}{T} \log_2 M \textit{ bits/sec} \end{aligned} \tag{13.22}$$

From (12.6) we have

$$P_{\mathrm{av}} = \frac{V^2}{2}$$

Therefore

$$\beta = \frac{TV^2}{N_0 \log_2 M} = \frac{2\mathcal{E}_b}{N_0} \tag{13.23}$$

and, from (12.29),

$$\lambda^2 = \frac{V^2 T}{N_0} = \beta \log_2 M = \beta T H \tag{13.24}$$

$P_{d_{\max}}$ can then be expressed in terms of β and M as

$$P_{d_{\max}} = 2 \int_0^\infty G(y)\, dy \int_{h(y)}^\infty G(x)\, dx \tag{13.25}$$

where

$$h(y) \triangleq \frac{y}{\tan(\pi/M)} - \sqrt{\beta \log_2 M}$$

The average mutual information of an additive gaussian continuous channel (Ref. [13.1]) is

$$R = \frac{1}{2} \int_{-\infty}^{\infty} \log_2 \frac{P_n(f) + P_s(f)}{P_n(f)}\, df = \int_0^\infty \log_2 \left[1 + \frac{P_s(f)}{P_n(f)} \right] df \tag{13.26}$$

where $P_s(f)$ and $P_n(f)$ are the signal and noise spectral densities, respectively. If the channel is band-limited to W Hz on either side of the carrier and the noise has flat spectral density, the total noise power is $N = 2WN_0$, and

$$R = 2\int_0^W \log_2\left[1 + \frac{P_s'(f)}{N_0/2}\right] df$$

where

$$P_s'(f) = P_s(f + f_c)$$

Channel capacity is the maximum attainable value of R under a given power restriction on the signal. If P_{av} is the maximum allowable signal power, R is maximized by a gaussian signal that also has a flat spectral density, say, S. Thus

$$P_{av} = 4WS$$

and

$$C_W = 2W \log_2\left(1 + \frac{P_{av}}{2WN_0}\right) \tag{13.27}$$

Since C_W is a monotonically increasing function in W, if there is no bandwidth restriction, we can allow $W \to \infty$, and

$$C_\infty = \frac{P_{av}}{N_0} \log_2 e > C_W \qquad \text{for all } W \tag{13.28}$$

By *Shannon's theorem* it is possible to transmit H bits/sec with zero error if and only if $H < C$. Suppose H is fixed at some value $m = (1/T) \log_2 M$. Hence

$$M = 2^{mT} = 2^{HT} \tag{13.29}$$

A method of obtaining an arbitrarily small probability of error is to allow T to become sufficiently large. In the limit as $T \to \infty$, for fixed H, $M \to \infty$. Similarly, $\lambda \to \infty$. β, however, remains constant. For a given coding scheme the probability of error is a function only of β and M.

When there is no restriction on bandwidth, Shannon's theorem states that

$$P_d \to \begin{cases} 1 & \text{as } M \to \infty \quad \text{if } H < C_\infty \\ 0 & \text{as } M \to \infty \quad \text{if } H > C_\infty \end{cases} \tag{13.30}$$

For $H < C_\infty$ we get

$$H < \frac{P_{av}}{N_0} \log_2 e \tag{13.31}$$

or

$$\beta \log_2 e > 2 \tag{13.32}$$

Thus, as $M \to \infty$,

$$P_d \to \begin{cases} 1 & \text{for } \beta \log_2 e > 2 \\ 0 & \text{for } \beta \log_2 e < 2 \end{cases} \tag{13.33}$$

So for a given H and given channel the minimum average signal power required for zero error is given by

$$P_{\text{av}_{\min}} = N_0 H \ln 2 \tag{13.34}$$

Consider the application of these results to the case where $D = 2$. In (13.25), for large M

$$\frac{y}{\tan(\pi/M)} - \sqrt{\beta \log_2 M} \approx \frac{yM}{\pi} - \sqrt{\beta \log_2 M} \to \infty \qquad \text{as } M \to \infty \qquad \text{for every } y > 0$$

Thus

$$\lim_{M\to\infty} P_{d_{\max}} = 0 \qquad \text{for every } \beta \tag{13.35}$$

Strictly speaking, the true bandwidth of M-phase-modulation systems is indefinable because the process is nonstationary.[1] If a substitute bandwidth is defined as the ratio of number of degrees of freedom D to the message length T, then

$$C_W = 2W \log_2\left[1 + \frac{P_{\text{av}}}{2W(N_0/2)}\right] \tag{13.36}$$

where now $W = D/T$. For D fixed at 2, as $T \to \infty$, $M \to \infty$ and

$$\lim_{T\to\infty} C_W = 0 \tag{13.37}$$

Since H is fixed at a value greater than zero, $H > C = 0$, and since

$$\lim_{T\to\infty} P_{d_{\max}} = 0$$

from (13.35), Shannon's theorem has been substantiated. T increasing and D fixed is an example of a fixed-bandwidth code. Regular simplex and orthogonal codes (to be discussed later) are examples of band-increasing codes.

Even though signal design for finite time T is our primary purpose, it is significant to show that the optimal results do satisfy *Shannon's limit theorems* for large T.

[1] We are speaking here of a stochastic process; in the previous chapter the definition of bandwidth was for deterministic waveforms and employed the percentage-of-energy concept.

13.3 PARTIAL ORDERING OF THE CLASS OF TWO-DIMENSIONAL SIGNAL SETS

In the previous chapter the dependence of the optimal signal choice on the signal-to-noise ratio λ was shown to be a significant part of the optimal-signal-design problem. In Sec. 13.1 we proved that in the two-dimensional case, at any rate, the optimal signal set is indeed independent of the signal-to-noise ratio. The question still remains, however, as to whether it is possible to have two suboptimal signal sets, say, $\{\mathbf{S}_i; i = 1, \ldots, M\}$ and $\{\mathbf{S}'_i; i = 1, \ldots, M\}$, such that $P_d(\lambda; \{\mathbf{S}_i\})$, the probability of detection for $\{\mathbf{S}_i\}$, is larger than that for $\{\mathbf{S}'_i\}$ for some λ, while the reverse is true for other λ.

This problem can be phrased in terms of a *partial ordering* of the class of signal sets (or, equivalently, a partial ordering of the class of nonnegative definite M-by-M matrices of given rank). We induce a partial ordering by saying that $\{\mathbf{S}_i\}$ is preferred over $\{\mathbf{S}'_i\}$ if and only if

$$P_d(\lambda;\{\mathbf{S}_i\}) > P_d(\lambda;\{\mathbf{S}'_i\}) \qquad \text{for all } \lambda > 0 \tag{13.38}$$

As in Sec. 13.1, let us represent a set of M signals in two dimensions by the angles between adjacent points on the unit circle; thus a set of M signals can be specified by the column vector

$$\boldsymbol{\varphi} \triangleq \begin{bmatrix} \varphi_1 \\ \cdot \\ \cdot \\ \cdot \\ \varphi_M \end{bmatrix}$$

where

$$\sum_{i=1}^{M} \varphi_i = 2\pi$$

The induced partial ordering is definitely nonempty because the optimal choice, equal spacing, is better than any other choice for all λ. However, we can say more. Consider the following statement.

DEFINITION 13.1 *One set of M signals $\boldsymbol{\varphi}'$ is said to be more equally spaced than another set of M signals $\boldsymbol{\varphi}$ if they can be related by*

$$\boldsymbol{\varphi}' = \mathbf{P}\boldsymbol{\varphi} \tag{13.39}$$

where the M-by-M matrix $\mathbf{P}$ is such that

$$\sum_{i=1}^{M} p_{ij} = \sum_{j=1}^{M} p_{ij} = 1 \qquad \text{for all } i, j = 1, \ldots, M$$

and

$$p_{ij} \geq 0 \qquad \text{for all } i, j = 1, \ldots, M$$

Then we can prove the following result.

THEOREM 13.3 *Suppose φ' is more equally spaced than φ. Then*

$$P_d(\lambda;\varphi') \geq P_d(\lambda;\varphi) \qquad \textit{for all } \lambda \geq 0$$

Proof: As always, assume all signals equally likely. Since

$$\varphi' = \mathbf{P}\varphi \qquad \varphi_i' = \sum_{j=1}^{M} p_{ij}\varphi_j$$

then, because of the convexity in β of $\int_0^\beta \exp(k \cos \alpha)\, d\alpha$, we have

$$I(\lambda r;\varphi') = \sum_{i=1}^{M} \int_0^{\varphi_i'/2} \exp(\lambda r \cos \alpha)\, d\alpha$$

$$= \sum_{i=1}^{M} \int_0^{\sum_{j=1}^{M} p_{ij}\varphi_j/2} \exp(\lambda r \cos \alpha)\, d\alpha$$

$$\geq \sum_i \sum_j p_{ij} \int_0^{\varphi_j/2} \exp(\lambda r \cos \alpha)\, d\alpha$$

$$= \sum_{j=1}^{M} \int_0^{\varphi_j/2} \exp(\lambda r \cos \alpha)\, d\alpha = I(\lambda r;\varphi) \tag{13.40}$$

This inequality holds for all $\lambda r \geq 0$, so

$$P_d(\lambda;\varphi') \geq P_d(\lambda;\varphi) \qquad \text{for all } \lambda \geq 0$$

13.4 THE DEPENDENCE OF SOME SUBOPTIMAL SIGNAL SETS ON THE SIGNAL-TO-NOISE RATIO

The subclass of signal sets that can be related as in Definition 13.1 is clearly totally ordered. However, not all sets of signals can be so related. The fact that the probability of detection is higher for some λ does not necessarily mean that it is higher for all λ. Equivalently, we state

THEOREM 13.4 *There exists sets φ and φ' such that*

$$P_d(\lambda;\varphi) > P_d(\lambda;\varphi') \qquad \textit{for small } \lambda$$

and

$$P_d(\lambda;\varphi) < P_d(\lambda;\varphi') \qquad \textit{for large } \lambda$$

Proof: Consider the following specific example which possesses the desired characteristics.

Example 13.1. Let

$$\boldsymbol{\varphi}' = \left(\varphi_1' = \frac{\pi}{4};\ \varphi_2' = \frac{\pi}{2};\ \varphi_3' = \frac{5\pi}{4}\right)$$

and

$$\boldsymbol{\varphi} = \left(\varphi_1 = \frac{\pi}{8};\ \varphi_2 = \frac{3\pi}{4};\ \varphi_3 = \frac{9\pi}{8}\right)$$

Remark: Since $\varphi_1 < \varphi_1', \varphi_2', \varphi_3'$, and since $\varphi_3' > \varphi_1, \varphi_2, \varphi_3$, $\boldsymbol{\varphi}'$ and $\boldsymbol{\varphi}$ cannot be related by a matrix $\mathbf{P}$ having the characteristics indicated above.

For small λ differentiating (13.13) with respect to λ and setting $\lambda = 0$ yields

$$\begin{aligned}\left.\frac{\partial P_d(\lambda;\boldsymbol{\varphi})}{\partial\lambda}\right|_{\lambda=0} &= \frac{1}{3\pi}\sum_{i=1}^{3}\sin\frac{\varphi_i}{2} \\ &= \frac{1}{3\pi}\left(\sin\frac{\pi}{16} + \sin\frac{6\pi}{16} + \sin\frac{9\pi}{16}\right) \\ &\approx \frac{1}{3\pi}(2.09976)\end{aligned}$$

and

$$\left.\frac{\partial P_d(\lambda;\boldsymbol{\varphi}')}{\partial\lambda}\right|_{\lambda=0} \approx \frac{1}{3\pi}(2.01367)$$

Therefore in the neighborhood of $\lambda = 0$

$$P_d(\lambda;\boldsymbol{\varphi}) > P_d(\lambda;\boldsymbol{\varphi}')$$

Conversely we must show that for λ sufficiently large

$$\begin{aligned}\Delta(\lambda) &= P_d(\lambda;\boldsymbol{\varphi}') - P_d(\lambda;\boldsymbol{\varphi}) \\ &= \frac{1}{3\pi}\exp\left(-\tfrac{1}{2}\lambda^2\right)\int_0^\infty r\,dr \\ &\exp\left(-\tfrac{1}{2}r^2\right)\left[\int_{\pi/16}^{2\pi/16} - \int_{4\pi/16}^{6\pi/16} + \int_{9\pi/16}^{10\pi/16}\exp(\lambda r\cos\alpha)\,d\alpha\right] > 0\end{aligned}$$

The integral from $9\pi/16$ to $10\pi/16$ is always positive. It is therefore sufficient to show that the difference between the first two integrals is positive. Since

$$\int_{\pi/16}^{2\pi/16} d\alpha\,\exp(\lambda r\cos\alpha) \geq \frac{\pi}{16}\exp\left(\lambda r\cos\frac{2\pi}{16}\right)$$

and

$$\int_{4\pi/16}^{6\pi/16} d\alpha\,\exp(\lambda r\cos\alpha) \leq \frac{2\pi}{16}\exp\left(\lambda r\cos\frac{4\pi}{16}\right)$$

for all $\lambda r \geq 0$, it is sufficient to show that for λ sufficiently large

$$\Delta(\lambda) > \tfrac{1}{48} \exp\left(-\tfrac{1}{2}\lambda^2\right) \int_0^\infty dr\, r \exp\left(-\tfrac{1}{2}r^2\right) \left[\exp\left(\lambda r \cos\frac{2\pi}{16}\right) - 2\exp\left(\lambda r \cos\frac{4\pi}{16}\right)\right] > 0 \quad (13.41)$$

From the identity

$$\int_0^\infty dr\, r \exp\left(-\tfrac{1}{2}r^2 + ar\right) = 1 + a\sqrt{2\pi}\,(1 - \operatorname{erfc} a) \exp\frac{a^2}{2} \quad (13.42)$$

and the inequality (see Ref. [13.6])

$$\frac{1}{\sqrt{2\pi}} \exp\left(-\tfrac{1}{2}x^2\right)\left(\frac{1}{x} - \frac{1}{x^3}\right) < \operatorname{erfc} x < \frac{1}{x\sqrt{2\pi}} \exp -\tfrac{1}{2}x^2 \qquad \text{for all } x > 0 \quad (13.43)$$

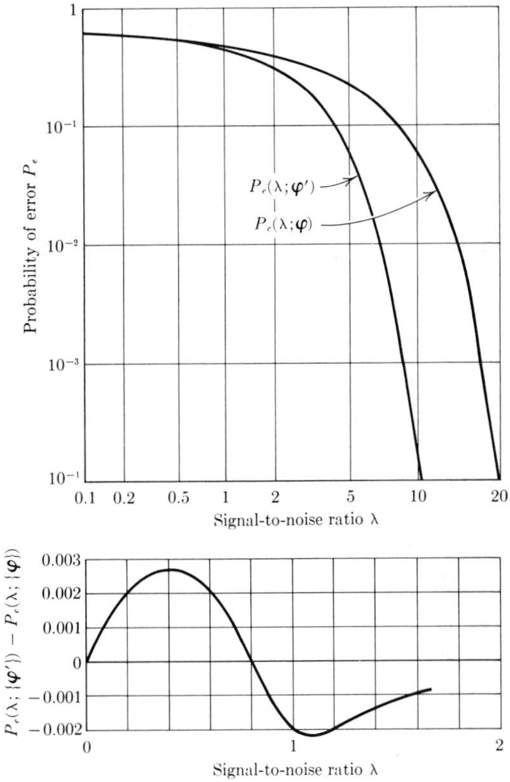

Fig. 13.5 Probability of error versus signal-to-noise ratio for the example in Theorem 13.4.

it follows that

$$\Delta(\lambda) > 0 \qquad \text{for all}$$

$$\lambda^2 > \frac{\ln 4 - \ln [\cos (4\pi/16)]}{\cos^2 (2\pi/16) - \cos^2 (4\pi/16)} = \lambda_0{}^2 > 1 \tag{13.44}$$

from which we conclude that

$$P_d(\lambda;\varphi) < P_d(\lambda;\varphi') \qquad \text{for all } \lambda > \lambda_0$$

This example therefore proves Theorem 13.4.

Probabilities of error are plotted against signal-to-noise ratios in Fig. 13.5 for φ and φ'. The results show, for example, that

$$P_e(\lambda = 10; \varphi) \approx 10^{-2}$$
$$P_e(\lambda = 10; \varphi') \approx 10^{-4}$$
$$P_e(\lambda = 20; \varphi) \approx 10^{-4}$$
$$P_e(\lambda = 20; \varphi') \approx 10^{-14}$$

indicating that φ' at $\lambda = 10$ is as good as φ at $\lambda = 20$.

PROBLEM

13.1 Complete the proof of Theorem 13.4 by verifying Eq. (13.44). This may be done by use of the identity in (13.42) and the inequality in (13.43).

REFERENCES

13.1 Gallager, R. G.: "Information Theory and Reliable Information," Wiley, New York, 1968.

13.2 Shannon C. E.: Probability of Error for Optimal Codes in Gaussian Channels, *Bell Systems Tech. J.*, vol. 38, no. 3, pp. 611–656.

13.3 Cahn, C. R.: Performance of Digital Phase Modulation Systems, *IRE Trans. Commun. Systems*, pp. 3–14, May, 1959.

13.4 Viterbi, A. J.: On Coded Phase Coherent Communications, *IRE Trans. Space Electron. Telemetry*, vol. SET-7, no. 1, March, 1961, pp. 3–14.

13.5 Weber, C. L.: On Optimal Signal Selection for *M*-ary Alphabets with Two Degrees of Freedom, *IEEE Trans. Inform. Theory*, April, 1965.

13.6 Feller, W.: "An Introduction to Probability Theory and Its Applications," Wiley, New York, 1950.

13.7 Scholtz, R. A., and I. S. Reed: *N*-orthogonal Phase Modulated Codes, *Proc. First IEEE Ann. Commun. Conv.*, Boulder, Colo., June, 1965.

13.8 Viterbi, A. J.: On a Class of Polyphase Codes for the Coherent Gaussian Channel, *IEEE Intern. Conv.*, New York, March, 1965.

13.9 Sanders, R. W.: Communication Efficiency of Several Communication Systems, *Proc. IRE*, vol. 48, no. 4, pp. 575–588, April, 1960.

14
General Theory

14.1 INTRODUCTION

In this chapter we begin the general theory of the signal-design problem as developed by Balakrishnan [14.1]. The main feature of this theory is the variational approach, that of reducing the problem to one of minimizing a functional on a convex set in euclidean space (or a subset thereof described by constraints). In this sense it opens up some new and general solutions to the signal-design problem.

Let us again begin with a precise statement of the signal-design problem, restated as we shall henceforth consider it: Let $\mathbf{Y}$ be a D-dimensional-vector random variable (real-valued) of the form

$$\mathbf{Y} = \lambda \mathbf{S}_j + \mathbf{Z} \tag{14.1}$$

where $\mathbf{Z}$ is a gaussian random vector with zero mean and covariance matrix the D-by-D identity matrix, where $\mathbf{S}_j$ is one of M equally likely signal vectors, each $\mathbf{S}_j$ being a known unit vector in E_D (D-dimensional euclidean space) with $D \leq M$, and where $\lambda > 0$ is the signal-to-noise ratio, as described in Chaps. 12 and 13. After observing $\mathbf{Y}$, we are asked to determine optimally (in the sense of maximizing the probability of detection) which $\mathbf{S}_j$ has been transmitted. We saw in Chap. 10 that for a fixed signal set $\{\mathbf{S}_j; j = 1, \ldots, M\}$ the probability of detection is maximized by the matched filter, which forms

$$E_i = \mathbf{Y}^T\mathbf{S}_i \qquad i = 1, \ldots, M$$

and decides $\mathbf{S}_k$ was transmitted if

$$E_k = \max_i E_i$$

The corresponding probability of detection is

$$P_d(\lambda;\{\mathbf{S}_j\}) = \frac{1}{M}\sum_{j=1}^{M} \Pr\,(E_j = \max_i E_i|\mathbf{Y} = \lambda\mathbf{S}_j + \mathbf{Z}) \qquad (14.2)$$

The optimal-signal-design problem is to find that set of vectors $\{\mathbf{S}_j\}$ which makes this probability a maximum for various M and D, and to determine the dependence or independence of the optimal signal set on the signal-to-noise ratio.

In this chapter we shall discuss some of the significant properties of the class of admissible signal sets (admissible in the sense that they satisfy all the designated restrictions) and find some subclasses of signal sets which contain the optimal sets. We shall define some characteristics of the optimal sets which reduce the size of the class containing them. In later chapters we shall show that certain sets are optimum under different dimensionality restrictions and indicate precisely in what sense they are optimum.

It will be convenient to denote the set of signal-vector inner products $\{\lambda_{ij}\}$ by the symmetric M-by-M matrix

$$\boldsymbol{\alpha} = \begin{bmatrix} 1 & & \lambda_{ij} \\ & \ddots & \\ \lambda_{ji} & & 1 \end{bmatrix} \qquad (14.3)$$

LEMMA 14.1 *$\boldsymbol{\alpha}$ is nonnegative-definite.*

Proof: Define $\mathbf{S} \triangleq (\mathbf{S}_1,\mathbf{S}_2, \ldots, \mathbf{S}_M)$, a row of column vectors which is D-by-M. Then $\boldsymbol{\alpha} = \mathbf{S}^T\mathbf{S}$ and is M-by-M. For any column vector $\mathbf{a}$

$$\mathbf{a}^T\boldsymbol{\alpha}\mathbf{a} = \mathbf{a}^T\mathbf{S}^T\mathbf{S}\mathbf{a} = (\mathbf{S}\mathbf{a})^T\mathbf{S}\mathbf{a} \geq 0$$

since the last quantity is a sum of squares. Hence $\boldsymbol{\alpha}$ is nonnegative-definite.

Note that the rank of the matrix $\boldsymbol{\alpha}$ is equal to the allowed degrees of freedom of the signal set. Thus $\boldsymbol{\alpha}$ is M-by-M and has rank D.

Also, $P_d(\lambda;\{\mathbf{S}_j\})$ is a nondecreasing function in λ with

$$P_d(0;\{\mathbf{S}_j\}) = \frac{1}{M} \qquad \text{for any set of } \{\mathbf{S}_j\}$$

$$\lim_{\lambda\to\infty} P_d(\lambda;\{\mathbf{S}_j\}) = 1 \qquad \text{if the } \mathbf{S}_j \text{ are all different}$$

From (14.2) we can write

$$\begin{aligned}\Pr\,(E_j &= \max_i E_i|\mathbf{Y} = \lambda\mathbf{S}_j + \mathbf{Z})\\ &= \int_{\Lambda_j} \frac{1}{(2\pi)^{D/2}} \exp\,(-\tfrac{1}{2}\|\mathbf{Y} - \lambda\mathbf{S}_j\|^2)\, d\mathbf{Y}\end{aligned} \tag{14.4}$$

where Λ_j is the region where $\mathbf{Y}^T\mathbf{S}_j \geq \mathbf{Y}^T\mathbf{S}_i$ for $i \neq j$.

The integrand in (14.4) is

$$\exp\,(-\tfrac{1}{2}\|\mathbf{Y} - \lambda\mathbf{S}_j\|^2) = \exp\,(-\tfrac{1}{2}\lambda^2)\exp\,(-\tfrac{1}{2}\|\mathbf{Y}\|^2 + \lambda\mathbf{Y}^T\mathbf{S}_j)$$

Substituting these into (14.2), we obtain

$$P_d(\lambda;\{\mathbf{S}_j\}) = \frac{1}{M}\exp\,(-\tfrac{1}{2}\lambda^2)\frac{1}{(2\pi)^{D/2}}\sum_{j=1}^{M}\int_{\Lambda_j} \exp\,(-\tfrac{1}{2}\|\mathbf{Y}\|^2 + \lambda\mathbf{Y}^T\mathbf{S}_j)\, d\mathbf{Y} \tag{14.5}$$

If we write $\mathbf{Y}^T\mathbf{S}_j$ as $\max_i \mathbf{Y}^T\mathbf{S}_i$ for all $\mathbf{Y}$ in Λ_j, the integrand (14.5) is then expressed such that it is independent of j. Since the union of the disjoint regions $\{\Lambda_j\}$ is the entire euclidean D-dimensional space, we can write

$$P_d(\lambda;\{\mathbf{S}_j\}) = \frac{1}{M}\exp\,(-\tfrac{1}{2}\lambda^2)\frac{1}{(2\pi)^{D/2}}\int_{E_D} \exp\,(-\tfrac{1}{2}\|\mathbf{Y}\|^2 + \lambda\max_j \mathbf{Y}^T\mathbf{S}_j)\, d\mathbf{Y} \tag{14.6}$$

where E_D is the entire D-dimensional space. This can be expressed as

$$P_d(\lambda;\{\mathbf{S}_j\}) = \frac{1}{M}\exp\,(-\tfrac{1}{2}\lambda^2)\, E(\exp\,(\lambda\max_j \mathbf{Y}^T\mathbf{S}_j)) \tag{14.7}$$

where $\mathbf{Y}$ may now be interpreted as a D-variate gaussian random vector with zero mean and covariance matrix equal to the D-by-D identity matrix. Interpreted in this way, $\mathbf{Y}$ is independent of $\mathbf{S}_j$. We shall adopt the notation

$$G(\mathbf{Y};\mathbf{m};\mathbf{C}) \tag{14.8}$$

for a gaussian variate $\mathbf{Y}$ with mean vector $\mathbf{m}$ and covariance matrix $\mathbf{C}$. Now let

$$\xi_j = \mathbf{Y}^T\mathbf{S}_j \qquad j = 1, \ldots, M \tag{14.9}$$

and

$$\boldsymbol{\xi} = \begin{bmatrix} \xi_1 \\ \cdot \\ \cdot \\ \cdot \\ \xi_M \end{bmatrix} \tag{14.10}$$

Then

$$\begin{aligned} \boldsymbol{\xi} &= \mathbf{S}^T\mathbf{Y} \\ E(\boldsymbol{\xi}) &= E(\mathbf{S}^T\mathbf{Y}) = \mathbf{S}^T E(\mathbf{Y}) = \mathbf{0} \\ \operatorname{Cov} \boldsymbol{\xi} &= E(\boldsymbol{\xi}\boldsymbol{\xi}^T) = E(\mathbf{S}^T\mathbf{Y}\mathbf{Y}^T\mathbf{S}) = \mathbf{S}^T\mathbf{S} = \boldsymbol{\alpha} \end{aligned}$$

Substituting, we have

$$\begin{aligned} P_d(\lambda;\{\mathbf{S}_j\}) &= \frac{1}{M} \exp\left(-\tfrac{1}{2}\lambda^2\right) E(\exp(\lambda \max_j \mathbf{Y}^T\mathbf{S}_j)) \\ &= \frac{1}{M} \exp\left(-\tfrac{1}{2}\lambda^2\right) E(\exp(\lambda \max_j \xi_j)) = P_d(\lambda;\boldsymbol{\alpha}) \end{aligned} \tag{14.11}$$

where $\boldsymbol{\xi}$ has a probability density function which is $G(\boldsymbol{\xi};\mathbf{0};\boldsymbol{\alpha})$. Hence *$P_d$ is a function only of λ and the matrix of signal-vector inner products $\boldsymbol{\alpha}$* and is therefore *invariant to any orthogonal transformation on the signal vectors.* Thus it is sufficient to specify a signal set by its set of inner products.

Also, if we define

$$\phi(\lambda;\boldsymbol{\alpha}) \triangleq E(\exp(\lambda \max_i \xi_i)) \tag{14.12}$$

then

$$P_d(\lambda;\boldsymbol{\alpha}) = \frac{1}{M} \exp\left(-\tfrac{1}{2}\lambda^2\right) \phi(\lambda;\boldsymbol{\alpha}) \tag{14.13}$$

and the optimization problem has been reduced to finding that $\boldsymbol{\alpha}$ matrix which maximizes $\phi(\lambda;\boldsymbol{\alpha})$ in (14.12) for various λ.

We now define $\mathcal{A}$, the class of admissible $\boldsymbol{\alpha}$, as those M-by-M symmetric nonnegative-definite matrices with 1s along the main diagonal and all off-diagonal elements less than or equal to 1 in magnitude. The remainder of this chapter is devoted to finding subclasses of $\mathcal{A}$ which contain the optimum $\boldsymbol{\alpha}$.

14.2 CONVEX-BODY CONSIDERATIONS: SMALL SIGNAL-TO-NOISE RATIOS

The maximization of $P_d(\lambda;\boldsymbol{\alpha})$ in the neighborhood of $\lambda = 0$ can be put in the context of *convex-body theory* (for the necessary theory of convex bodies see Refs. [14.4, 14.5]). Let

$$H_S(\mathbf{Y}) = \max_i \mathbf{Y}^T\mathbf{S}_i \tag{14.14}$$

$H_S(\mathbf{Y})$ is the *support function* of the polytope formed by the set of vectors $\{\mathbf{S}_i\}$. The polytope in this case is the set

$$\left\{\mathbf{Y} | \mathbf{Y} = \sum_{i=1}^{M} \gamma_i \mathbf{S}_i;\ \sum_{i=1}^{M} \gamma_i = 1;\ \gamma_i \geqq 0;\ i = 1, \ . \ . \ . \ , M\right\}$$

which is the convex hull generated by the $\{\mathbf{S}_i\}$.

Substitution into (14.6) yields

$$P_d(\lambda;\boldsymbol{\alpha}) = \frac{1}{M} \exp\left(-\tfrac{1}{2}\lambda^2\right) \int_{E_D} \exp\left(\lambda H_S(\mathbf{Y}) - \frac{\|Y\|^2}{2}\right) d\mathbf{Y} \tag{14.15}$$

By noting that for any function f

$$\int_{E_D} f(\mathbf{Y})\, d\mathbf{Y} = \int_0^\infty dr\, r^{D-1} \int_{\Omega_D} f(\mathbf{Y})\, d\boldsymbol{\Omega}$$

where

$$r^2 = \sum_{i=1}^{D} y_i^2$$

Ω_D = surface of the D-dimensional unit sphere

$d\boldsymbol{\Omega}$ = surface element on Ω_D

we can write $\phi(\lambda;\boldsymbol{\alpha})$ as

$$\phi(\lambda;\boldsymbol{\alpha}) = \frac{1}{(2\pi)^{D/2}} \int_0^\infty dr\, r^{D-1} \exp\left(-\tfrac{1}{2}r^2\right) \int_{\Omega_D} \exp\left(\lambda r H_S(\mathbf{Y}_n)\right) d\boldsymbol{\Omega} \tag{14.16}$$

where $\mathbf{Y}_n$ is now of unit magnitude.

Since $\phi(0;\boldsymbol{\alpha}) = 1$ for every $\boldsymbol{\alpha}$, if there is a choice of $\boldsymbol{\alpha}$ which maximizes $\phi(\lambda;\boldsymbol{\alpha})$ for small λ (or if there is a choice of $\boldsymbol{\alpha}$ which maximizes $\phi(\lambda;\boldsymbol{\alpha})$ independent of λ), it necessarily must maximize the derivative of $\phi(\lambda;\boldsymbol{\alpha})$ with respect to λ at the origin. Thus, from (14.16),

$$\left.\frac{\partial\phi(\lambda;\boldsymbol{\alpha})}{\partial\lambda}\right|_{\lambda=0} = \frac{B}{(2\pi)^{D/2}} \int_0^\infty dr\, r^D \exp -\tfrac{1}{2}r^2 = \sqrt{2}\, \frac{\Gamma[(D+1)/2]}{\Gamma(D/2)}\, \bar{B} \tag{14.17}$$

where

$$B = \int_{\Omega_D} H_S(\mathbf{Y}_n)\, d\boldsymbol{\Omega} \tag{14.18}$$

The *mean width* of the convex body $\bar{B}$ is now defined as

$$\bar{B} = \frac{B}{\omega_D} \tag{14.19}$$

where ω_D is the surface area of the D-dimensional unit sphere. That is,

$$\omega_D = \int_{\Omega_D} d\boldsymbol{\Omega}$$

which becomes

$$\omega_D = \frac{2\pi^{D/2}}{\Gamma(D/2)}$$

where Γ in (14.17) is the *gamma function.*

Thus for a given D and M a necessary condition for the optimum $\boldsymbol{\alpha}$ is that it maximize the mean width. This formulation is independent of coordinate rotations. As defined here, the *mean width* is an average radial distance, averaged uniformly over the D-dimensional unit sphere. It is not an average diameter, as the name might imply.

With the theory of convex bodies (see Ref. [14.4]) it can be proved that when $D = M - 1$, the polytope which maximizes the mean width is the regular simplex, which consists of M vectors having an inner-product structure given by

$$\boldsymbol{\alpha}_R = \begin{bmatrix} 1 & & \dfrac{-1}{M-1} \\ & \ddots & \\ \dfrac{-1}{M-1} & & 1 \end{bmatrix} \tag{14.20}$$

that is, $\lambda_{ij} = -1/(M-1)$ for all $i \neq j$. From further geometrical considerations it can be shown that $\boldsymbol{\alpha}_R$ is the only polytope which maximizes the mean width. Therefore for small λ (small in the sense that a first-order approximation is sufficient) and $D = M - 1$ the regular simplex is the optimal signal set. The maximum mean width for M points and $D < M - 1$ will be less than that of the regular simplex. In the next section we shall see that the mean width is not increased by allowing D to equal M. Thus, if D is left unspecified, the dimensionality in which the largest mean width is attained is $D \doteq M - 1$, and the corresponding signal set is the regular simplex. Hence, if the optimal set is independent of λ, it must be $\boldsymbol{\alpha}_R$.

Example 14.1 Mean width in two dimensions For agreement with the two-dimensional results of the previous chapter, we must be able to show the following.

THEOREM 14.1 *The mean width for $D = 2$ and M any integer is maximized by equally spacing the M points on the unit circle.*

Proof: First we show that in two dimensions the mean width $\bar{B}_2$ is defined as

$$\bar{B}_2 \triangleq \frac{1}{2\pi} P \tag{14.21}$$

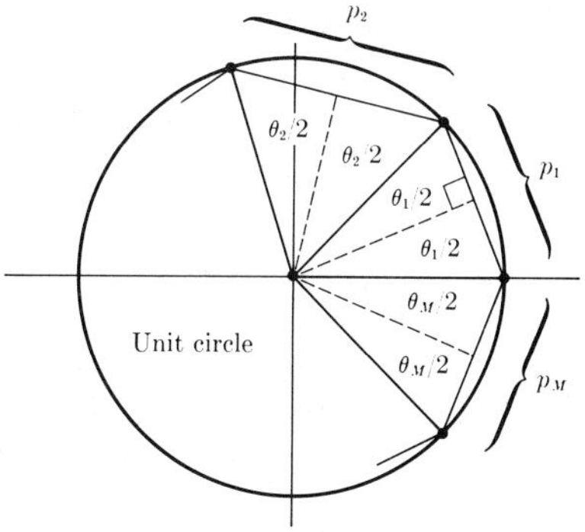

Fig. 14.1 Mean width of a polygon formed from the tips of M unit vectors in two dimensions.

where P is the perimeter. Given any M points on the unit circle with corresponding angles $\{\theta_i\}$, as indicated in Fig. 14.1, then

$$P = \sum_{i=1}^{M} p_i \qquad \text{where } p_i = 2 \sin \frac{\theta_i}{2}$$

Hence

$$P = 2 \sum_{i=1}^{M} \sin \frac{\theta_i}{2} \tag{14.22}$$

$\bar{B}_2$ can be written as

$$\bar{B}_2 = \frac{1}{2\pi} \int_0^{2\pi} L(\zeta)\, d\zeta \tag{14.23}$$

where for any angle ζ, $L(\zeta)$ is the radial distance at that angle, from the origin to the point where lines drawn perpendicular to the radial line first intersect the perimeter. This is illustrated in Fig. 14.2.

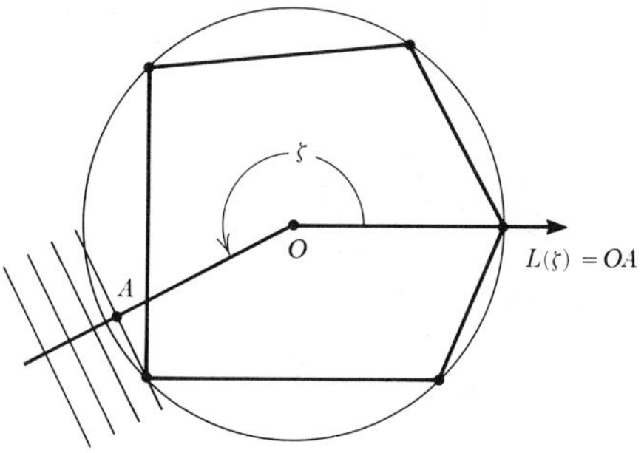

Fig. 14.2 Evaluation of the mean width of a polygon in two dimensions.

Now, from Figs. 14.1 and 14.2,

$$\bar{B}_2 = \frac{2}{2\pi} \sum_{i=1}^{M} \int_0^{\theta_i/2} \cos \zeta \, d\zeta = 2 \frac{1}{2\pi} \sum_{i=1}^{M} \sin \frac{\theta_i}{2} = \frac{P}{2\pi} \tag{14.24}$$

Thus

$$\bar{B}_2 = \frac{1}{2\pi} P$$

for any arbitrary spacing of M points.

Now we maximize the perimeter. The function

$$\sin \frac{\theta}{2} \qquad 0 \leq \theta \leq 2\pi$$

is convex downward. Because of its convexity, we can write

$$\sin \sum_{i=1}^{M} \gamma_i x_i \geq \sum_{i=1}^{M} \gamma_i \sin x_i$$

where

$$\sum_{i=1}^{M} \gamma_i = 1 \qquad \gamma_i \geq 0$$

and

$$x_i \in [0,\pi]$$

Let

$$x_i = \frac{\theta_i}{2}$$

therefore

$$x_i \in [0,\pi] \qquad \text{for all } i$$

Let $\gamma_i = 1/M$. Thus $\gamma_i > 0$ and $\sum_{i=1}^{M} \gamma_i = 1$. Substituting, we have

$$\sin \sum_{i=1}^{M} \frac{1}{M} \frac{\theta_i}{2} = \sin \frac{\pi}{M} \geq \sum_{i=1}^{M} \frac{1}{M} \sin \frac{\theta_i}{2}$$

since

$$\sum_{i=1}^{M} \frac{1}{M} \frac{\theta_i}{2} = \frac{\pi}{M}$$

Thus

$$2M \sin \frac{\pi}{M} \geq \sum_{i=1}^{M} 2 \sin \frac{\theta_i}{2} = P$$

This proves that the perimeter is reduced if the θ are chosen different from $\theta_i = 2\pi/M$ for $i = 1, \ldots, M$.

We have shown that finding optimal $\boldsymbol{\alpha}$ for small λ is equivalent to maximizing the mean width. It is worth noting that the converse is also true; that is, for a given M and D maximizing the mean width is equivalent to maximizing $P_d(\lambda;\boldsymbol{\alpha})$ for small λ. This is particularly significant because the geometrical problem of maximizing the mean width for arbitrary M and D is in general still open.

It should be emphasized that if it were known that the optimal $\boldsymbol{\alpha}$ were independent of the signal-to-noise ratio for all M and D, then the problem would be reduced to maximizing the mean width for various M and D, and the optimal-signal-design problem would be strictly a geometrical one. However, to date the optimal $\boldsymbol{\alpha}$ has been shown to be independent of λ only for $D = 2$ and $M - 1$. Also, the counterexample in the previous chapter, indicating that the preference of suboptimal signal sets can indeed depend on λ, puts all the more emphasis on being able to demonstrate the dependence or lack of dependence of the optimal set on λ. However, all the local optimal results that exist at present are independent of λ. In later chapters we shall discuss these results in detail.

We now find other characteristics which the optimal $\boldsymbol{\alpha}$ must possess, thereby further reducing the size of the class containing the optimal sets.

14.3 LINEARLY DEPENDENT VERSUS LINEARLY INDEPENDENT SIGNAL SETS

THEOREM 14.2 *For each set of linearly independent vectors* $\{\mathbf{S}_i;\ i = 1, \ldots, M\}$ *there exists a set of linearly dependent vectors* $\{\mathbf{S}'_i;\ i = 1, \ldots, M\}$ *with a greater probability of detection at all signal-to-noise ratios.*

Remark: This is not to say that all dependent sets are preferred over all independent sets. This theorem does state, however, that all optimal signal sets lie in the class of linearly dependent sets of vectors.

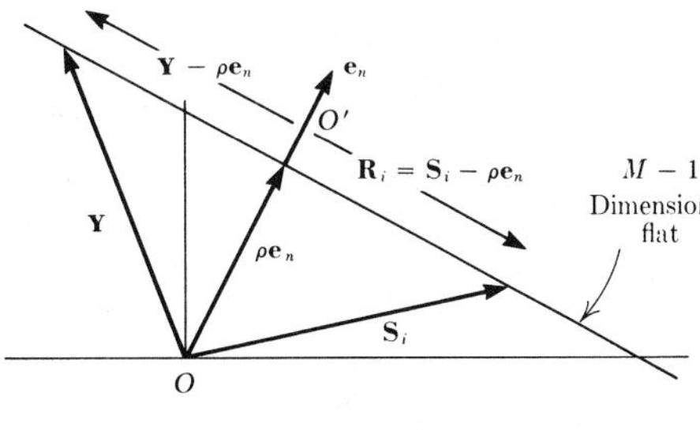

Fig. 14.3 An $M - 1$ flat containing M linearly independent signal vectors.

Proof: Since the dimensionality of the linearly independent signal set is M, there exists an $M - 1$ flat through the tips of the $\{\mathbf{S}_i\}$, defined by the M equations (see Fig. 14.3):

$$(\mathbf{S}_i - \rho \mathbf{e}_n)^T \mathbf{e}_n = 0 \qquad i = 1, \ldots, M$$

where $\mathbf{e}_n$ is the unit normal to the $M - 1$ flat and ρ is the projected distance from $\mathbf{O}$ to the flat, which is the distance OO' in Fig. 14.3. These defining equations can be rewritten as

$$\mathbf{S}_i^T \mathbf{e}_n = \rho \qquad i = 1, \ldots, M$$

where $\rho \neq 0$ ($\rho = 0$ implies $\mathbf{S}_i^T \mathbf{e}_n = 0$, which implies that each $\mathbf{S}_i$ is orthogonal to $\mathbf{e}_n$, implying that the $\{\mathbf{S}_i\}$ occupy only $M - 1$ dimensions and are therefore linearly dependent).

Let $\mathbf{R}_i$ be such that

$$\mathbf{S}_i = \mathbf{R}_i + \rho \mathbf{e}_n$$

Then

$$\|\mathbf{S}_i\| = 1 = \sqrt{\mathbf{R}_i^T \mathbf{R}_i + \rho^2}$$

and

$$\|\mathbf{R}_i\| = \sqrt{1 - \rho^2} < 1$$

Now let

$$\mathbf{S}'_i \triangleq \frac{\mathbf{R}_i}{\sqrt{1 - \rho^2}} \qquad i = 1, \ldots, M$$

The $\mathbf{S}'_i$ are M unit vectors in the $M - 1$ flat; hence they are linearly dependent.

As in (14.9), let

$$\xi'_i \triangleq \mathbf{Y}^T \mathbf{S}'_i = \frac{\mathbf{Y}^T \mathbf{R}_i}{\sqrt{1 - \rho^2}}$$

Then

$$E(\xi'_i \xi'_j) = \frac{\mathbf{R}_i^T \mathbf{R}_j}{1 - \rho^2}$$

and

$$\begin{aligned} \xi_i = \mathbf{Y}^T \mathbf{S}_i &= \sqrt{1 - \rho^2}\, \frac{\mathbf{Y}^T \mathbf{R}_i}{\sqrt{1 - \rho^2}} + \rho \mathbf{Y}^T \mathbf{e}_n \\ &= \sqrt{1 - \rho^2}\, \xi'_i + \rho \mathbf{Y}^T \mathbf{e}_n \end{aligned}$$

Therefore

$$\begin{aligned} \phi(\lambda;\alpha) &= E(\exp(\lambda \max_i \xi_i)) \\ &= E(\exp(\lambda \rho \mathbf{Y}^T \mathbf{e}_n) \exp(\lambda \sqrt{1 - \rho^2} \max_i \xi'_i)) \end{aligned}$$

Let

$$u = \mathbf{Y}^T\mathbf{e}_n$$

The vector $\mathbf{Y}$ is a gaussian variate, so u is a gaussian random variable with

$$E(u) = 0$$
$$E(u^2) = E(\mathbf{e}_n{}^T\mathbf{Y}\mathbf{Y}^T\mathbf{e}_n) = \mathbf{e}_n{}^T\mathbf{e}_n = 1$$

and

$$E(u\xi_i') = E\left(\mathbf{e}_n{}^T\mathbf{Y}\frac{\mathbf{Y}^T\mathbf{R}_i}{\sqrt{1-\rho^2}}\right) = \frac{\mathbf{e}_n{}^T\mathbf{R}_i}{\sqrt{1-\rho^2}} = 0 \qquad i = 1, \ldots, M$$

Thus u is independent of the ξ_i'. Since

$$E(\exp \lambda\rho u) = \exp\frac{\lambda^2\rho^2}{2}$$

then

$$\phi(\lambda;\alpha) = \exp\left(\frac{\lambda^2\rho^2}{2}\right)E(\exp(\lambda\sqrt{1-\rho^2}\max_i \xi_i'))$$

and

$$P_d(\lambda;\alpha) = \frac{1}{M}\exp\left[\frac{-\lambda^2}{2}(1-\rho^2)\right]E(\exp(\lambda\sqrt{1-\rho^2}\max_i \xi_i'))$$

Define

$$P_d(\lambda;\alpha') = \frac{1}{M}\exp(-\tfrac{1}{2}\lambda^2)\,E(\exp(\lambda \max_i \xi_i'))$$

where

$$\alpha' \triangleq \begin{bmatrix} 1 & & \lambda_{ij}' \\ & \ddots & \\ \lambda_{ji}' & & 1 \end{bmatrix}$$

with

$$\lambda_{ij}' = \mathbf{S}_i'^T\mathbf{S}_j'$$

is the inner-product matrix of the dependent vectors. Therefore

$$P_d(\lambda\sqrt{1-\rho^2};\,\alpha') = P_d(\lambda;\alpha) \tag{14.25}$$

Since $\sqrt{1-\rho^2} < 1$ and P_d is a monotonically increasing function in λ, we have

$$P_d(\lambda;\alpha') > P_d(\lambda;\alpha) \qquad \text{for all } \lambda \tag{14.26}$$

Stated another way, (14.25) indicates that the signal-to-noise ratio required to attain a given error rate by the linearly dependent signal set is not as great as that required for the linearly independent signal set. We now use this result to get a precise comparison between certain signal sets, as demonstrated in the following two important examples.

Example 14.2 Orthogonal signal set versus the regular simplex The orthogonal signal set is characterized by an inner-product matrix equal to the M-by-M identity matrix, which we shall denote by $\boldsymbol{\alpha}_0$; that for the regular simplex $\boldsymbol{\alpha}_R$ is given by (14.20). The orthogonal signal set is a linearly independent signal set. We assume that the vectors of this set align themselves with the coordinate axes in E_M. If they do not, we can use an orthogonal transformation to align them. This does not alter P_d, since P_d is invariant to orthogonal transformations on the signal set.

To find ρ for $\boldsymbol{\alpha}_0$ we follow the same procedure as in the preceding proof, that is,

$$\mathbf{S}_i^T\mathbf{e}_n = \rho, \qquad i = 1, \ldots, M$$

We define

$$\boldsymbol{\varrho} \triangleq \begin{bmatrix} \rho \\ \cdot \\ \cdot \\ \cdot \\ \rho \end{bmatrix}$$

Then

$$\mathbf{S}^T\mathbf{e}_n = \boldsymbol{\varrho}$$

where, as before, $\mathbf{S} \triangleq (\mathbf{S}_1, \ldots, \mathbf{S}_M)$.

Now

$$\boldsymbol{\varrho}^T\boldsymbol{\varrho} = M\rho^2 = (\mathbf{S}^T\mathbf{e}_n)^T\mathbf{S}^T\mathbf{e}_n = \mathbf{e}_n^T\mathbf{e}_n = 1$$

and

$$\rho^2 = \frac{1}{M}$$

As before,

$$\mathbf{S}_i^T\mathbf{S}_j = \delta_{ij} = \mathbf{R}_i^T\mathbf{R}_j + \rho^2$$

Thus

$$\mathbf{R}_i^T\mathbf{R}_j = \begin{cases} 1 - \rho^2 & \text{if } i = j \\ -\rho^2 & \text{if } i \neq j \end{cases}$$

and

$$\mathbf{S}_i'^T\mathbf{S}_j' = \begin{cases} 1 & \text{if } i = j \\ \dfrac{-\rho^2}{1-\rho^2} = \dfrac{-1}{M-1} & \text{if } i \neq j \end{cases}$$

which corresponds to the regular simplex signal structure, where

$$\boldsymbol{\alpha}_R \triangleq \begin{bmatrix} 1 & & \dfrac{-1}{M-1} \\ & \ddots & \\ \dfrac{-1}{M-1} & & 1 \end{bmatrix}$$

By substitution, we have

$$P_d(\lambda;\boldsymbol{\alpha}_0) = P_d\left(\lambda\sqrt{\frac{M-1}{M}};\boldsymbol{\alpha}_R\right)$$

or [rescaling λ, that is, $\lambda\sqrt{(M-1)/M} \to \lambda$]

$$P_d\left(\lambda\sqrt{\frac{M}{M-1}};\boldsymbol{\alpha}_0\right) = P_d(\lambda;\boldsymbol{\alpha}_R) \tag{14.27}$$

Note that for small M the improvement of $\boldsymbol{\alpha}_R$ over $\boldsymbol{\alpha}_0$ is greater than for large M. As $M \to \infty$, $\boldsymbol{\alpha}_R \to \boldsymbol{\alpha}_0$, and the signal sets become identical.

In the orthogonal signal set the ξ_i are all independent gaussian random variables with zero mean and unit variance. Hence $P_d(\lambda;\boldsymbol{\alpha}_0)$ can be written directly as

$$\begin{aligned} P_d(\lambda;\boldsymbol{\alpha}_0) &= \frac{1}{M}\exp\left(-\tfrac{1}{2}\lambda^2\right) E(\exp(\lambda \max_i \xi_i)) \\ &= \frac{1}{M}\exp\left(-\tfrac{1}{2}\lambda^2\right)\sum_{j=1}^{M}\int_{-\infty}^{\infty}\exp\lambda\xi_j\,\frac{\exp -\frac{1}{2}\xi_j^2}{\sqrt{2\pi}}\,d\xi_j \\ &\qquad \underbrace{\int\cdots\int_{-\infty}^{\xi_j}}_{(M-1)\text{-fold}} \frac{\exp\left(-\frac{1}{2}\sum_{i\neq j}\xi_i^2\right)}{(\sqrt{2\pi})^{M-1}}\,d\xi_1\cdots d\xi_{j-1}\,d\xi_{j+1}\cdots d\xi_M \\ &= \int_{-\infty}^{\infty} G(x-\lambda)\,[\phi(x)]^{M-1}\,dx \end{aligned} \tag{14.28}$$

where

$$\phi(x) = \int_{-\infty}^{x} G(y)\,dy \tag{14.29}$$

and

$$G(x) = \frac{1}{\sqrt{2\pi}}\exp -\tfrac{1}{2}x^2 \tag{14.30}$$

With (14.27) we also have a direct way of evaluating $P_d(\lambda;\boldsymbol{\alpha}_R)$:

$$P_d(\lambda;\boldsymbol{\alpha}_R) = \int_{-\infty}^{\infty} G\left(x - \lambda\sqrt{\frac{M}{M-1}}\right)[\phi(x)]^{M-1}\,dx \qquad (14.31)$$

From (13.24), this can be expressed in terms of *communication efficiency* β and M as

$$P_d(\beta;\boldsymbol{\alpha}_R) = \int_{-\infty}^{\infty} G\left(x - \sqrt{\frac{\beta M\log_2 M}{M-1}}\right)[\phi(x)]^{M-1}\,dx \qquad (14.32)$$

Plots of $P_d(\lambda;\boldsymbol{\alpha}_R)$ versus λ^2 for various M are given in Fig. 15.2.

Example 14.3 Equicorrelated signal set As a slight generalization of the previous example we take

$$\mathbf{S}_i^T\mathbf{S}_j = \gamma \qquad \text{for all } i \neq j$$

which is the *equicorrelated signal set*, with inner-product matrix

$$\boldsymbol{\alpha}_\gamma = \begin{bmatrix} 1 & & \gamma \\ & \ddots & \\ \gamma & & 1 \end{bmatrix} \qquad (14.33)$$

Note that $\gamma \geq -1/(M-1)$, because

$$0 \leq \left(\sum_{i=1}^{M}\mathbf{S}_i\right)^T\sum_{j=1}^{M}\mathbf{S}_j = M + M(M-1)\gamma$$

Also, $\boldsymbol{\alpha}_\gamma$ characterizes linearly independent signal sets for all γ such that $1 > \gamma > -1/(M-1)$.

LEMMA 14.2 *The distance ρ from the origin to the $M-1$ flat generated by any set of M vector tips which are linearly independent is*

$$\rho^2 = \frac{D(\boldsymbol{\alpha})}{\sum_{i=1}^{M}\sum_{j=1}^{M}C_{ij}(\boldsymbol{\alpha})} \qquad (14.34)$$

where $D(\boldsymbol{\alpha})$ is the determinant of $\boldsymbol{\alpha}$ and $C_{ij}(\boldsymbol{\alpha})$ is the cofactor of the ijth element of $\boldsymbol{\alpha}$.

Proof:

$$(\mathbf{S}^T\mathbf{e}_N)^T = \mathbf{e}_N{}^T\mathbf{S} = \boldsymbol{\varrho}^T \triangleq (1,1,\ldots,1)\rho$$

Since $\mathbf{S}$ is an M-by-M matrix in this case, with $D(\mathbf{S}) \neq 0$, we can write

$$\mathbf{e}_n^T = \boldsymbol{\rho}^T\mathbf{S}^{-1}$$

from which

$$\mathbf{e}_n = (\mathbf{S}^{-1})^T\boldsymbol{\rho}$$

Then

$$\mathbf{e}_n^T\mathbf{e}_n = 1 = \boldsymbol{\rho}^T\mathbf{S}^{-1}(\mathbf{S}^{-1})^T\boldsymbol{\rho} = \rho^2(1,1, \ldots ,1)\mathbf{S}^{-1}(\mathbf{S}^{-1})^T(1, \ldots ,1)^T$$

But

$$\mathbf{S}^{-1}(\mathbf{S}^{-1})^T = (\mathbf{S}^T\mathbf{S})^{-1} = \boldsymbol{\alpha}^{-1}$$

Hence

$$\rho^2 = \frac{1}{(1, \ldots ,1)(\boldsymbol{\alpha})^{-1}(1, \ldots ,1)^T} = \frac{1}{\sum_{i=1}^{M}\sum_{j=1}^{M} \alpha_{ij}^{-1}} = \frac{D(\boldsymbol{\alpha})}{\sum_{i=1}^{M}\sum_{j=1}^{M} C_{ij}(\boldsymbol{\alpha})}$$

In our example

$$D(\boldsymbol{\alpha}_\gamma) = (1 - \gamma)^{M-1}[1 + (M - 1)\gamma] \qquad (14.35)$$

$$C_{ii} = [1 + (M - 2)\gamma](1 - \gamma)^{M-2} \qquad (14.36)$$

$$C_{ij} = -\gamma(1 - \gamma)^{M-2} \qquad \text{for } i \neq j \qquad (14.37)$$

from which

$$\rho^2 = \frac{1}{M}[1 + (M - 1)\gamma]$$

Again using the same procedure as in the previous example, we have

$$\mathbf{S}_i'^T\mathbf{S}_j' = \frac{\mathbf{R}_i^T\mathbf{R}_j}{1 - \rho^2} = \frac{\gamma - \rho^2}{1 - \rho^2} = \frac{-1}{M - 1} \qquad \text{for all } i \neq j \qquad (14.38)$$

which again agrees with the regular simplex. Substituting and using

$$1 - \rho^2 = \frac{M - 1}{M}(1 - \gamma)$$

yields

$$P_d(\lambda;\boldsymbol{\alpha}_\gamma) = P_d\left(\lambda\sqrt{\frac{M - 1}{M}(1 - \gamma)};\, \boldsymbol{\alpha}_R\right) \qquad (14.39)$$

or, equivalently, from Example 14.2,

$$P_d(\lambda;\boldsymbol{\alpha}_\gamma) = P_d(\lambda\sqrt{1 - \gamma};\, \boldsymbol{\alpha}_0) \qquad (14.40)$$

14.4 GRADIENT OF THE PROBABILITY OF DETECTION

Let us now determine the gradient of $P_d(\lambda;\boldsymbol{\alpha})$, which is a fundamental quantity in the analysis. This gradient will be used to determine necessary conditions for signal sets to be locally optimal. We obtain a general result here that further reduces the size of the class of signal sets which contain the optimal sets.

From (14.13) we have

$$P_d(\lambda;\boldsymbol{\alpha}) = \frac{1}{M} \exp\left(-\tfrac{1}{2}\lambda^2\right) E(\exp(\lambda \max_i \xi_i))$$

Let

$$x = \max_i \xi_i$$

and let $p_M(x;\boldsymbol{\alpha})$ be the probability density function of x. Then

$$P_d(\lambda;\boldsymbol{\alpha}) = \frac{1}{M} \exp\left(-\tfrac{1}{2}\lambda^2\right) \int_{-\infty}^{\infty} \exp(\lambda x)\, p_M(x;\boldsymbol{\alpha})\, dx \tag{14.41}$$

We define

$$\begin{aligned}\Phi(x;\boldsymbol{\alpha}) &\triangleq \int_{-\infty}^{x} p_M(y;\boldsymbol{\alpha})\, dy = \Pr(\max_i \xi_i = y \le x) \\ &= \underbrace{\int_{-\infty}^{x} \cdots \int}_{M\text{-fold}} G(\boldsymbol{\xi};\mathbf{0};\boldsymbol{\alpha})\, d\boldsymbol{\xi}\end{aligned} \tag{14.42}$$

Then, after substitution, we have

$$\begin{aligned}\phi(\lambda;\boldsymbol{\alpha}) &= MP_d(\lambda;\boldsymbol{\alpha}) \exp \tfrac{1}{2}\lambda^2 = \int_{-\infty}^{\infty} \exp(\lambda x) \frac{d}{dx}[\Phi(x;\boldsymbol{\alpha})]\, dx \\ &= \int_{-\infty}^{\infty} \exp(\lambda x) \frac{d}{dx}\{\Phi(x;\boldsymbol{\alpha}) - [\phi(x)]^M\}\, dx \\ &\quad + M \int_{-\infty}^{\infty} \exp(\lambda x)\, [\phi(x)]^{M-1} G(x)\, dx\end{aligned} \tag{14.43}$$

The second integral is not a function of $\boldsymbol{\alpha}$. Integrating the first integral by parts gives

$$\begin{aligned}\phi(\lambda;\boldsymbol{\alpha}) &= \exp(\lambda x)\, \{\Phi(x;\boldsymbol{\alpha}) - [\phi(x)]^M\}_{-\infty}^{\infty} - \lambda \int_{-\infty}^{\infty} \exp(\lambda x)\, \{\Phi(x;\boldsymbol{\alpha}) \\ &\quad - [\phi(x)]^M\}\, dx + M \int_{-\infty}^{\infty} \exp(\lambda x)\, G(x) [\phi(x)]^{M-1}\, dx\end{aligned} \tag{14.44}$$

As $x \to \pm\infty$, $\Phi(x;\boldsymbol{\alpha}) - [\phi(x)]^M \to 0$ as $\exp -\tfrac{1}{2}x^2$. Thus the first term in (14.44) vanishes.

Now we take the derivative of $\phi(\lambda;\boldsymbol{\alpha})$ with respect to λ_{12}:

$$\frac{\partial\phi(\lambda;\boldsymbol{\alpha})}{\partial\lambda_{12}} = -\lambda\int_{-\infty}^{\infty}\exp(\lambda x)\,\frac{\partial\Phi(x;\boldsymbol{\alpha})}{\partial\lambda_{12}}\,dx$$

$$= -\lambda\int_{-\infty}^{\infty}\exp(\lambda x)\,\frac{\partial}{\partial\lambda_{12}}\left[\int\overset{x}{\underset{-\infty}{\cdots}}\int G(\boldsymbol{\xi};\mathbf{0};\boldsymbol{\alpha})\,d\boldsymbol{\xi}\right]dx$$

$$= -\lambda\int_{-\infty}^{\infty}\exp(\lambda x)\left\{\int\overset{x}{\underset{-\infty}{\cdots}}\int\left[\frac{\partial}{\partial\lambda_{12}}G(\boldsymbol{\xi};\mathbf{0};\boldsymbol{\alpha})\right]d\boldsymbol{\xi}\right\}dx \tag{14.45}$$

We denote the *characteristic function* of $G(\boldsymbol{\xi};\mathbf{0};\boldsymbol{\alpha})$ by $C(\mathbf{t};\boldsymbol{\alpha})$, where $\mathbf{t}$ is an M-by-1 column vector. Then

$$C(\mathbf{t};\boldsymbol{\alpha}) = \exp -\tfrac{1}{2}\mathbf{t}^T\boldsymbol{\alpha}\mathbf{t} \tag{14.46}$$

and

$$G(\boldsymbol{\xi};\mathbf{0};\boldsymbol{\alpha}) = \mathfrak{F}[C(\mathbf{t};\boldsymbol{\alpha})]$$

where $\mathfrak{F}$ indicates the M-dimensional Fourier transform. Thus

$$G(\boldsymbol{\xi};\mathbf{0};\boldsymbol{\alpha}) = \int\overset{\infty}{\underset{-\infty}{\cdots}}\int \exp i\mathbf{t}^T\boldsymbol{\xi}\,\frac{\exp -\frac{1}{2}\mathbf{t}^T\boldsymbol{\alpha}\mathbf{t}}{(2\pi)^M}\,d\mathbf{t} \tag{14.47}$$

Hence

$$\frac{\partial G(\boldsymbol{\xi};\mathbf{0};\boldsymbol{\alpha})}{\partial\lambda_{12}} = -\int\overset{\infty}{\underset{-\infty}{\cdots}}\int t_1t_2\exp(-i\mathbf{t}^T\boldsymbol{\xi})\,\frac{\exp -\frac{1}{2}\mathbf{t}^T\boldsymbol{\alpha}\mathbf{t}}{(2\pi)^M}\,d\mathbf{t}$$

$$= \frac{\partial^2}{\partial\xi_1\,\partial\xi_2}\left\{\int\overset{\infty}{\underset{-\infty}{\cdots}}\int\exp(-i\mathbf{t}^T\boldsymbol{\xi})\,\frac{\exp -\frac{1}{2}\mathbf{t}^T\boldsymbol{\alpha}\mathbf{t}}{(2\pi)^M}\,d\mathbf{t}\right\}$$

$$= \frac{\partial^2}{\partial\xi_1\,\partial\xi_2}[G(\boldsymbol{\xi};\mathbf{0};\boldsymbol{\alpha})]$$

or, with slightly different notation,

$$\frac{\partial G(\boldsymbol{\xi};\mathbf{0};\boldsymbol{\alpha})}{\partial\lambda_{12}} = \frac{\partial^2 G(\xi_1,\ \ldots\ ,\xi_M;\ \mathbf{0};\ \boldsymbol{\alpha})}{\partial\xi_1\,\partial\xi_2} \tag{14.48}$$

Substitution gives

$$\frac{\partial\phi(\lambda;\boldsymbol{\alpha})}{\partial\lambda_{12}} = -\lambda\int_{-\infty}^{\infty}dx\exp\lambda x\underbrace{\int\overset{x}{\underset{-\infty}{\cdots}}\int}_{M\text{-fold}}\frac{\partial^2 G(\xi_1,\ \ldots\ ,\xi_M;\ \mathbf{0};\ \boldsymbol{\alpha})}{\partial\xi_1\,\partial\xi_2}\,d\boldsymbol{\xi}$$

$$= -\lambda\int_{-\infty}^{\infty}dx\exp\lambda x\underbrace{\int\overset{x}{\underset{-\infty}{\cdots}}\int}_{(M-2)\text{-fold}}G(x,x,\xi_3,\ \ldots\ ,\xi_M;\ \mathbf{0};\ \boldsymbol{\alpha})\,d\xi_3\cdots d\xi_M \tag{14.49}$$

The integral is greater than zero for all λ. Therefore

$$\frac{\partial\phi(\lambda;\boldsymbol{\alpha})}{\partial\lambda_{12}} < 0 \qquad \text{for all } \lambda > 0 \tag{14.50}$$

Similarly,

$$\frac{\partial\phi(\lambda;\boldsymbol{\alpha})}{\partial\lambda_{ij}} < 0 \qquad \text{for all } \lambda > 0, \text{ all } i \neq j \tag{14.51}$$

Hence decreasing λ_{ij} increases $\phi(\lambda;\boldsymbol{\alpha})$, and we have proved the following theorem.

THEOREM 14.3 *If*

$$\lambda'_{ij} \leq \lambda_{ij} \qquad \textit{for all } i \neq j \tag{14.52}$$

then

$$P_d(\lambda;\boldsymbol{\alpha}') \geq P_d(\lambda;\boldsymbol{\alpha}) \qquad \textit{for all } \lambda > 0 \tag{14.53}$$

This proves that in determining the optimum $\boldsymbol{\alpha}$ we should make the $\{\lambda_{ij}\}$ as small as possible, which corresponds to placing the set of signal vectors as far apart from one another as possible within restrictions imposed by the covariance matrix. This is what we would expect. It gives a partial ordering of the class of admissible signal sets, which clearly is not a complete ordering. Hence this result reduces the class of sets which contains the optimal sets but does not tell us precisely which sets are optimal. There is also no dimensionality restriction in this result.

In particular, if

$$\rho_0 = \max_{i \neq j} \{\lambda_{ij}\}$$

then, with the corresponding inner-product matrix denoted by $\boldsymbol{\alpha}_{\rho_0}$,

$$\boldsymbol{\alpha}_{\rho_0} = \begin{bmatrix} 1 & & \rho_0 \\ & \ddots & \\ \rho_0 & & 1 \end{bmatrix}$$

we have

$$P_d(\lambda;\boldsymbol{\alpha}) > P_d(\lambda;\boldsymbol{\alpha}_{\rho_0}) \qquad \text{for all } \lambda$$

and hence minimizing ρ_0, which corresponds to maximizing the minimum distance within the admissible class, provides a lower bound in this class and is independent of λ.

14.5 SIGNAL SETS WHOSE CONVEX HULL DOES NOT INCLUDE THE ORIGIN

DEFINITION 14.1 *The convex hull of a set of M vectors $\{\mathbf{S}_i\}$ is the set*

$$\Pi = \left\{\mathbf{Y} \middle| \mathbf{Y} = \sum_{i=1}^{M} \gamma_i \mathbf{S}_i, \gamma_i \geq 0, \sum_{i=1}^{M} \gamma_i = 1\right\} \tag{14.54}$$

We must first prove the following theorem, which will be needed in what follows.

THEOREM 14.4 *Let $x = \max_i \xi_i$, where $\xi_i = \mathbf{Y}^T\mathbf{S}_i$ for $i = 1, \ldots, M$ and*

$$\boldsymbol{\xi} = \begin{bmatrix} \xi_1 \\ \cdot \\ \cdot \\ \cdot \\ \xi_M \end{bmatrix}$$

is a gaussian variate with zero mean and covariance matrix equal to $\boldsymbol{\alpha}$. Let $p_M(x;\boldsymbol{\alpha})$ be the probability density function of x. Then

$$p_M(x;\boldsymbol{\alpha}) = 0 \qquad \text{for } x < 0 \tag{14.55}$$

if and only if Π contains the origin.

Proof: $\mathbf{0} \in \Pi$ implies that there exists a set of $\{a_i\}$ such that

$$\sum_{i=1}^{M} a_i \mathbf{S}_i = \mathbf{0} \qquad a_i \geq 0 \qquad \sum_{i=1}^{M} a_i = 1$$

First, suppose $p_M(x;\boldsymbol{\alpha}) = 0$ for all $x < 0$. Consider the minimum

$$\min \left\{\left\|\sum_{i=1}^{M} a_i \mathbf{S}_i\right\| \,\middle|\, \sum_{i=1}^{M} a_i = 1, a_i \geq 0\right\}$$

over all possible sets of $\{a_i\}$. That is, consider the point closest to the origin of the closed convex hull generated by the $\{\mathbf{S}_i\}$. Let

$$\delta = \min \left\|\sum_{i=1}^{M} a_i \mathbf{S}_i\right\|$$

Suppose $\delta > 0$. We have then satisfied all the hypotheses of the following lemma from convex-body theory (illustrated in Fig. 14.4 and proven in [14.5]).

LEMMA 14.3 *Let $\mathbf{S}$ equal a closed convex set in E_D, with $\mathbf{0} \notin \mathbf{S}$. Then there exists a vector $\mathbf{p} \in E_D$ such that*

$$\mathbf{p}^T\mathbf{x} > 0 \qquad \text{for all } \mathbf{x} \in \mathbf{S}$$

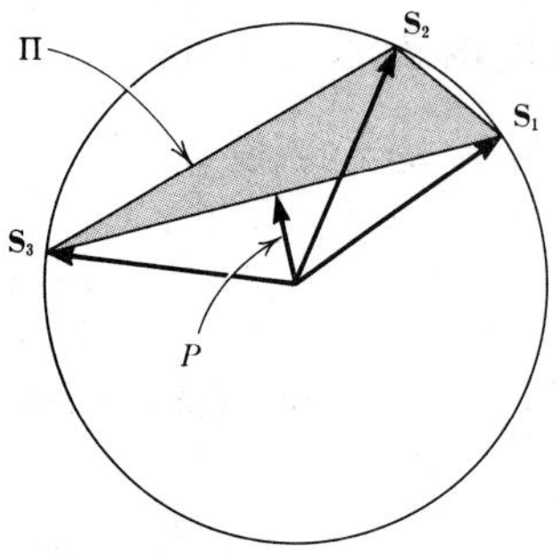

Fig. 14.4 Example of a signal structure whose convex hull does not include the origin.

Applying this lemma, call the vector which satisfies the lemma $\mathbf{V}_0$ and adjust its magnitude so that

$$\|\mathbf{V}_0\| = \delta$$

Then, by the lemma,

$$\mathbf{V}_0{}^T\mathbf{S}_i > 0 \qquad i = 1, \ldots, M$$

Now take

$$\mathbf{\Delta} = \{\mathbf{Y} | \mathbf{Y} \in E_D;\, x = \max_i \mathbf{Y}^T\mathbf{S}_i < 0\}$$

$\mathbf{\Delta}$ is not empty because $-\mathbf{V}_0 \in \mathbf{\Delta}$. Because of the strict inequality in the lemma, $-\mathbf{V}_0$ is in the interior of $\mathbf{\Delta}$ and thus has nonzero probability, which is a contradiction. Therefore δ must be zero.

Conversely, assume $\delta = 0$, which implies that there exists $a_i \geq 0$ with $\sum_{i=1}^{M} a_i = 1$, such that

$$\sum_{i=1}^{M} a_i\mathbf{S}_i = \mathbf{0}$$

This implies that for every $\mathbf{Y} \in E_D$ not all the $\mathbf{Y}^T\mathbf{S}_i$ can have the same sign, because

$$\sum_{i=1}^{M} a_i\mathbf{Y}^T\mathbf{S}_i = \mathbf{Y}^T \sum_{i=1}^{M} a_i\mathbf{S}_i = 0$$

Hence at least one of the $\mathbf{Y}^T\mathbf{S}_i$ must be positive; that is

$$x = \max_i \mathbf{Y}^T\mathbf{S}_i \geq 0 \qquad \text{for all } \mathbf{Y}$$

Therefore

$$p_M(x;\boldsymbol{\alpha}) = 0 \qquad \text{for all } x < 0$$

Note that this proof of Theorem 14.4 is true for any M and any D.

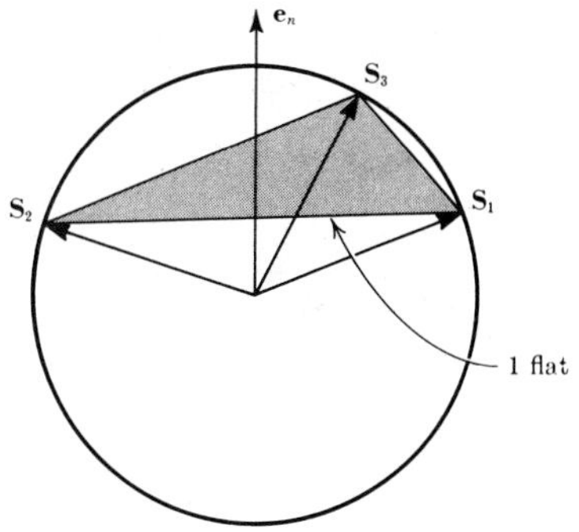

Fig. 14.5 Example of a convex hull which does not include the origin.

We now prove the main theorem of this section.

THEOREM 14.5 *If the convex hull generated by a set of* $\{\mathbf{S}_i; i = 1, \ldots, M\}$ *does not include the origin, there exists a signal set whose convex hull does include the origin with a higher probability of detection for all signal-to-noise ratios.*

Proof: Let $\boldsymbol{\alpha}$ correspond to a set of M linearly dependent vectors $\{\mathbf{S}_i\}$. From Theorem 14.2 we know that all optimal signal sets are contained in the class which is linearly dependent, and we can thus restrict our attention to this class. Assume that the convex hull generated by the $\{\mathbf{S}_i\}$ does not contain the origin.

Assume that the M vectors span E_{M-1}. Then, since the convex hull generated by the $\{\mathbf{S}_i\}$ does not contain the origin, from Lemma 14.3, there exists an $M - 2$ flat through the tips of $M - 1$ of the $\{\mathbf{S}_i\}$, say, $\mathbf{S}_1, \ldots, \mathbf{S}_{M-1}$, which separates the origin from the remaining vector $\mathbf{S}_M$, as illustrated for three signals in two-dimensional space in Fig. 14.5. Let the flat have equation

$$\mathbf{Y}^T\mathbf{e}_n = \rho$$

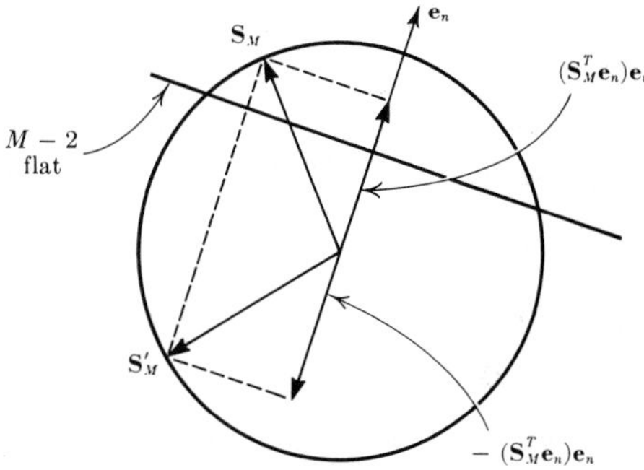

Fig. 14.6 Example of an $M - 2$ flat separating the origin from one signal vector.

where $\rho > 0$ is the distance from the origin to the flat and $\mathbf{e}_n$ is the unit normal to the flat. Then

$$\mathbf{S}_i^T\mathbf{e}_n = \rho \qquad i = 1, \ldots, M-1$$

and

$$\mathbf{S}_M^T\mathbf{e}_n > \rho$$

Now define a new set of vectors $\{\mathbf{S}_i'\}$ such that

$$\mathbf{S}_M' = \mathbf{S}_M - 2(\mathbf{e}_n^T\mathbf{S}_M)\mathbf{e}_n$$

and

$$\mathbf{S}_i' = \mathbf{S}_i \qquad i = 1, \ldots, M-1 \tag{14.56}$$

The $\mathbf{S}_i'$ are generated by choosing the $\mathbf{S}_i'$ the same as the $\mathbf{S}_i$ for those $\mathbf{S}_i$ whose tips generate the $M-2$ flat and forming $\mathbf{S}_M'$ from $\mathbf{S}_M$ by choosing the component of $\mathbf{S}_M$ perpendicular to the flat, taking its negative, and leaving the components of $\mathbf{S}_M$ in the flat unchanged, as illustrated in Fig. 14.6. We have

$$\mathbf{S}_i'^T\mathbf{S}_i' = 1 \qquad i = 1, \ldots, M$$

Also,

$$\begin{aligned}\mathbf{S}_i'^T\mathbf{S}_M' &= \mathbf{S}_i^T(\mathbf{S}_M - 2\mathbf{e}_n(\mathbf{e}_n^T\mathbf{S}_M)) \\ &= \mathbf{S}_i^T\mathbf{S}_M - 2(\mathbf{S}_i^T\mathbf{e}_n)(\mathbf{e}_n^T\mathbf{S}_M)\end{aligned}$$

Since

$$\mathbf{S}_i^T\mathbf{e}_n = \rho > 0$$

and

$$\mathbf{e}_n^T\mathbf{S}_M > \rho > 0$$

then

$$\lambda_{iM}' = \mathbf{S}_i'^T\mathbf{S}_M' < \lambda_{iM} - 2\rho^2 < \lambda_{iM} \qquad i = 1, \ldots, M-1$$

Since

$$\lambda_{ij}' = \lambda_{ij} \qquad \text{for all } i \neq j;\ i, j < M$$

we can conclude from Theorem 14.3 that

$$P_d(\lambda;\alpha') \geq P_d(\lambda;\alpha) \qquad \text{for all } \lambda > 0$$

If the dimensionality of the $\{\mathbf{S}_i\}$ is $D < M-1$, and the convex hull generated by the $\{\mathbf{S}_i\}$ does not include the origin, there exists a $D - 1$ flat through the tips of D of the $\{\mathbf{S}_i\}$, say, $\mathbf{S}_1, \ldots,$ $\mathbf{S}_D$, such that

$$\mathbf{e}_n\mathbf{S}_i = \rho \qquad i = 1, \ldots, D$$

and

$$\mathbf{e}_n\mathbf{S}_i > \rho \qquad i = D + 1, \ldots, M$$

where again ρ is the distance from the origin to the flat and $\mathbf{e}_n$ is the unit normal to the flat.

We define $\{\mathbf{S}'_i\}$ such that

$$\mathbf{S}'_i = \mathbf{S}_i \qquad i = 1, \ldots, D$$

and

$$\mathbf{S}'_i = \mathbf{S}_i - 2\mathbf{e}_n(\mathbf{e}_n{}^T\mathbf{S}_i) \qquad i = D + 1, \ldots, M$$

and again we can conclude that

$$P_d(\lambda;\boldsymbol{\alpha}') \geq P_d(\lambda;\boldsymbol{\alpha}) \qquad \text{for all } \lambda > 0$$

If the convex hull generated by the $\{\mathbf{S}'_i\}$ does not contain the origin, we repeat the procedure. Successive repetitions will result in a signal set whose convex hull does contain the origin; a maximum of $D - 1$ iterations will be needed. Since the above inequality is true for each iteration, the proof is complete.

Let us examine at this point the following lemma concerning the behavior of the components of the gradient vector, which will be used extensively in later chapters. It contains an important property which the gradient of the optimal set must possess.

LEMMA 14.4 *Define*

$$\phi_{ij}(\lambda;\boldsymbol{\alpha}) \triangleq -\frac{M}{\lambda}\frac{\partial\phi(\lambda;\boldsymbol{\alpha})}{\partial\lambda_{ij}} \tag{14.57}$$

For those $\boldsymbol{\alpha}$ *whose corresponding convex hull contains the origin,*

$$\phi_{ij}(\lambda;\boldsymbol{\alpha}) = 0 \tag{14.58}$$

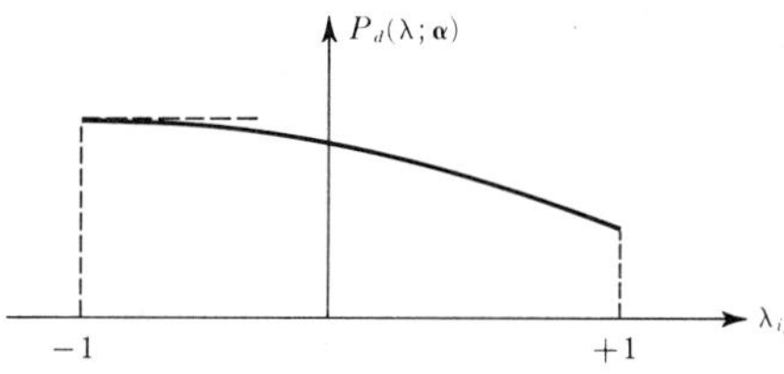

Fig. 14.7 $P_d(\lambda;\boldsymbol{\alpha})$ versus λ_{ij}.

if and only if

$$\lambda_{ij} = -1 \tag{14.59}$$

Remark: With the fact that $\partial\phi(\lambda;\alpha)/\partial\lambda_{ij} \leq 0$ for any λ_{ij} and this lemma, we can conclude that, as a function of λ_{ij}, the behavior of $P_d(\lambda;\alpha)$ is as shown in Fig. 14.7, where the derivative with respect to λ_{ij} is zero only at $\lambda_{ij} = -1$.

Proof: It is sufficient to show that $\phi_{12}(\lambda;\alpha) = 0$ if and only if $\lambda_{12} = -1$. From (14.57) and (14.49),

$$\frac{\phi_{12}(\lambda;\alpha)}{M} = \int_{-\infty}^{\infty} dx \exp \lambda x \underbrace{\int \cdots \int_{-\infty}^{x}}_{(M-2)\text{-fold}} G(x, x, \xi_3, \ldots, \xi_M; \mathbf{0}; \boldsymbol{\alpha})\, d\xi_3 \cdots d\xi_M \tag{14.60}$$

From Theorem 14.2 we know that the gaussian density function in (14.60) is concentrated in an r flat, where $r \leq M - 1$. Also, from Theorem 14.4, $p_M(x;\alpha) = 0$ for $x < 0$, from which we can conclude that the r flat cannot be located in the negative orthant, where the ξ_i are all less than zero. From this it follows immediately that the r flat also cannot be in the positive orthant, where the ξ_i are all greater than zero. Thus the r flat must be located as in Fig. 14.8, where it necessarily intersects the origin, since the means of all the ξ_i are zero.

From Theorems 14.4 and 14.5, for optimal sets the integral over negative x in (14.60) vanishes and can be written as

$$\frac{\phi_{12}(\lambda;\alpha)}{M} = \int_{0}^{\infty} dx \exp \lambda x \underbrace{\int \cdots \int_{-\infty}^{x}}_{(M-2)\text{-fold}} G(x, x, \xi_3, \ldots, \xi_M; \mathbf{0}; \boldsymbol{\alpha})\, d\xi_3 \cdots d\xi_M \tag{14.61}$$

Assume first that $\lambda_{12} = -1$. This implies that $\xi_1 = -\xi_2$ with

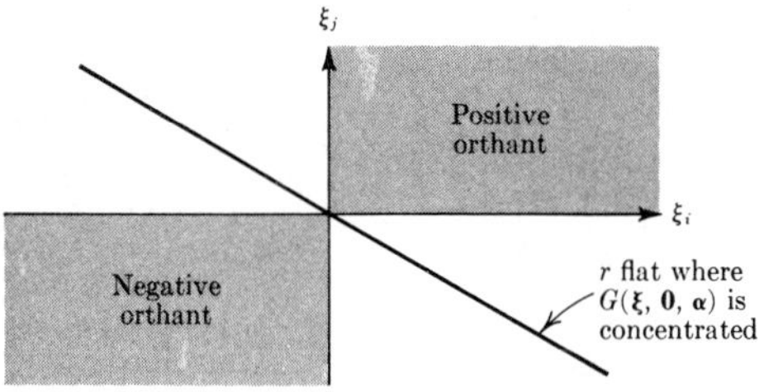

Fig. 14.8 Position of the r flat containing $G(\boldsymbol{\xi};\mathbf{0};\boldsymbol{\alpha})$.

probability 1. But in the integration over x in (14.61) we are integrating only over points where $\xi_2 = \xi_1 = x$. Hence (see Fig. 14.8) at every $x > 0$ the gaussian density is exactly zero, and we can conclude that

$$\phi_{12}(\lambda;\boldsymbol{\alpha}) = 0$$

Conversely, assume that $\phi_{12}(\lambda;\boldsymbol{\alpha}) = 0$. Further assume that $\lambda_{12} > -1$, from which we want to arrive at a contradiction. To do this we write the gaussian density in (14.61) as a conditional density,

$$\begin{aligned} G(\xi_1 = x,\ \xi_2 = x,\ \xi_3,\ \ldots\ ,\xi_M;\ \mathbf{0};\ \boldsymbol{\alpha}) \\ = G(\xi_3,\ \ldots\ ,\xi_M | \xi_1 = x,\ \xi_2 = x) G(\xi_1 = x,\ \xi_2 = x) \end{aligned} \tag{14.62}$$

where we call

$$G(\xi_1 = x,\ \xi_2 = x) = G_{12}(x) = \frac{\exp\,[-x^2/(1 + \lambda_{12})]}{2\pi\sqrt{1 - \lambda_{12}{}^2}} \tag{14.63}$$

and the conditional gaussian density has mean

$$E(\xi_3,\ \ldots\ ,\xi_M | \xi_1 = x,\ \xi_2 = x) = \left(\frac{\lambda_{13} + \lambda_{23}}{1 + \lambda_{12}}\, x,\ \ldots\ ,\frac{\lambda_{1M} + \lambda_{2M}}{1 + \lambda_{12}}\, x \right) \tag{14.64}$$

and covariance matrix[1] equal to

$$\begin{bmatrix} 1\ \lambda_{34} & \cdots & \lambda_{3M} \\ \lambda_{34} \ \ddots & \cdots & \cdots \\ \cdots\cdots & \ddots & \cdots\cdots \\ \cdots & \ddots & \lambda_{M-1,M} \\ \lambda_{3M} & \cdots\ \lambda_{M-1,M} \ \ddots & 1 \end{bmatrix} - \begin{bmatrix} \lambda_{13} & \lambda_{23} \\ \cdot & \cdot \\ \cdot & \cdot \\ \cdot & \cdot \\ \lambda_{1M} & \lambda_{2M} \end{bmatrix} \frac{\begin{bmatrix} 1 & -\lambda_{12} \\ -\lambda_{12} & 1 \end{bmatrix}}{1 - \lambda_{12}{}^2} \begin{bmatrix} \lambda_{13} & \cdots & \lambda_{1M} \\ \lambda_{23} & \cdots & \lambda_{2M} \end{bmatrix} \tag{14.65}$$

Substitution gives

$$\frac{\phi_{12}(\lambda;\boldsymbol{\alpha})}{M} = \int_0^\infty \exp\,(\lambda x)\, G_{12}(x) F_{12}(x)\, dx \tag{14.66}$$

[1] A summary of conditional gaussian densities is given in Appendix A.

where

$$F_{12}(x) = \int_{-\infty}^{x} \cdots \int G(\xi_3, \ldots, \xi_M | \xi_1 = x, \ \xi_2 = x) \, d\xi_3 \cdots d\xi_M \tag{14.67}$$

When $\lambda_{12} > -1$, $G_{12}(x)$ has positive measure for all x. Also, $F_{12}(x) \geq 0$ for all x. Therefore, if we can show that there exists some x for which $F_{12}(x)$ has positive measure, then we can conclude that

$\phi_{12}(\lambda;\boldsymbol{\alpha}) > 0$

in contradiction of our original assumption.

Let

$$\hat{\xi}_j = \xi_j - E(\xi_j | \xi_1 = x, \ \xi_2 = x)$$

$$= \xi_j - \frac{\lambda_{1j} + \lambda_{2j}}{1 + \lambda_{12}} x \qquad j = 3, \ldots, M \tag{14.68}$$

Then

$$F_{12}(x) = \int_{-\infty}^{\alpha_3 x} \cdots \int^{\alpha_M x} \hat{G}(\hat{\xi}_3, \ldots, \hat{\xi}_M) \, d\hat{\xi}_3 \cdots d\hat{\xi}_M \tag{14.69}$$

where the $\hat{\xi}_j$ have zero mean and covariance matrix given by (14.65), and

$$\alpha_j = 1 - \frac{\lambda_{1j} + \lambda_{2j}}{1 + \lambda_{12}} \qquad j = 3, \ldots, M$$

If the probability density function $\hat{G}(\hat{\xi}_3, \ldots, \hat{\xi}_M)$ is nonsingular, then $F_{12}(x)$ has positive measure for all x, and the proof is complete. We know that this is not true, however, because $F_{12}(x)$ would then have positive measure for negative x, implying that $p_M(x;\boldsymbol{\alpha}) > 0$ for $x < 0$, which is a contradiction. Therefore the density is singular, and as before, it is concentrated in an r flat through the origin, where $r < M - 2$. Similarly, the r flat cannot be in the negative or positive orthant; thus it is again in a position as indicated in Fig. 14.8 for the $\{\xi_j\}$.

In this case to show that $F_{12}(x)$ has positive measure for some x we must show that the direction taken by the line segment $(\alpha_3 x, \ldots, \alpha_M x)$ as x varies over $(0, \infty)$ is such that the negative orthant consisting of all $\hat{\xi}_i < \alpha_i x$ for $i = 3, \ldots, M$ intersects the r flat for some x. We claim that this is always the case. Suppose there is no x, as x varies over $(0, \infty)$, such that the negative orthant below $(\alpha_3 x \ \ldots, \alpha_M x)$ intersects the r flat containing the singular gaussian

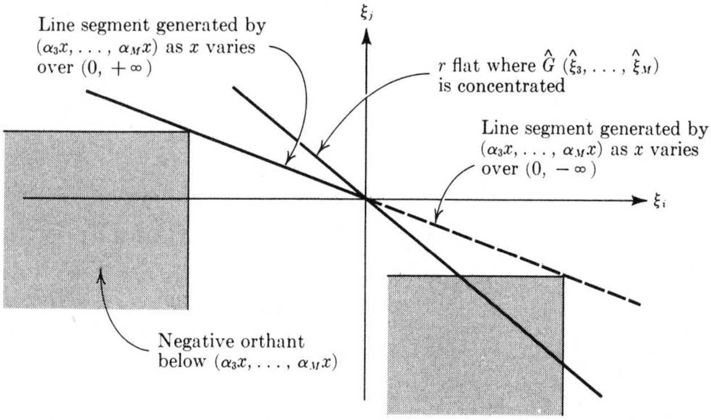

Fig. 14.9 Position of the r flat in which $\hat{G}(\hat{\xi}_3, \ldots, \hat{\xi}_M)$ is concentrated.

distribution. This situation is illustrated in Fig. 14.9. In this case, if we then consider the integral over negative x (the dotted line in Fig. 14.9), it is immediate that the negative orthant below points on this line intersects the r flat, which implies that $p_M(x;\boldsymbol{\alpha}) > 0$ for $x < 0$, contradicting the assumption that the convex hull of the corresponding signal set contains the origin. Therefore we can conclude that $F_{12}(x)$ necessarily has positive measure for $x > 0$ and $\phi_{12}(\lambda;\boldsymbol{\alpha}) > 0$, in contradiction of our original assumption. Thus $\phi_{12}(\lambda;\boldsymbol{\alpha}) = 0$ implies $\lambda_{12} = -1$.

14.6 THE ADMISSIBLE α SPACE

We have already seen that the admissible $\boldsymbol{\alpha}$ space $\mathcal{A}$ consists of those M-by-M symmetric nonnegative-definite matrices with 1s along the main diagonal and all off-diagonal elements such that $|\lambda_{ij}| \leq 1$. Let us now discuss some of the pertinent properties of the $\boldsymbol{\alpha}$ space.

First, note that $\mathcal{A}$ is bounded, for if we define the magnitude of a matrix in the usual way, by $|\text{tr}\ \boldsymbol{\alpha}^T\boldsymbol{\alpha}|$, where tr means *the trace of*, then

$$\|\boldsymbol{\alpha}\|^2 = |\text{tr}\ \boldsymbol{\alpha}^T\boldsymbol{\alpha}| \leq M^2 \tag{14.70}$$

$\mathcal{A}$ is also convex, for if $\boldsymbol{\alpha}_1$ and $\boldsymbol{\alpha}_2$ are in $\mathcal{A}$, then

$$\boldsymbol{\alpha}^0 = \beta\boldsymbol{\alpha}_1 + (1 - \beta)\boldsymbol{\alpha}_2 \qquad 0 \leq \beta \leq 1 \tag{14.71}$$

is of the form

$$\begin{bmatrix} 1 & & \lambda_{ij}{}^0 \\ & \ddots & \\ \lambda_{ji}{}^0 & & 1 \end{bmatrix}$$

and is nonnegative-definite, since for any M-by-1 vector $\mathbf{t}$

$$\mathbf{t}^T\boldsymbol{\alpha}^0\mathbf{t} = \beta\mathbf{t}^T\boldsymbol{\alpha}_1\mathbf{t} + (1-\beta)\mathbf{t}^T\boldsymbol{\alpha}_2\mathbf{t} \geq 0 \tag{14.72}$$

Hence

$$\boldsymbol{\alpha}^0 \in \mathcal{A}$$

It can similarly be shown that the class of positive-definite $\boldsymbol{\alpha}$ is also convex. Therefore the interior of $\mathcal{A}$ consists of $\boldsymbol{\alpha}$ which are positive-definite, for which $D(\boldsymbol{\alpha}) > 0$, and the boundary consists of those admissible $\boldsymbol{\alpha}$ for which $D(\boldsymbol{\alpha}) = 0$. Therefore we know that all the optimal $\boldsymbol{\alpha}$ lie on the boundary of $\mathcal{A}$.

It can also be shown that $\mathcal{A}$ is closed. Further, its surface is smooth, since $D(\boldsymbol{\alpha}) = 0$ is a polynomial in the set of $\{\lambda_{ij}\}$.

LEMMA 14.5 *$\sum_{i=1}^{M} \mathbf{S}_i = \mathbf{0}$ if and only if $\sum_{i=1}^{M} \lambda_{ij} = 0$ for $j = 1, \ldots, M$. In addition,*

$$\sum_{i=1}^{M} \lambda_{ij} = 0 \text{ for } j = 1, \ldots, M \text{ if and only if } \sum_{i=1}^{M}\sum_{j=1}^{M} \lambda_{ij} = 0$$

Proof: Assume $\sum_{i=1}^{M} \mathbf{S}_i = \mathbf{0}$. Then

$$\sum_{i=1}^{M} \mathbf{S}_i^T\mathbf{S}_j = \sum_{i=1}^{M} \lambda_{ij} = 0 \qquad j = 1, \ldots, M \tag{14.73}$$

and hence

$$\sum_{i=1}^{M}\sum_{j=1}^{M} \lambda_{ij} = 0$$

Conversely,

$$\sum_{i=1}^{M}\sum_{j=1}^{M} \lambda_{ij} = \sum_{i=1}^{M}\sum_{j=1}^{M} \mathbf{S}_i^T\mathbf{S}_j = \left(\sum_{i=1}^{M} \mathbf{S}_i\right)^T \sum_{j=1}^{M} \mathbf{S}_j \tag{14.74}$$

If $\sum_{i=1}^{M} \mathbf{S}_i = \mathbf{A}$, where $\mathbf{A}$ is not the zero vector, then $\mathbf{A}^T\mathbf{A} > 0$ and $\sum_{i=1}^{M}\sum_{j=1}^{M} \lambda_{ij} > 0$, in contradiction of the original assumption.

Therefore the class of signals for which

$$\sum_{i=1}^{M} \mathbf{S}_i = \mathbf{0} \tag{14.75}$$

is identical to the class for which

$$\sum_{i=1}^{M}\sum_{j=1}^{M}\lambda_{ij} = 0$$

Note also that this class is a convex set, and that the $\boldsymbol{\alpha}$ for which

$$\sum_{i=1}^{M}\sum_{j=1}^{M}\lambda_{ij} = 0$$

is a hyperplane in the $M(M-1)/2$ euclidean space containing $\mathcal{A}$.

If $\sum_{i=1}^{M}\lambda_{ij} = 0$ for $j = 1, \ldots, M$, then the column vector

$$\mathbf{1} = (1, \ldots, 1)^T$$

is an eigenvector with corresponding eigenvalue equal to zero, because

$$\boldsymbol{\alpha}\mathbf{1} = \mathbf{0} \tag{14.76}$$

Since $D(\boldsymbol{\alpha})$ is the product of the eigenvalues of $\boldsymbol{\alpha}$, we know that all admissible signal sets for which

$$\sum_{i=1}^{M}\sum_{j=1}^{M}\lambda_{ij} = 0 \tag{14.77}$$

all lie on the surface of $\mathcal{A}$ because they have an eigenvalue equal to zero. Note that this hyperplane defined by (14.76) contains nonadmissible as well as admissible $\boldsymbol{\alpha}$. The regular simplex signal structure $\boldsymbol{\alpha}_R$ and the optimal signal set when $D = 2$ (equal spacing) both lie in this hyperplane. In fact, we shall see that all our solutions satisfy (14.75) and (14.77) (see Prob. 14.4). Fig. 14.10 is an illustration of the $\boldsymbol{\alpha}$ space.

The following relationship will be used extensively in the next two chapters.

LEMMA 14.6 *For any symmetric matrix with independent off-diagonal elements,*

$$\frac{\partial D(\boldsymbol{\alpha})}{\partial \lambda_{ij}} = 2C_{ij}$$

where C_{ij} is the cofactor including sign of λ_{ij}.

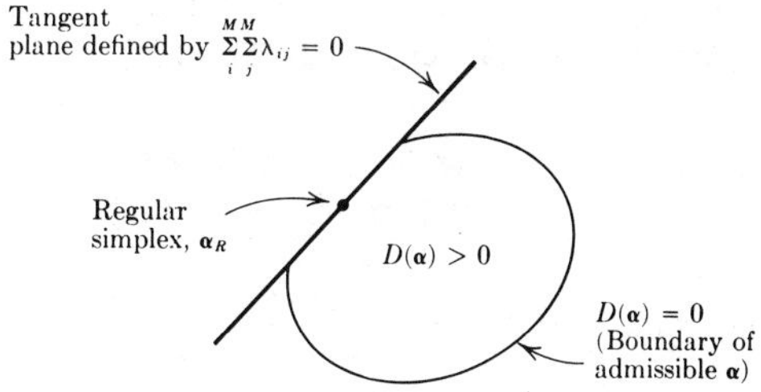

Fig. 14.10 The $[M(M-1)/2]$-dimensional $\boldsymbol{\alpha}$ space.

Proof: Expanding $D(\boldsymbol{\alpha})$ in terms of the jth column yields

$$D(\boldsymbol{\alpha}) = \sum_{k=1}^{M} \lambda_{kj} C_{kj}$$

Then

$$\frac{\partial D(\boldsymbol{\alpha})}{\partial \lambda_{ij}} = C_{ij} + \sum_{\substack{k=1 \\ k \neq i}}^{M} \lambda_{kj} \frac{\partial C_{kj}}{\partial \lambda_{ij}}$$

Now

$$\left.\frac{\partial C_{kj}}{\partial \lambda_{ij}}\right|_{k \neq i} = \frac{\partial}{\partial \lambda_{ij}} \sum_{\substack{l=1 \\ l \neq i}}^{M} \lambda_{li} C_{li}{}^{kj} = C_{ji}{}^{kj} \tag{14.78}$$

where $C_{li}{}^{kj}$ is the cofactor of the cofactor, that is, the determinant resulting after the kth and lth row and the ith and jth column have been removed from $\boldsymbol{\alpha}$. Substitution yields

$$\frac{\partial D(\boldsymbol{\alpha})}{\partial \lambda_{ij}} = C_{ij} + \sum_{\substack{k=1 \\ k \neq i}}^{M} \lambda_{kj} C_{ji}{}^{kj}$$

Using the fact that $C_{ji}{}^{kj} = C_{jk}{}^{ij}$, which results from the symmetry ot $\boldsymbol{\alpha}$, the summation becomes C_{ij}.

With this lemma, and noting that for the regular simplex the C_{ij} are independent of i and j, we can conclude that the plane tangent to the surface defined by $D(\boldsymbol{\alpha}) = 0$ at $\boldsymbol{\alpha}_R$ is given by

$$\sum_{i=1}^{M} \sum_{j=1}^{M} \lambda_{ij} = 0$$

14.7 SERIES EXPANSIONS AND ASYMPTOTIC APPROXIMATIONS

In this section we use *Hermite polynomial expansions* for the probability of detection for large M and large λ. A summary of these expansions, known as *tetrachoric series*, is given in Appendix E. For further details on tetrachoric series see Ref. [14.7]. The asymptotic expansion that is of most interest to us can be stated in the following manner.

THEOREM 14.6 *For large λ and fixed M the probability of error can be asymptotically approximated by*

$$P_e(\lambda;\boldsymbol{\alpha}) \approx \frac{1}{M} \sum_{i \neq j}^{M} \sum \frac{\exp -\frac{1}{2}\gamma_{ij}{}^2\lambda^2}{\sqrt{2\pi}\,\gamma_{ij}\lambda} \tag{14.79}$$

where

$$\gamma_{ij} = \sqrt{\frac{1 - \lambda_{ij}}{1 + \lambda_{ij}}} \tag{14.80}$$

Proof: From (14.41) we have

$$P_d(\lambda;\boldsymbol{\alpha}) = \frac{1}{M} \exp\left(-\tfrac{1}{2}\lambda^2\right) \int_{-\infty}^{\infty} \exp(\lambda x)\, p_M(x;\boldsymbol{\alpha})\, dx$$

For $p_M(x;\boldsymbol{\alpha})$ we can write

$$\begin{aligned} p_M(x;\boldsymbol{\alpha}) &= \sum_{i=1}^{M} p(\xi_i = x) \Pr(\xi_j < \xi_i \text{ for all } j \neq i | \xi_i = x) \\ &= \sum_{i=1}^{M} G(x) \Pr(\xi_j < x \text{ for all } j \neq i | \xi_i = x) \end{aligned} \tag{14.81}$$

Now,

$$\Pr(\xi_j < x \text{ for all } j \neq i | \xi_i = x) = \underbrace{\int \cdots \int_{-\infty}^{x}}_{(M-1)\text{-fold}} p_{M-1}(\xi_1, \ldots, \xi_{i-1}, \xi_{i+1}, \ldots, \xi_M | \xi_i = x)\, d\xi_1 \cdots d\xi_{i-1}\, d\xi_{i+1} \cdots d\xi_M \tag{14.82}$$

We let

$$\boldsymbol{\xi}_i = \begin{bmatrix} \xi_1 \\ \cdot \\ \cdot \\ \cdot \\ \xi_{i-1} \\ \xi_{i+1} \\ \cdot \\ \cdot \\ \cdot \\ \xi_M \end{bmatrix} \qquad \mathbf{m}_i = E(\boldsymbol{\xi}_i | \xi_i = x) \qquad \mathbf{R}_i = E(\boldsymbol{\xi}_i \boldsymbol{\xi}_i^T | \xi_i = x)$$

Then

$$p_{M-1}(\xi_1, \ldots, \xi_{i-1}, \xi_{i+1}, \ldots, \xi_M | \xi_i = x) = G(\boldsymbol{\xi}_i;\mathbf{m}_i;\mathbf{R}_i) \tag{14.83}$$

and

$$\Pr(\xi_j < x \text{ for all } j \neq i | \xi_i = x) = \underbrace{\int \cdots \int_{-\infty}^{x}}_{(M-1)\text{-fold}} G(\boldsymbol{\xi}_i;\mathbf{m}_i;\mathbf{R}_i)\, d\boldsymbol{\xi}_i \tag{14.84}$$

Now we let

$$\boldsymbol{\rho}_i = \begin{bmatrix} \lambda_{1i} \\ \cdot \\ \cdot \\ \cdot \\ \lambda_{i-1,i} \\ \lambda_{i+1,i} \\ \cdot \\ \cdot \\ \cdot \\ \lambda_{Mi} \end{bmatrix}$$

$\boldsymbol{\rho}_i$ is the ith column of $\boldsymbol{\alpha}$ with $\lambda_{ii} = 1$ removed. Thus $\boldsymbol{\rho}_i$ is an $(M - 1)$-by-1 column vector. Also, we define $\boldsymbol{\alpha}_i$ as $\boldsymbol{\alpha}$ with the ith row and ith column removed. $\boldsymbol{\alpha}_i$ is the unconditional covariance matrix of $\boldsymbol{\xi}_i$. Then, from the conditional gaussian distribution, we have

$$\mathbf{m}_i = \boldsymbol{\rho}_i x \tag{14.85}$$

and

$$\mathbf{R}_i = \boldsymbol{\alpha}_i - \boldsymbol{\rho}_i \boldsymbol{\rho}_i^T \tag{14.86}$$

If we define the jkth element of $\mathbf{R}_i$ as $r_{i_{jk}}$, then

$$r_{i_{jk}} = \lambda_{jk} - \lambda_{ij}\lambda_{ik} \qquad j \neq i,\ k \neq i \tag{14.87}$$

Now we define

$$\hat{\boldsymbol{\xi}}_i \triangleq \boldsymbol{\xi}_i - \mathbf{m}_i = \boldsymbol{\xi}_i - \boldsymbol{\rho}_i x \tag{14.88}$$

Then

$$\Pr(\xi_j < x \text{ for all } j \neq i \mid \xi_i = x) = \int_{-\infty}^{x(1-\lambda_{1i})} \int_{-\infty}^{x(1-\lambda_{i-1,i})} \int_{-\infty}^{x(1-\lambda_{i+1,i})} \int_{-\infty}^{x(1-\lambda_{Mi})} G(\hat{\boldsymbol{\xi}}_i;\mathbf{0};\mathbf{R}_i)\, d\hat{\boldsymbol{\xi}}_i \tag{14.89}$$

If we now substitute

$$\eta_j \triangleq \frac{\hat{\xi}_j}{\sqrt{1 - \lambda_{ij}^2}} \qquad j = 1, \ldots, i-1, i+1, \ldots, M \tag{14.90}$$

in (14.89), where $\hat{\xi}_j \triangleq \xi_j - \lambda_{ji} x$, then

$$E(\eta_j) = 0$$

and

$$E(\eta_j \eta_k) = \frac{\lambda_{jk} - \lambda_{ij}\lambda_{ik}}{\sqrt{1 - \lambda_{ij}^2}\sqrt{1 - \lambda_{ik}^2}} \tag{14.91}$$

Therefore

$$\Pr(\xi_j < x \text{ for all } j \neq i | \xi_i = x) = \Phi_i(x\gamma_{i1}, \ldots, x\gamma_{i,i-1}, x\gamma_{i,i+1}, \ldots, x\gamma_{iM}) \quad (14.92)$$

where Φ_i is an $(M - 1)$-variate cumulative gaussian distribution with zero means, unit variances, and covariances given by

$$\mu_{jk} = \frac{\lambda_{jk} - \lambda_{ij}\lambda_{ik}}{\sqrt{1 - \lambda_{ij}^2}\sqrt{1 - \lambda_{ik}^2}} \qquad j \neq i,\ k \neq i \quad (14.93)$$

and where

$$\gamma_{ij} = \sqrt{\frac{1 - \lambda_{ij}}{1 + \lambda_{ij}}} \quad (14.94)$$

Substituting into (14.81), we have

$$P_d(\lambda;\alpha) = \frac{1}{M}\int_{-\infty}^{\infty} dx\, G(x - \lambda) \sum_{i=1}^{M} \Phi_i(\gamma x_{i1}, \ldots, x\gamma_{i,i-1}, x\gamma_{i,i+1}, \ldots, x\gamma_{iM}) \quad (14.95)$$

Note that λ appears only in $G(x - \lambda)$. Expansion of Φ_i in a tetrachoric series yields

$$\Phi_i(x\gamma_{i1}, \ldots, x\gamma_{i,i-1}, x\gamma_{i,i+1}, \ldots, x\gamma_{iM}) = \prod_{\substack{j=1 \\ j\neq i}}^{M} \phi(\gamma_{ij}x) + \sum_{r=1}^{\infty} (\textit{tetrachoric series}) \quad (14.96)$$

Substitution then gives

$$P_d(\lambda;\alpha) = \frac{1}{M}\int_{-\infty}^{\infty} G(x - \lambda) \sum_{i=1}^{M} \prod_{\substack{j=1 \\ j\neq i}}^{M} \phi(\gamma_{ij}x)\, dx + \frac{1}{M} R(\lambda;\alpha) \quad (14.97)$$

where

$$R(\lambda;\alpha) = \exp(-\tfrac{1}{2}\lambda^2) \int_{-\infty}^{\infty} \exp \lambda x \exp -\tfrac{1}{2}x^2 \sum_{i=1}^{M} \left[\sum_{r=1}^{\infty} (\textit{tetrachoric series})\right] dx \quad (14.98)$$

The integral here is bounded, and therefore $R(\lambda;\alpha)$ is of order $\exp -\frac{1}{2}\lambda^2$ for large λ.

We rewrite the first integral in (14.97) as

$$\frac{1}{M}\int_{-\infty}^{\infty} G(x)\,dx \sum_{i=1}^{M} \prod_{\substack{j=1\\ j\neq i}}^{M} \phi(\gamma_{ij}(x+\lambda)) \tag{14.99}$$

and approximate $\phi(t)$ by

$$1 - \frac{\exp -\frac{1}{2}t^2}{t\sqrt{2\pi}} \qquad \text{for large } t \tag{14.100}$$

Upon substitution, for large λ we then have the approximation

$$P_d(\lambda;\boldsymbol{\alpha}) \approx \frac{1}{M}\int_{-\infty}^{\infty} G(x)\,dx \sum_{i=1}^{M} \prod_{\substack{j=1\\ j\neq i}}^{M} \left(1 - \frac{\exp\{-\frac{1}{2}[\gamma_{ij}(x+\lambda)]^2\}}{\gamma_{ij}(x+\lambda)\sqrt{2\pi}}\right) \tag{14.101}$$

For large λ we can approximate the product expansion by the first two terms,

$$\prod_{\substack{j=1\\ j\neq i}}^{M} \left\{1 - \frac{\exp[-\frac{1}{2}\gamma_{ij}^2(x+\lambda)^2]}{\gamma_{ij}(x+\lambda)\sqrt{2\pi}}\right\} \approx 1 - \sum_{\substack{j=1\\ j\neq i}}^{M} \frac{\exp[-\frac{1}{2}\gamma_{ij}^2(x+\lambda)^2]}{\gamma_{ij}(x+\lambda)\sqrt{2\pi}} \tag{14.102}$$

which, when substituted into (14.101), gives

$$P_d(\lambda;\boldsymbol{\alpha}) \approx 1 - \frac{1}{M}\sum\sum_{i\neq j}^{M} \int_{-\infty}^{\infty} dx\, G(x)\,\frac{\exp[-\frac{1}{2}\gamma_{ij}^2(x+\lambda)^2]}{\gamma_{ij}(x+\lambda)\sqrt{2\pi}} \tag{14.103}$$

which for large λ can be finally approximated by

$$P_d(\lambda;\boldsymbol{\alpha}) \approx 1 - \frac{1}{M}\sum\sum_{i\neq j}^{M} \frac{\exp -\frac{1}{2}\gamma_{ij}^2\lambda^2}{\gamma_{ij}\lambda\sqrt{2\pi}} \tag{14.104}$$

With the equality (14.92) we can obtain a recursion formula for the probability density corresponding to the regular simplex. In this case

$$\lambda_{ij} = \frac{-1}{M-1} \tag{14.105}$$

from which

$$\mu_{jk} = \frac{-1}{M-2} \qquad \text{for all } j \neq k \tag{14.106}$$

which corresponds to the regular simplex in $M - 2$ dimensions.

Also,

$$\gamma_{ij} = \sqrt{\frac{M}{M-2}} \tag{14.107}$$

for the regular simplex. Substituting, we have

$$\Phi_i(x\gamma_{1i}, \ldots ,x\gamma_{i-1,i},x\gamma_{i+1,i}, \ldots ,x\gamma_{Mi}) \Big|_{\alpha_R} = \underbrace{\int_{-\infty}^{x\sqrt{M/(M-2)}} \cdots \int}_{(M-1)\text{-fold}} G(\mathbf{Y};\mathbf{0};\boldsymbol{\alpha}_{R_{M-1}})\, d\mathbf{Y} \qquad (14.108)$$

where $\boldsymbol{\alpha}_{R_{M-1}}$ is the inner-product matrix for the regular simplex signal structure having $M - 1$ signals. Since

$$p_M(x;\boldsymbol{\alpha}_R) = MG(x)\Phi_i(x\gamma_{1i}, \ldots ,x\gamma_{i-1,i}, x\gamma_{i+1,i}, \ldots ,x\gamma_{Mi}) \Big|_{\alpha_R} \qquad (14.109)$$

we obtain the recursion formula

$$p_M(x;\boldsymbol{\alpha}_R) = MG(x) \underbrace{\int_{-\infty}^{x\sqrt{M/(M-2)}} \cdots \int}_{(M-1)\text{-fold}} G(\mathbf{Y};\mathbf{0};\boldsymbol{\alpha}_{R_{M-1}})\, d\mathbf{Y} \qquad (14.110)$$

for the density of an M-variate gaussian random vector with covariance matrix given by $\boldsymbol{\alpha}_R$. By repeated use of this relationship, $P_d(\lambda;\boldsymbol{\alpha}_R)$ can be expressed in terms of successive integrations of $G(x)$. This result then can be used for numerical evaluation of $P_d(\lambda;\boldsymbol{\alpha}_R)$.

For any $\boldsymbol{\alpha}$ we now determine an upper bound on the magnitude of

$$\frac{\partial P_d(\lambda;\boldsymbol{\alpha})}{\partial \lambda_{ij}}$$

LEMMA 14.7 *Define*

$$P_{ij}(\lambda;\boldsymbol{\alpha}) \triangleq \frac{\partial P_d(\lambda;\boldsymbol{\alpha})}{\partial \lambda_{ij}} \qquad (14.111)$$

Then

$$|P_{ij}(\lambda;\boldsymbol{\alpha})| \leq \frac{\lambda \exp -\frac{1}{2}\lambda^2}{M^2\, 2\sqrt{\pi}\sqrt{1-\lambda_{ij}}} \exp \frac{\lambda^2(1+\lambda_{ij})}{4} \qquad (14.112)$$

Proof: It is sufficient to prove (14.112) for λ_{12}. From (14.57),

$$P_{12}(\lambda;\boldsymbol{\alpha}) = \frac{-\lambda}{M^2} \exp(-\tfrac{1}{2}\lambda^2)\, \phi_{12}(\lambda;\boldsymbol{\alpha}) \qquad (14.113)$$

and from (14.66),

$$\frac{\phi_{12}(\lambda;\boldsymbol{\alpha})}{M} = \int_0^\infty \exp(\lambda x)\, G_{12}(x)F_{12}(x)\, dx \qquad (14.114)$$

where $G_{12}(x)$ is given by (14.63) and $F_{12}(x)$ by (14.69).

For $\phi_{12}(\lambda;\alpha)$, let

$$x = u_{12}y \tag{14.115}$$

where

$$u_{12} \triangleq \sqrt{\frac{1 + \lambda_{12}}{2}} \tag{14.116}$$

Then (14.114) becomes

$$\frac{\phi_{12}(\lambda;\alpha)}{M} = \frac{1}{2\sqrt{\pi}\sqrt{1 - \lambda_{12}}} \exp\left(\frac{\lambda^2(1 + \lambda_{12})}{4}\right) R_{12}(\lambda) \tag{14.117}$$

where

$$R_{12}(\lambda) = \int_0^\infty G(y - \lambda u_{12})F_{12}(yu_{12})\,dy \tag{14.118}$$

Since F_{12} is a cumulative density function, it is clear that

$$0 \le R_{12}(\lambda) \le 1$$

Therefore

$$|P_{12}(\lambda;\alpha)| \le \frac{\lambda}{M^2} \exp\left(-\tfrac{1}{2}\lambda^2\right) \frac{1}{2\sqrt{\pi}\sqrt{1 - \lambda_{12}}} \exp \frac{\lambda^2(1 + \lambda_{12})}{4} \tag{14.119}$$

As in the special case for $D = 2$, we now prove the following result, also a special case of Shannon's coding theorems, but again by a more direct derivation.

THEOREM 14.7 *For the orthogonal signal structure, as $\lambda \to \infty$ and $M \to \infty$, so that the communication efficiency $\beta = \lambda^2/\log_2 M$ [from (13.24)] remains fixed,*

$$\lim_{\substack{\lambda \to \infty \\ M \to \infty}} P_d(\lambda;\alpha_0) = \begin{cases} 1 & \text{if } \beta \log_2 e > 2 \\ 0 & \text{if } \beta \log_2 e < 2 \end{cases} \tag{14.120}$$

Proof: From (14.28)

$$P_d(\lambda;\alpha_0) = \int_{-\infty}^{\infty} G(x)[\phi(x + \lambda)]^{M-1}\,dx$$

Now

$$[\phi(x)]^{M-1} = \exp\,[(M - 1)\ln \phi(x)]$$

and

$$\ln \phi(x) = \ln \{1 - [1 - \phi(x)]\}$$

But

$$1 - \phi(x) \approx \frac{1}{x\sqrt{2\pi}} \exp -\tfrac{1}{2}x^2 \qquad \text{for large } x$$

Thus

$$\ln \{1 - [1 - \phi(x)]\} \approx \ln \left(1 - \frac{\exp -\frac{1}{2}x^2}{x\sqrt{2\pi}}\right)$$

Since

$$\ln (1 - a) \approx -a \qquad \text{for small } a$$

we have

$$\ln \phi(x + \lambda) \approx \frac{-1}{(x + \lambda)\sqrt{2\pi}} \exp [-\tfrac{1}{2}(x + \lambda)^2]$$
$$\approx \frac{-1}{\lambda\sqrt{2\pi}} \exp -\tfrac{1}{2}\lambda^2 \qquad \text{for large } \lambda$$

Hence for large λ

$$[\phi(x + \lambda)]^{M-1} \approx \exp \left[\frac{-(M - 1)}{\lambda\sqrt{2\pi}} \exp (-\tfrac{1}{2}\lambda^2)\right]$$
$$\approx \exp \left[\frac{-(M - 1)}{\sqrt{2\pi}\sqrt{\beta \log_2 M}} \exp (-\tfrac{1}{2}\beta \log_2 M)\right]$$

By noting that

$$\log_2 M = \log_2 e \ln M$$

we have

$$[\phi(x + \lambda)]^{M-1} \approx \exp \frac{-(M - 1) \exp (-\frac{1}{2}\beta \log_2 e \ln M)}{\sqrt{2\pi}\sqrt{\log_2 e \ln M}}$$
$$\approx \exp \frac{-(M - 1)M^{-\frac{1}{2}\beta \log_2 e}}{\sqrt{2\pi}\sqrt{\log_2 e \ln M}} \tag{14.121}$$

Thus, as $M \to \infty$, if $\frac{1}{2}\beta \log_2 e > 1$, then Eq. (14.121) goes to 1, and $P_d(\lambda \to \infty; \boldsymbol{\alpha}_0) = 1$, and if $\frac{1}{2}\beta \log_2 e < 1$, $P_d(\lambda \to \infty; \boldsymbol{\alpha}_0) = 0$.

The *biorthogonal signal structure* can also be shown to satisfy Theorem 14.7. An extension of this result is the following:

COROLLARY[1] *For each M let*

$$\gamma_M = \max_{i \neq j} \lambda_{ij} \tag{14.122}$$

[1] Due to Viterbi.

Then, as $\lambda \to \infty$ *and* $M \to \infty$,

$$P_d(\lambda;\boldsymbol{\alpha}) \to 1 \tag{14.123}$$

for β *such that*

$$\lim_{M} \beta(1 - \gamma_M) > 2 \log_2 e \tag{14.124}$$

Remark: This is a sufficient condition for β, but not a necessary condition.

Proof: Define

$$\boldsymbol{\alpha}_{\gamma_M} \triangleq \begin{bmatrix} 1 & & \gamma_M \\ & \ddots & \\ \gamma_M & & 1 \end{bmatrix} \tag{14.125}$$

Then, from Theorem 14.3,

$$P_d(\lambda;\boldsymbol{\alpha}) \geq P_d(\lambda;\boldsymbol{\alpha}_{\gamma_M})$$

and from (14.40),

$$P_d(\lambda;\boldsymbol{\alpha}_{\gamma_M}) = P_d(\lambda\sqrt{1 - \gamma_M};\, \boldsymbol{\alpha}_0)$$

or, equivalently,

$$P_d(\lambda;\boldsymbol{\alpha}) \geq P_d(\lambda\sqrt{1 - \gamma_M};\, \boldsymbol{\alpha}_0) \tag{14.126}$$

Now let

$$\bar{\lambda} = \lambda\sqrt{1 - \gamma_M} \tag{14.127}$$

and

$$\bar{\beta} = \frac{\bar{\lambda}^2}{\log_2 M} \tag{14.128}$$

Then,

$$\lim_{\substack{\lambda \to \infty \\ M \to \infty}} P_d(\lambda;\boldsymbol{\alpha}_0) = 1 \qquad \text{if } \bar{\beta}\log_2 e > 2$$

from which we get

$$\lim_{\substack{\lambda \to \infty \\ M \to \infty}} P_d(\lambda;\boldsymbol{\alpha}) = 1 \tag{14.129}$$

if

$$\lim_{M} \beta\sqrt{1 - \gamma_M} > \frac{2}{\log_2 e} \tag{14.130}$$

Equation (14.125) provides an upper bound for the probability of error for any α which can be evaluated numerically.

Again it should be emphasized that these asymptotic results agree with Shannon's limit theorems, but they do not provide us with optimal signal sets.

PROBLEMS

14.1 Show that the biorthogonal signal structure satisfies Theorem 14.7.

14.2 (*a*) Which of the results derived in this chapter are still valid when the amplitude of the transmitted signal is assumed to be a nonnegative random variable which is unknown to the receiver?

(*b*) What if the amplitude may take on negative as well as positive values?

14.3 For an arbitrary signal structure, show that the probability of detection is a monotonically increasing function of the signal-to-noise ratio.

14.4 Given any signal set which is such that

$$\sum_{k=1}^{M} \mathbf{S}_k = \mathbf{C} \neq \mathbf{0}$$

find a signal set $\{\mathbf{S}'_i\}$ such that

$$\sum_{k=1}^{M} \mathbf{S}'_k = \mathbf{0}$$

with the same probability of detection. Assume matched-filter receivers are to be used, and note that the restriction that all signals be of unit magnitude has not been imposed.

14.5 Show that the probability of detection is invariant under orthogonal transformations of the signal vectors.

14.6 (*a*) From the representation of the probability of detection in Sec. 14.2, show that the probability of detection for equally likely and equal-energy signal vectors can be expressed as

$$P_d = \int_0^\infty dr\, r^{D-1} \frac{1}{M} \sum_{k=1}^{M} \int_{\Omega_k} g_{r,\lambda}(\|\mathbf{Y}_n - \mathbf{S}_k\|)\, d\boldsymbol{\Omega}$$

where $\mathbf{Y}_n$ is of unit magnitude, Ω_k is that region of the surface of the D-dimensional unit sphere (with surface elements $d\boldsymbol{\Omega}$), where

$$\mathbf{Y}_n{}^T\mathbf{S}_k \geq \mathbf{Y}_n{}^T\mathbf{S}_j \qquad \text{for all } j$$

and where

$$g_{r,\lambda}(x) = (2\pi)^{-D/2} \exp\left\{-\tfrac{1}{2}\lambda^2[(1-r)^2 + rx^2]\right\}$$

(*b*) Show that this result can be extended to cases where the additive noise vector $\mathbf{N}$ has probability density function $f(\|\mathbf{N}\|)$, when f is a monotonically decreasing function of $\|\mathbf{N}\|$, and the resulting $g_{r,\lambda}(x)$ is monotonically decreasing in $\|x\|$ for all $\lambda > 0$ and $r > 0$ (Landau and Slepian [14.8]).

REFERENCES

14.1 Balakrishnan, A. V.: A Contribution to the Sphere Packing Problem of Communication Theory, *J. Math. Anal. Appl.*, vol. 3, December, 1961, pp. 485–506.

14.2 Max, J.: Signals Sets with Uniform Correlation Properties, *J. SIAM*, vol. 10, March, 1962, pp. 113–118.

14.3 Slepian, D.: Signaling Systems, *Bell Lab. Tech. Memo.*, May 7, 1951.

14.4 Bonnesen, T., and Fenchel, W.: "Theorie der Konvexen Korper," Chelsea, New York, 1948.

14.5 Hadley, G.: "Linear Algebra," Addison-Wesley, Reading, Mass., 1961.

14.6 Balakrishnan, A. V., and J. E. Taber: Error Rates in Coherent Communication Systems, *IRE Trans. Commun. Systems*, vol. CS-10, no. 1, March, 1962, pp. 86–89.

14.7 Kendall, M. G.: Proof of Relations Connected with Tetra-choric Series and Its Generalizations, *Biometrica*, vol. 32, 1941, p. 196.

14.8 Landau, H. J., and D. Slepian: On the Optimality of the Regular Simplex Code, *Bell Sys. Tech. J.*, vol. 45, no. 8, October, 1966, pp. 1247–1272.

15

Optimality for Coherent Systems When Dimensionality Is Not Specified: Regular Simplex Coding

The variational approach to the signal-design problem described in the previous chapter was developed and exploited by Balakrishnan in obtaining the first known results on the optimality of the regular simplex signal structure when no restrictions are placed on the dimensionality of the signal space. We now develop these results for their own interests, as well as their applicability to the general problem with constraints on the dimensionality D. As shown in the previous chapter, the class of $\boldsymbol{\alpha}$ for which $D(\boldsymbol{\alpha}) = 0$ contains all of the optimal $\boldsymbol{\alpha}$. Also, we concluded that the maximum value necessary for D is $M - 1$. Therefore, even in this case of no bandwidth restriction, only a finite bandwidth is required for the optimal signal set. This again does not violate Shannon's channel-capacity theorem, since M and T can vary there but are fixed and finite here.

15.1 NECESSARY (FIRST-ORDER) CONSIDERATIONS FOR OPTIMALITY

If we fix λ, then, since $P_d(\lambda;\boldsymbol{\alpha})$ is a continuous function in $\boldsymbol{\alpha}$, and since $\mathcal{A}$ is closed and bounded, it follows that the optimal $\boldsymbol{\alpha}$ is actually attained at some point $\boldsymbol{\alpha}_0$ in $\mathcal{A}$. However, $\boldsymbol{\alpha}_0$ may depend on λ. Since $P_d(\lambda;\boldsymbol{\alpha})$ is analytic, from the fact that it is expandable in a tetrachoric series, it is, of course, differentiable. Therefore the directional derivative in $\mathcal{A}$ space directed away from $\boldsymbol{\alpha}_0$ toward any other admissible $\boldsymbol{\alpha}$ must be nonpositive. Since $\mathcal{A}$ is convex, the adjoining line segment $\boldsymbol{\alpha}_0$ to $\boldsymbol{\alpha}$ will also be in $\mathcal{A}$.

Now, for any other admissible choice, say, $\boldsymbol{\alpha}'$, we can expand $P_d(\lambda;\boldsymbol{\alpha}')$ about $P_d(\lambda;\boldsymbol{\alpha}_0)$ in the Taylor's series expansion

$$P_d(\lambda;\boldsymbol{\alpha}') = P_d(\lambda;\boldsymbol{\alpha}_0) + \sum_{i>j}\sum (\lambda'_{ij} - \lambda^\circ_{ij}) \frac{\partial P_d(\lambda;\boldsymbol{\alpha}_0)}{\partial \lambda_{ij}} + \frac{1}{2}\sum_{i>j}\sum \sum_{k>l}\sum (\lambda'_{ij} - \lambda^\circ_{ij})(\lambda'_{kl} - \lambda^\circ_{kl}) \frac{\partial^2 P_d(\lambda;\boldsymbol{\alpha}_0)}{\partial \lambda_{ij}\,\partial \lambda_{kl}} + \cdots \tag{15.1}$$

If $\boldsymbol{\alpha}_0$ is truly the optimal choice, then the first-order term in (15.1) must be nonpositive for any $\boldsymbol{\alpha}'$ in the neighborhood of $\boldsymbol{\alpha}_0$.

THEOREM 15.1 *For the regular simplex signal structure $\boldsymbol{\alpha}_R$ the first-order variation in* (15.1) *is nonpositive for any admissible $\boldsymbol{\alpha}'$ at all signal-to-noise ratios.*

Proof: At $\boldsymbol{\alpha}_R$, $\partial P_d(\lambda;\boldsymbol{\alpha}_R)/\partial\lambda_{ij}$ is independent of i and j and is strictly negative. Thus it is sufficient to show that

$$\sum_{i>j}\sum (\lambda'_{ij} - \lambda_{ij}{}^R) \geq 0 \tag{15.2}$$

For this we have only to note that

$$\left(\sum_{i=1}^{M} \mathbf{S}'_i\right)^T \sum_{j=1}^{M} \mathbf{S}'_j = M + 2\sum_{i>j}\sum \lambda'_{ij} \geq 0 \tag{15.3}$$

or, equivalently,

$$\sum_{i>j}\sum \lambda'_{ij} \geq -\frac{M}{2} \tag{15.4}$$

However, for $\boldsymbol{\alpha}_R$

$$\sum_{i>j}\sum \lambda_{ij}{}^R = -\frac{M}{2} \tag{15.5}$$

which validates (15.2) and proves the theorem.

Note, however, that equality exists in (15.2) for any $\boldsymbol{\alpha}'$ such that $\sum_{i=1}^{M} \mathbf{S}'_i = \mathbf{0}$. Also, from Lemma 14.5, there is strict inequality in (15.2) for all $\{\mathbf{S}'_i\}$ such that $\sum_{i=1}^{M} \mathbf{S}'_i \neq \mathbf{0}$, and this class of sets is off the tangent plane defined by

$$\sum_{i=1}^{M} \sum_{j=1}^{M} \lambda_{ij} = 0 \tag{15.6}$$

Therefore Theorem 15.1 proves that $\boldsymbol{\alpha}_R$ is a local maximum in every admissible direction in $\mathcal{A}$ except those in the tangent plane given by (15.6). In these directions Theorem 15.1 shows that $\boldsymbol{\alpha}_R$ is a local extremum. To prove that $\boldsymbol{\alpha}_R$ is also a local maximum in these directions, the second-order variation in the Taylor's series expansion must be examined (see Sec. 15.4).

There are certain properties that $\boldsymbol{\alpha}_0$ must possess. Consider an $\boldsymbol{\alpha}'$ on the surface of $\mathcal{A}$, but not on the tangent plane defined by (15.6), and consider points on the adjoining line segment

$$\boldsymbol{\alpha} = (1 - \theta)\boldsymbol{\alpha}_0 + \theta\boldsymbol{\alpha}' \qquad 0 \leq \theta \leq 1 \tag{15.7}$$

The eigenvector corresponding to zero eigenvalue for $\boldsymbol{\alpha}'$ is not $(1, \ldots, 1)^T$, as it is for $\boldsymbol{\alpha}_0$, since $\boldsymbol{\alpha}'$ is not in the tangent plane defined by (15.6). This statement requires that $\boldsymbol{\alpha}_0$ be in this tangent plane and therefore have eigenvector $(1, \ldots, 1)^T$ corresponding to zero eigenvalue, which we assume for the present. Therefore, for $0 < \theta < 1$, $\boldsymbol{\alpha}$ is in the interior of $\mathcal{A}$, for which $D(\boldsymbol{\alpha}) > 0$, and, from Theorem 14.2, $\boldsymbol{\alpha}$ can be projected onto the boundary of $\mathcal{A}$ at a point $\tilde{\boldsymbol{\alpha}}$ to improve the probability of detection (as shown in Fig. 15.1). Now let us look at $P_d(\lambda;\tilde{\boldsymbol{\alpha}})$ as $\tilde{\boldsymbol{\alpha}} \rightarrow \boldsymbol{\alpha}_0$. That is, we are examining the probability of detection along a specific path in the neighborhood of $\boldsymbol{\alpha}_0$ which is entirely on the surface of $\mathcal{A}$. The necessary characteristic of $\boldsymbol{\alpha}_0$ which results from this analysis can be stated in the following way.

THEOREM 15.2 *As* $\tilde{\boldsymbol{\alpha}} \rightarrow \boldsymbol{\alpha}_0$,

$$P_d(\lambda;\tilde{\boldsymbol{\alpha}}) \leq P_d(\lambda;\boldsymbol{\alpha}_0) \tag{15.8}$$

for θ *sufficiently small and any* $\boldsymbol{\alpha}'$ *if*

$$P_{ij}(\lambda;\boldsymbol{\alpha}_0) = kC^{\circ}_{ij} \qquad \text{for all } i \neq j \tag{15.9}$$

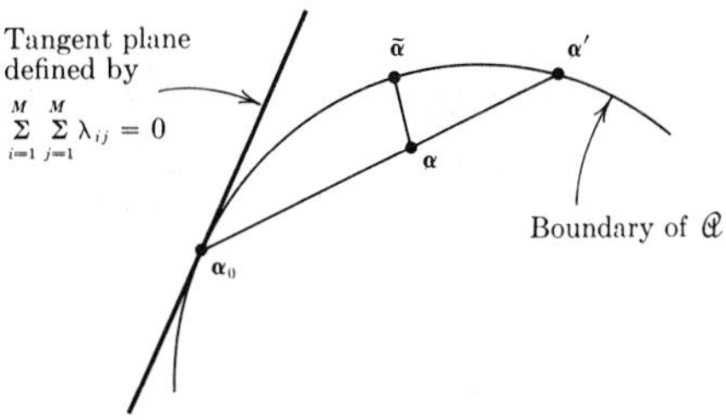

Fig. 15.1 The position of the tangent plane defined by $\sum_{i=1}^{N} \sum_{j=1}^{N} \lambda_{ij} = 0$ and the boundary of $\mathcal{A}$.

where

$$P_{ij}(\lambda;\boldsymbol{\alpha}_0) = \frac{\partial P_d(\lambda;\boldsymbol{\alpha}_0)}{\partial \lambda_{ij}} \tag{15.10}$$

and C_{ij}° *is the cofactor (including sign) of* λ_{ij}° *in* $\boldsymbol{\alpha}_0$.

Proof: For $0 < \theta < 1$ we have $\boldsymbol{\alpha}$ in the interior of $\mathcal{A}$. Define C_{ij} as the cofactor of λ_{ij} in $\boldsymbol{\alpha}$. For a fixed θ project $\boldsymbol{\alpha}$ onto the boundary of $\mathcal{A}$ as in Theorem 14.2, which results in

$$\tilde{\boldsymbol{\alpha}} = \frac{\boldsymbol{\alpha} - \rho^2(\mathbf{1})}{1 - \rho^2} = \frac{(1-\theta)\boldsymbol{\alpha}_0 + \theta\boldsymbol{\alpha}' - \rho^2(\mathbf{1})}{1 - \rho^2} \tag{15.11}$$

where $\mathbf{1}$ is the M-by-M matrix of all 1s, and, from Lemma 14.2,

$$\rho^2 = \frac{D}{C} \tag{15.12}$$

where

$$D = D(\boldsymbol{\alpha})$$

and

$$C = \sum_{i=1}^{M}\sum_{j=1}^{M} C_{ij} \tag{15.13}$$

Then $\tilde{\boldsymbol{\alpha}}$ can be expressed in terms of D and C as

$$\tilde{\boldsymbol{\alpha}} = \frac{C[(1-\theta)\boldsymbol{\alpha}_0 + \theta\boldsymbol{\alpha}'] - D(\mathbf{1})}{C - D} \tag{15.14a}$$

or

$$\tilde{\lambda}_{ij} = \frac{C\lambda_{ij} - D}{C - D} \tag{15.14b}$$

The directional derivative at $\boldsymbol{\alpha}_0$ along this prescribed path is

$$\left.\frac{\partial P_d(\lambda;\tilde{\boldsymbol{\alpha}})}{\partial \theta}\right|_{\theta=0} = \sum\sum_{i>j} P_{ij}(\lambda;\boldsymbol{\alpha}_0) \left.\frac{\partial \tilde{\lambda}_{ij}}{\partial \theta}\right|_{\theta=0} \tag{15.15}$$

Now,

$$\left.\frac{\partial \tilde{\lambda}_{ij}}{\partial \theta}\right|_{\theta=0} = (\lambda'_{ij} - \lambda_{ij}^{\circ}) - \frac{1}{C}(1 - \lambda_{ij}^{\circ}) \left.\frac{\partial D}{\partial \theta}\right|_{\theta=0} \tag{15.16}$$

and $\left.\frac{\partial D}{\partial \theta}\right|_{\theta=0}$ can be determined by writing

$$D = D(\boldsymbol{\alpha}_0) + \left[\sum\sum_{i>j} \left.\frac{\partial D(\boldsymbol{\alpha})}{\partial \lambda_{ij}}\right|_{\theta=0} \left.\frac{\partial \lambda_{ij}}{\partial \theta}\right|_{\theta=0}\right]\theta + \cdots \tag{15.17}$$

From Lemma 14.6,

$$\left.\frac{\partial D(\boldsymbol{\alpha})}{\partial \lambda_{ij}}\right|_{\theta=0} = 2C_{ij}\Big|_{\theta=0} = 2C_{ij}^{\circ} \tag{15.18}$$

Also,

$$\frac{\partial \lambda_{ij}}{\partial \theta} = \lambda'_{ij} - \lambda_{ij}^{\circ} \tag{15.19}$$

from which

$$\left.\frac{\partial D}{\partial \theta}\right|_{\theta=0} = 2\sum_{i>j}\sum (\lambda'_{ij} - \lambda_{ij}^{\circ})C_{ij}^{\circ} \tag{15.20}$$

If we substitute (15.16) and (15.20) into (15.15) we have

$$\left.\frac{\partial P_d(\lambda;\tilde{\boldsymbol{\alpha}})}{\partial \theta}\right|_{\theta=0} = \sum_{i>j}\sum (\lambda'_{ij} - \lambda_{ij}^{\circ})P_{ij}(\lambda;\boldsymbol{\alpha}_0)$$
$$- 2\left[\sum_{i>j}\sum (\lambda'_{ij} - \lambda_{ij}^{\circ})C_{ij}^{\circ}\right] \cdot \frac{\sum_{k>l}\sum (1 - \lambda_{kl}^{\circ})P_{kl}(\lambda;\boldsymbol{\alpha}_0)}{C} \tag{15.21}$$

For $\boldsymbol{\alpha}_0$ to be a maximum independent of $\boldsymbol{\alpha}'$ (15.21) must vanish, which occurs only when

$$P_{ij}(\lambda;\boldsymbol{\alpha}_0) = kC_{ij}^{\circ} \qquad \text{for all } i \neq j \tag{15.9}$$

This result can be formulated in another way by use of the *Lagrange variational technique*. Since we have seen that $D(\boldsymbol{\alpha}_0) = 0$, we form the lagrangian functional

$$L(\lambda;\boldsymbol{\alpha}) = P_d(\lambda;\boldsymbol{\alpha}) + \nu D(\boldsymbol{\alpha}) \tag{15.22}$$

where ν is the lagrangian multiplier. Differentiating with respect to λ_{ij} and equating to zero at $\boldsymbol{\alpha}_0$ yields

$$P_{ij}(\lambda;\boldsymbol{\alpha}_0) = kC_{ij}^{\circ} \tag{15.9}$$

as the necessary property for $\boldsymbol{\alpha}_0$ to be a local extremum. The regular simplex satisfies (15.9), since the $C_{ij}\big|_{\boldsymbol{\alpha}_R}$ are all equal for all $i \neq j$, as also are all the $P_{ij}(\lambda;\boldsymbol{\alpha}_R)$.

Since $C_{ij}\big|_{\boldsymbol{\alpha}_R} > 0$ and $P_{ij}(\lambda;\boldsymbol{\alpha}_R) < 0$, we require $k < 0$ for $\boldsymbol{\alpha}_R$ to satisfy (15.9). Note that Theorem 15.2 required that we choose $\boldsymbol{\alpha}'$ off the tangent plane defined by

$$\sum_{i=1}\sum_{j=1} \lambda_{ij} = 0$$

Thus this theorem provides a necessary property for $\boldsymbol{\alpha}_0$ but still does not answer the question of how $P_d(\lambda;\boldsymbol{\alpha})$ behaves in the tangent plane in the neighborhood of $\boldsymbol{\alpha}_R$.

15.2 UNIQUENESS OF THE REGULAR SIMPLEX SATISFYING NECESSARY CONDITIONS FOR ALL SIGNAL-TO-NOISE RATIOS

We shall show now that in all of $\mathcal{A}$, $\boldsymbol{\alpha}_R$ is the only signal structure that satisfies the above first-order conditions at all signal-to-noise ratios. With this fact, then, if $\boldsymbol{\alpha}_R$ is not the global optimum at all signal-to-noise ratios, then the global optimum, whatever it may be, must necessarily depend in some way on the signal-to-noise ratio. Or, phrased in another way, we can say that with this fact the only remaining fact necessary for us to conclude that $\boldsymbol{\alpha}_R$ is the global maximum at all signal-to-noise ratios is that the global optimum itself be independent of the signal-to-noise ratio. To date, however, for arbitrary but fixed M and D this question of the global optimum's independence of the signal-to-noise ratio is still open.

Before stating and proving this uniqueness theorem, however, we must first derive several properties that $\boldsymbol{\alpha}_0$ must possess. To begin with, again fix λ, and let $\boldsymbol{\alpha}_0$ correspond to one of the local extrema that is independent of λ (any one, if there are several). The relationship in (15.9) must hold for each of these; or, equivalently,

$$\phi_{ij}(\lambda;\boldsymbol{\alpha}_0) = KC_{ij}^{\circ} \qquad i \neq j \tag{15.23}$$

where $\phi_{ij}(\lambda;\boldsymbol{\alpha})$ is defined by (14.57) and is nonnegative for all $i \neq j$ and any $\boldsymbol{\alpha}$. Since the directional derivative of $P_d(\lambda;\boldsymbol{\alpha}_0)$ along any admissible path directed away from $\boldsymbol{\alpha}_0$ must be nonpositive, we have

$$\sum_{i>j}\sum (\lambda_{ij} - \lambda_{ij}^{\circ})\phi_{ij}(\lambda;\boldsymbol{\alpha}_0) \geq 0 \tag{15.24}$$

Substituting (15.23) into (15.24) gives

$$K\sum_{i>j}\sum (\lambda_{ij} - \lambda_{ij}^{\circ})C_{ij}^{\circ} \geq 0 \tag{15.25}$$

In particular, setting

$$\lambda_{ij} = 0 \qquad i \neq j \tag{15.26}$$

which corresponds to the line segment from $\boldsymbol{\alpha}_0$ toward the orthogonal signal set, we get

$$K\sum_{i<j}\sum \lambda_{ij}^{\circ}C_{ij}^{\circ} \leq 0 \tag{15.27}$$

This leads to our next result.

LEMMA 15.1

$$K > 0 \tag{15.28}$$

Proof: We know that $D(\boldsymbol{\alpha}_0) = 0$. Hence

$$MD(\boldsymbol{\alpha}_0) = \sum_{i=1}^{M} C_{ii}^{\circ} + \sum\sum_{i \neq j}^{M} \lambda_{ij}^{\circ} C_{ij}^{\circ} = 0$$

and $\boldsymbol{\alpha}_0$ is nonnegative-definite. This implies

$$C_{ii}^{\circ} \geq 0 \qquad i = 1, \ldots, M$$

from which we get

$$\sum\sum_{i \neq j} \lambda_{ij}^{\circ} C_{ij}^{\circ} \leq 0$$

Therefore, from (15.27),

$$K \geq 0$$

If $K = 0$, however, then $\phi_{ij}^{\circ} = 0$ for all $i \neq j$, which, from Lemma 14.4, implies $\lambda_{ij}^{\circ} = -1$ for all $i \neq j$. This is impossible for $M > 2$. But since the $M = 2$ result is already well known, we consider only $M > 2$ and conclude that $K > 0$.

With this lemma and the realization that $\phi_{ij} \geq 0$ for all $i \neq j$, we have that

$$C_{ij}^{\circ} \geq 0 \qquad \text{for all } i \neq j \tag{15.29}$$

Actually, we can write this as follows.

LEMMA 15.2 *For any local extremum in* $\mathcal{A}$

$$C_{ij}^{\circ} > 0 \qquad \textit{for all } i \neq j \tag{15.30}$$

Proof: Suppose $C_{12}^{\circ} = 0$. Then, from (15.23) and Lemma 14.4, $\lambda_{12}^{\circ} = -1$. Hence ξ_1 and ξ_2 are linearly dependent, and any covariance matrix involving ξ_1 and ξ_2 will have its corresponding determinant equal to zero. Therefore

$$C_{ii}^{\circ} = 0 \qquad i = 3, \ldots, M$$

Now, since $\boldsymbol{\alpha}_0$ is nonnegative-definite, the matrix of cofactors $\{C_{ij}^{\circ}\}$ is also nonnegative-definite, which implies

$$C_{ii}^{\circ} C_{jj}^{\circ} - (C_{ij}^{\circ})^2 \geq 0$$

Hence

$$C_{ij}^{\circ} = 0 \qquad \begin{aligned} &\text{for } i, j = 3, \ldots, M \\ &i = 1, j = 3, \ldots, M \\ &i = 2, j = 3, \ldots, M \end{aligned}$$

which implies

$$D(\boldsymbol{\alpha}_0) = C_{11}^{\circ} + C_{12}^{\circ}\lambda_{12}^{\circ} = 0$$

However, since we assumed $C_{12}^{\circ} = 0$, we have $C_{11}^{\circ} = 0$. Similarily, $C_{22}^{\circ} = 0$.

Therefore

$$C_{ij}^{\circ} = 0 \qquad \text{for all } i, j$$

and hence

$$\phi_{ij}^{\circ} = 0 \qquad \text{for all } i, j$$

which implies

$$\lambda_{ij}^{\circ} = -1 \qquad \text{for all } i, j$$

which is impossible (again for $M > 2$).

Actually, we can say more about these cofactors.

LEMMA 15.3

$$C_{ii}^{\circ}C_{jj}^{\circ} = (C_{ij}^{\circ})^2 \tag{15.31}$$

Proof: Since $D(\boldsymbol{\alpha}_0) = 0$,

$$\sum_{i=1}^{M} \lambda_{ij}^{\circ}C_{ik}^{\circ} = 0 \qquad j, k = 1, \ldots, M$$

For $j \neq k$, this is zero for any matrix. For $j = k$ it is zero because $D(\boldsymbol{\alpha}_0) = 0$. Now, if we define $\mathbf{C}^{\circ}$ as the matrix of cofactors $\{C_{ij}^{\circ}\}$, then

$$\boldsymbol{\alpha}_0\mathbf{C}^{\circ} = \mathbf{0}$$

where $\mathbf{0}$ is the matrix of all zeros. Now, $\boldsymbol{\alpha}_0$ can have rank as high as $M - 1$, and for $\boldsymbol{\alpha}_0 = \boldsymbol{\alpha}_R$ the rank is $M - 1$. Thus the rank of $\mathbf{C}^{\circ}$ is at most 1, implying that all 2-by-2 minors of $\mathbf{C}^{\circ}$ are zero. Therefore

$$C_{ii}^{\circ}C_{jj}^{\circ} = (C_{ij}^{\circ})^2$$

Another significant property of $\boldsymbol{\alpha}_0$ can be stated as follows.

LEMMA 15.4 *For $\boldsymbol{\alpha}_0$, the flat in which the probability density is concentrated is given by*

$$\sum_{i=1}^{M} \sqrt{C_{ii}^{\circ}}\, \xi_i = 0$$

Proof: It is immediate that

$$E\left(\sum_{i=1}^{M} \sqrt{C_{ii}^{\circ}}\, \xi_i\right) = 0$$

Also,

$$E\left(\sum_{i=1}^{M} \sqrt{C_{ii}^{\circ}}\, \xi_i\right)^2 = \sum_{i=1}^{M} \sum_{j=1}^{M} \sqrt{C_{ii}^{\circ}} \sqrt{C_{jj}^{\circ}}\, \lambda_{ij}^{\circ}$$

Applying Lemma 15.3 and using the fact that $D(\boldsymbol{\alpha}_0) = 0$, we conclude that the variance of

$$\sum_{i=1}^{M} \sqrt{C_{ii}^{\circ}}\, \xi_i$$

is zero. Hence it is zero with probability 1.

We come now to the main result of this section.

THEOREM 15.3 *The regular simplex signal structure is the only signal set which is a local extremum in the class of all admissible signal sets $\mathcal{A}$ at all signal-to-noise ratios.*

Proof: For any of the local extrema we have from (15.9)

$$\phi_{ij}(\lambda;\boldsymbol{\alpha}_0) = KC_{ij}^{\circ} = \int_0^{\infty} \exp(\lambda x) G_{ij}(x) F_{ij}(x)\, dx \tag{15.32}$$

where $F_{ij}(x)$ is given by (14.67) and $G_{ij}(x)$ is given by (14.63). From this we obtain that

$$K \int_0^{\infty} \exp(\lambda x)[C_{ij}^{\circ} G_{kl}(x) F_{kl}(x) - C_{kl}^{\circ} G_{ij}(x) F_{ij}(x)]\, dx = 0 \tag{15.33}$$

From Lemma 15.3, $K > 0$; hence the integral must vanish. The integral is over $(0, \infty)$ instead of $(-\infty, \infty)$, as a result of Theorems 14.4 and 14.5. Because of the uniqueness and analytic properties of the Laplace transform, equality can be attained in (15.33) if and only if

$$C_{ij}^{\circ} G_{kl}(x) F_{kl}(x) = C_{kl}^{\circ} G_{ij}(x) F_{ij}(x) \qquad \text{for all } x \tag{15.34}$$

Note that we are able to associate the integral in (15.32) with the Laplace transform and conclude that (15.34) must be satisfied only if we make the assumption that $G_{ij}(x)$ and $F_{ij}(x)$ are independent of λ, or, equivalently, that $\boldsymbol{\alpha}_0$ is independent of λ.

It is immediate that $\boldsymbol{\alpha}_R$ satisfies (15.34), since

$$C_{ij}\Big|_{\boldsymbol{\alpha}_R} = C_{kl}\Big|_{\boldsymbol{\alpha}_R}$$

$$G_{ij}(x)\Big|_{\boldsymbol{\alpha}_R} = G_{kl}(x)\Big|_{\boldsymbol{\alpha}_R} \tag{15.35}$$

and

$$F_{ij}(x)\Big|_{\boldsymbol{\alpha}_R} = F_{kl}(x)\Big|_{\boldsymbol{\alpha}_R} \qquad \text{for all } x$$

To prove that $\boldsymbol{\alpha}_R$ is the only signal structure that satisfies (15.34) for all x, suppose $\boldsymbol{\alpha}_0 \neq \boldsymbol{\alpha}_R$ also satisfies (15.34) and let

$$\lambda_{12}^\circ = \max_{i \neq j} \lambda_{ij}^\circ \tag{15.36}$$

Since $\boldsymbol{\alpha}_0 \neq \boldsymbol{\alpha}_R$, $\lambda_{12}^\circ > \lambda_{kl}^\circ$ for some k and l. This implies that for $F_{12}(x)$ as expressed in (14.69),

$$\alpha_j > 0 \qquad j = 3, \ldots, M$$

and therefore

$$\lim_{x \to \infty} F_{12}(x) = 1 \tag{15.37}$$

Now, if

$$\lim_{x \to \infty} F_{kl}(x) = a > 0 \tag{15.38}$$

by using the tetrachoric series we can show that for large x, $F_{kl}(x)$ is of the form

$$a - x^{-d} \exp -bx^2 \tag{15.39}$$

where $d > 0$ and $b > 0$. Using the same approximation for $F_{12}(x)$ for large x and substituting into (15.34), we require

$$C_{kl}^\circ \frac{\exp[-x^2/(1 + \lambda_{12}^\circ)]}{2\pi\sqrt{1 - (\lambda_{12}^\circ)^2}} (a - x^{-d} \exp -bx^2) \approx C_{12}^\circ \frac{\exp[-x^2/(1 + \lambda_{kl}^\circ)]}{2\pi\sqrt{1 - (\lambda_{kl}^\circ)^2}} (1 - x^{-d'} \exp -b'x^2) \tag{15.40}$$

which for large x is of the form

$$d_1 \exp \frac{-x^2}{1 + \lambda_{12}^\circ} \approx d_2 \exp \frac{-x^2}{1 + \lambda_{kl}^\circ} \tag{15.41}$$

and it is immediate that equality can be attained only if

$$\lambda_{12}^\circ = \lambda_{kl}^\circ \tag{15.42}$$

However, if

$$\lim_{x \to \infty} F_{kl}(x) = 0 \tag{15.43}$$

then

$$F_{kl}(x) = o(x^{-d} \exp -bx^2) \tag{15.44}$$

and no value for λ_{kl} will give equality for large x in (15.34). Hence all the λ_{ij}° must be equal.

Thus far we have that the regular simplex is a local extremum at every signal-to-noise ratio and is the only such signal structure in $\mathcal{A}$. However, there remains the major problem of showing whether it is the global or absolute maximum, and moreover, that the global optimum is independent of the signal-to-noise ratio. So far we have that if there exists a global maximum independent of λ, it necessarily must be the regular simplex.

15.3 GLOBAL OPTIMALITY OF THE REGULAR SIMPLEX FOR LARGE SIGNAL-TO-NOISE RATIOS

THEOREM 15.4 *In the class of all admissible signal sets $\mathcal{A}$ the regular simplex signal structure is the global optimum for large signal-to-noise ratios.*

Proof: By substituting (14.117) into (14.113), we obtain for any $\boldsymbol{\alpha}$

$$-\frac{M}{\lambda} \exp\left(\tfrac{1}{2}\lambda^2\right) P_{12}(\lambda;\boldsymbol{\alpha}) = \frac{1}{2\sqrt{\pi}\sqrt{1-\lambda_{12}}} \exp\left(\frac{\lambda^2(1+\lambda_{12})}{4}\right) R_{12}(\lambda) \tag{15.45}$$

where $R_{12}(\lambda)$ is defined by (14.118). Consider now a sequence $\{\lambda_n\}$ such that

$$\lambda_n \longrightarrow \infty$$

and define

$$\boldsymbol{\alpha}_n \sim \{\lambda_{ij}{}^n\} \tag{15.46}$$

as the global optimum at λ_n. At each n let us renumber so that

$$\lambda_{12}{}^n = \max_{i \neq j} \lambda_{ij}{}^n \tag{15.47}$$

Then, from (15.37),

$$F_{12}(x) \rightarrow 1 \qquad \text{as } x \longrightarrow \infty$$

for each n, and consequently,

$$R_{12}(x) \rightarrow 1 \qquad \text{as } \lambda_n \longrightarrow \infty \tag{15.48}$$

But for the optimal choice we already require that

$$\left.\frac{\partial P_d(\lambda_n;\boldsymbol{\alpha}_\theta)}{\partial \theta}\right|_{\theta=0} = \sum_{i>j}\sum (\lambda'_{ij} - \lambda_{ij}{}^n) P_{ij}(\lambda_n;\boldsymbol{\alpha}_n) \leq 0 \tag{15.49}$$

for any other admissible $\boldsymbol{\alpha}'$, where

$$\boldsymbol{\alpha}_\theta = (1-\theta)\boldsymbol{\alpha}_n + \theta\boldsymbol{\alpha}'$$

Substituting (15.45) into (15.49), we require that α_n satisfy

$$\sum_{i>j}\sum \frac{\lambda'_{ij} - \lambda_{ij}{}^n}{\sqrt{1 - \lambda_{ij}{}^n}} \exp\left(\frac{\lambda_n{}^2\lambda_{ij}{}^n}{4}\right) R_{ij}(\lambda_n) \geq 0 \tag{15.50}$$

for any admissible $\{\lambda'_{ij}\}$ and every n. For n sufficiently large, however,

$$\lambda^\circ_{ij} = \text{constant} \qquad i \neq j$$

is the only signal structure which satisfies (15.50) for all admissible α'. For if $\alpha_n = \alpha_R$, then by dividing through by $\exp(\lambda_n{}^2\lambda_{12}{}^R/4)$ and letting n become large, (15.50) reduces to

$$\sum_{i>j}\sum (\lambda'_{ij} - \lambda_{ij}{}^R) \geq 0 \tag{15.51}$$

which we already know to be the case.

If, however, we assume that the global maximum is something other than α_R and again divide through by $\exp(\lambda_n{}^2\lambda_{12}{}^n/4)$ we obtain

$$\sum_{s}\sum \frac{(\lambda'_{ij} - \lambda_{ij}{}^n)R_{ij}(\lambda_n)}{\sqrt{1 - \lambda_{ij}{}^n}} + \sum_{\bar{s}}\sum \frac{\lambda'_{ij} - \lambda_{ij}{}^n}{\sqrt{1 - \lambda_{ij}{}^n}} \exp\left(\frac{\lambda_n{}^2(\lambda_{ij}{}^n - \lambda_{12}{}^n)}{4}\right) R_{ij}(\lambda_n) \geq 0 \tag{15.52}$$

where

$$s \triangleq \{(i,j) | \lambda_{ij}{}^n = \lambda_{12}{}^n\} \qquad \bar{s} \triangleq \{(i,j) | \lambda_{ij}{}^n < \lambda_{12}{}^n\}$$

For large n the condition to be satisfied is

$$\sum_{s}\sum \frac{\lambda'_{ij} - \lambda_{ij}{}^n}{\sqrt{1 - \lambda_{ij}{}^n}} \geq 0$$

This inequality will not be satisfied for arbitrary admissible α' in the tangent plane described by $\sum_{i=1}^{n}\sum_{j=1}^{n} \lambda_{ij} = 0$ unless the sum is over all $i \neq j$, from which we conclude that only α_R satisfies (15.50) for large λ.

A different, but equivalent, proof of this theorem has recently been given by Ziv [15.2].

15.4 SUFFICIENT (SECOND-ORDER) CONDITIONS FOR OPTIMALITY

As indicated in Sec. 15.1, a study of first-order variations was not enough to conclude that the regular simplex is a local maximum in all admissible directions of $\mathfrak{A}$. These first-order variations about the regular simplex are exactly zero when the direction is in the tangent plane defined by

$$\sum_{i=1}^{M} \sum_{j=1}^{M} \lambda_{ij} = 0 \tag{15.6}$$

Hence the possibility exists that the regular simplex is a saddle point. In order to conclude that $\boldsymbol{\alpha}_R$ indeed is a local maximum in these directions also, it is necessary to examine the second-order variations in the neighborhood of the regular simplex.

Let $\boldsymbol{\alpha}'$ be any other admissible choice such that

$$\|\boldsymbol{\alpha}' - \boldsymbol{\alpha}_R\| < \delta \tag{15.53}$$

for some fixed $\delta > 0$. Let

$$\lambda_{ij}(\theta) = (1 - \theta)\lambda_{ij}{}^R + \theta\lambda'_{ij} \tag{15.54}$$

Then the Taylor's series expansion in θ about $\boldsymbol{\alpha}_R$ for the cumulative density function $\Phi(x;\boldsymbol{\alpha}')$ is

$$\Phi(x;\boldsymbol{\alpha}') = \Phi(x;\boldsymbol{\alpha}_R) + a_1(x)\theta + a_2(x)\frac{\theta^2}{2} + \sum_{k=3}^{\infty} a_k(x)\frac{\theta^k}{k!} \tag{15.55}$$

where

$$a_1(x) = \sum_{i>j}\sum \frac{\partial\Phi(x;\boldsymbol{\alpha}_R)}{\partial\lambda_{ij}}(\lambda'_{ij} - \lambda_{ij}{}^R) \tag{15.56}$$

and

$$a_2(x) = \sum_{i>j}\sum \sum_{k>l}\sum a_{ijkl}(\lambda'_{ij} - \lambda_{ij}{}^R)(\lambda'_{kl} - \lambda_{kl}{}^R) \tag{15.57}$$

where for convenience we adopt the notation

$$a_{ijkl} = \frac{\partial^2\Phi(x;\boldsymbol{\alpha}_R)}{\partial\lambda_{ij}\,\partial\lambda_{kl}} \tag{15.58}$$

To proceed further we must partially compute a_{ijkl}, which we do in the following three lemmas. The method used in this computation consists in first considering

$$\boldsymbol{\alpha}_\rho = \begin{bmatrix} 1 & & & \\ & \ddots & & \rho \\ & & \ddots & \\ \rho & & & \ddots \\ & & & 1 \end{bmatrix} \tag{15.59}$$

and then taking limits as

$$\rho \downarrow \frac{-1}{M-1} \tag{15.60}$$

LEMMA 15.5 *For $i \neq j \neq k \neq l$ define*

$$r = a_{ijkl} = a_{1234} \tag{15.61}$$

Then

$$r = \underbrace{\int_{-\infty}^{x} \cdots \int}_{(M-4)\text{-fold}} G(x, x, x, x, \xi_5, \ldots, \xi_M; \mathbf{0}; \boldsymbol{\alpha}_R)\, d\xi_5 \cdots d\xi_M$$

$$= G_4(x)F_4(x) \tag{15.62}$$

where

$$G_4(x) = \left(\frac{1}{2\pi}\right)^2 \sqrt{\frac{(M-1)^4}{M^3(M-4)}} \exp\left(-2\,\frac{M-1}{M-4}\,x^2\right) \tag{15.63}$$

and

$$F_4(x) = \Pr\left[\psi_i \leq x\sqrt{\frac{M(M-1)}{(M-4)(M-5)}};\ i = 5, \ldots, M\right] \tag{15.64}$$

where the ψ_i are gaussian with zero means, unit variances, and

$$E(\psi_i\psi_j) = \frac{-1}{M-5} \qquad i \neq j \tag{15.65}$$

Proof: Because of the symmetry of $\boldsymbol{\alpha}_R$, it is immediate that for any $i \neq j \neq k \neq l$

$$a_{ijkl} = a_{1234}$$

By the same technique used in Sec. 14.4 we can show that

$$\frac{\partial^2\Phi(x;\boldsymbol{\alpha}_\rho)}{\partial\lambda_{12}\,\partial\lambda_{34}} = \underbrace{\int_{-\infty}^{x} \cdots \int}_{(M-4)\text{-fold}} G(x, x, x, x, \xi_5, \ldots, \xi_M; \mathbf{0}; \boldsymbol{\alpha}_\rho)\, d\xi_5 \cdots d\xi_M$$

$$= G(\xi_1 = x, \xi_2 = x, \xi_3 = x, \xi_4 = x) \int_{-\infty}^{x} \cdots \int G(\xi_5, \ldots, \xi_M | \xi_1 = x, \xi_2 = x, \xi_3 = x, \xi_4 = x)\, d\xi_5 \cdots d\xi_M$$

$G(\xi_1, \xi_2, \xi_3, \xi_4)$ is a nonsingular density and can therefore be written down explicitly. If this is done, after taking limits we have

$$G_4(x) = G(\xi_1 = x, \xi_2 = x, \xi_3 = x, \xi_4 = x)$$
$$= \left(\frac{1}{2\pi}\right)^2 \sqrt{\frac{(M-1)^4}{M^3(M-4)}} \exp\left(-2\,\frac{M-1}{M-4}\,x^2\right)$$

Hence

$$F_4(x) = \int_{-\infty}^{x} \cdots \int G(\xi_5, \ldots, \xi_M | \xi_1 = x, \xi_2 = x, \xi_3 = x, \xi_4 = x)\, d\xi_5 \cdots d\xi_M$$

which can be written as

$$F_4(x) = \underbrace{\int_{-\infty}^{x} \cdots \int}_{(M-4)\text{-fold}} G(\xi_5, \ldots, \xi_M; \mathbf{m}_4; \mathbf{C}_4)\, d\xi_5 \cdots d\xi_M$$

where

$$\mathbf{m}_4 \triangleq \begin{bmatrix} \dfrac{-4x}{M-4} \\ \cdot \\ \cdot \\ \cdot \\ \dfrac{-4x}{M-4} \end{bmatrix}$$

and the elements of $\mathbf{C}_4$ are

$$\frac{M(M-5)}{(M-1)(M-4)}$$

along the diagonal and

$$\frac{-M}{(M-1)(M-4)}$$

off the diagonal. Finally, we express $F_4(x)$ in terms of normalized random variables by setting

$$\psi_i = \frac{\xi_i + [4x/(M-4)]}{\sqrt{M(M-5)/[(M-4)(M-1)]}} \qquad i = 5, \ldots, M$$

The ψ_i have zero mean, unit variance, and covariances equal to $-1/(M-5)$, from which

$$F_4(x) = \Pr\left[\psi_i \le x\sqrt{\frac{M(M-1)}{(M-4)(M-5)}} : i = 5, \ldots, M\right]$$

LEMMA 15.6 *For $i = k$ and $j \neq l$ define*

$$q = a_{kjkl} = a_{1213} \tag{15.66}$$

Then

$$q = G_4(x)F_4(x) - \frac{M-1}{M-3}\, xG_3(x)F_3(x) \tag{15.67}$$

where

$$G_3(x) = \left(\frac{1}{2\pi}\right)^{\frac{3}{2}} \sqrt{\frac{(M-1)^3}{M^2(M-3)}} \exp\left(-\frac{3}{2}\frac{M-1}{M-3}x^2\right) \tag{15.68}$$

and

$$F_3(x) = \Pr\left[\psi_i \le x\sqrt{\frac{M(M-1)}{(M-3)(M-4)}};\, i = 4, \ldots, M\right] \tag{15.69}$$

where the ψ_i *are gaussian with zero mean, unit variance, and*

$$E(\psi_i\psi_j) = \frac{-1}{M-4} \qquad i \ne j \tag{15.70}$$

Proof: Again, because of the symmetry in $\boldsymbol{\alpha}_\rho$ and $\boldsymbol{\alpha}_R$,

$$a_{kjkl} = a_{1213} \qquad j \ne k \ne l$$

By the same technique,

$$\frac{\partial^2\Phi(x;\boldsymbol{\alpha}_R)}{\partial\lambda_{12}\,\partial\lambda_{13}} = \lim_{\rho\to -1/(M-1)} \underbrace{\int_{-\infty}^{x}\cdots\int}_{M\text{-fold}} \frac{\partial^4}{\partial\xi_1{}^2\,\partial\xi_2\,\partial\xi_3} G(\boldsymbol{\xi};\mathbf{0};\boldsymbol{\alpha}_\rho)\,d\boldsymbol{\xi}$$

$$= -\lim_{\rho\to -1/(M-1)} \underbrace{\int_{-\infty}^{x}\cdots\int}_{M\text{-fold}} \frac{\partial^3}{\partial\xi_1\,\partial\xi_2\,\partial\xi_3} \frac{\sum_{i=1}^{M} C_{1i}\xi_i}{D} G(\boldsymbol{\xi};\mathbf{0};\boldsymbol{\alpha}_\rho)\,d\boldsymbol{\xi}$$

$$= -\lim_{\rho\to -1/(M-1)} \underbrace{\int_{-\infty}^{x}\cdots\int}_{(M-3)\text{-fold}} \left[\frac{(1+2\epsilon)x}{\sigma_1{}^2} + \frac{\epsilon}{\sigma_1{}^2}\sum_{i=4}^{M}\xi_i\right] G(x, x, x, \xi_4, \ldots, \xi_M; \mathbf{0}; \boldsymbol{\alpha}_\rho)\,d\xi_4\cdots d\xi_M$$

where

$$\epsilon = \frac{C_{1i}}{C_{11}} = \frac{-\rho}{1+(M-2)\rho} \qquad i \geqq 2$$

and

$$\sigma_1{}^2 = \frac{D}{C_{11}} = \frac{(1-\rho)[1+(M-1)\rho]}{1+(M-2)\rho}$$

By the same procedure as for Lemma 15.5 it can be shown that

$$\lim_{\rho\to -1/(M-1)} \int_{-\infty}^{x}\cdots\int G(x, x, x, \xi_4, \ldots, \xi_M; \mathbf{0}; \boldsymbol{\alpha}_\rho)\,d\xi_4\cdots d\xi_M = G_3(x)F_3(x)$$

Substitution gives

$$q = -\lim_{\rho\to -1/(M-1)} \left[\frac{(1+2\epsilon)x}{\sigma_1{}^2} G_3(x)F_3(x) + \frac{\epsilon}{\sigma_1{}^2}\sum_{j=4}^{M} \underbrace{\int_{-\infty}^{x}\cdots\int}_{(M-3)\text{-fold}} \xi_j G(x, x, x, \xi_4, \ldots, \xi_M; \mathbf{0}; \boldsymbol{\alpha}_\rho)\,d\xi_4\cdots d\xi_M\right] \tag{15.71}$$

Now we define

$$a_3 \triangleq \underbrace{\int_{-\infty}^{x} \cdots \int}_{(M-3)\text{-fold}} \xi_j G(x, x, x, \xi_4, \ldots, \xi_M; \mathbf{0}; \boldsymbol{\alpha}_\rho)\, d\xi_4 \cdots d\xi_M$$

$$j = 4, \ldots, M \qquad (15.72)$$

Here also, a_3 does not depend on j because of the symmetry of $\boldsymbol{\alpha}_\rho$. Next we note that

$$G_4(x)F_4(x) = \underbrace{\int_{-\infty}^{x} \cdots \int}_{(M-4)\text{-fold}} G(x, x, x, x, \xi_5, \ldots, \xi_M; \mathbf{0}; \boldsymbol{\alpha}_R)\, d\xi_5 \cdots d\xi_M$$

$$= \underbrace{\int_{-\infty}^{x} \cdots \int}_{(M-3)\text{-fold}} \frac{\partial}{\partial \xi_4} G(x, x, x, \xi_4, \ldots, \xi_M; \mathbf{0}; \boldsymbol{\alpha}_R)\, d\xi_4 \cdots d\xi_M$$

$$= -\lim_{\rho \to -1/(M-1)} \int_{-\infty}^{x} \cdots \int \left[\frac{3\epsilon x}{\sigma_1^2} + \frac{\sum_{j=4}^{M} C_{4j}\xi_j}{D}\right] G(x, x, x, \xi_4, \ldots, \xi_M; \mathbf{0}; \boldsymbol{\alpha}_\rho)\, d\xi_4 \cdots d\xi_M$$

$$= -\lim_{\rho \to -1/(M-1)} \left\{\frac{3\epsilon x}{\sigma_1^2} G_3(x)F_3(x) + \frac{1 + (M-4)\epsilon}{\sigma_1^2} a_3\right\}$$

from which

$$a_3 = -\lim_{\rho \to -1/(M-1)} \frac{\sigma_1^2 G_4(x)F_4(x) + 3\epsilon x G_3(x)F_3(x)}{1 + (M-4)\epsilon} \qquad (15.73)$$

Substituting into (15.71), we have

$$q = \lim_{\rho \to -1/(M-1)} \frac{(M-3)\epsilon}{1 + (M-4)\epsilon} G_4(x)F_4(x) - \left\{\frac{1 + 2\epsilon}{\sigma_1^2} - \frac{3\epsilon^2(M-3)}{[1 + (M-4)\epsilon]\sigma_1^2}\right\} xG_3(x)F_3(x)$$

$$= G_4(x)F_4(x) - \frac{M-1}{M-3} xG_3(x)F_3(x)$$

LEMMA 15.7 *For $i = k$ and $j = l$, define*

$$p = a_{klkl} = a_{1212} \tag{15.74}$$

Then

$$p = \left(\frac{M-1}{M-2}\right)^2 x^2 G_2(x)F_2(x) + [C(x) - 2F_2(x)]\frac{M-1}{M(M-2)}G_2(x) - 2\frac{M-1}{M-2}xG_3(x)F_3(x) + G_4(x)F_4(x) \tag{15.75}$$

where

$$G_2(x) = \frac{1}{2\pi}\sqrt{\frac{(M-1)^2}{M(M-2)}}\exp\left(\frac{-(M-1)}{M-2}x^2\right) \tag{15.76}$$

$$F_2(x) = \Pr(\psi_i \leq x\beta; i = 3, \ldots, M) \tag{15.77}$$

and

$$C(x) = \underbrace{\int_{-\infty}^{x\beta} \cdots \int}_{(M-2)\text{-fold}} \psi_i^2 G(\psi_3, \ldots, \psi_M; \mathbf{0}; \mathbf{C}_3)\, d\psi_3 \cdots d\psi_M \qquad i = 3, \ldots, M \tag{15.78}$$

where

$$\beta = \sqrt{\frac{M(M-1)}{(M-2)(M-3)}} \tag{15.79}$$

and

$$\mathbf{C}_3 = \begin{bmatrix} 1 & & \frac{-1}{M-3} \\ & \ddots & \\ \frac{-1}{M-3} & & 1 \end{bmatrix} \tag{15.80}$$

Proof: As in the previous lemmas, it can be shown that

$$G_2(x)F_2(x) = \lim_{\rho \to -1/(M-1)} \int_{-\infty}^{x} \cdots \int G(x, x, \xi_3, \ldots, \xi_M; \mathbf{0}; \boldsymbol{\alpha}_\rho)\, d\xi_3 \cdots d\xi_M \tag{15.81}$$

Now

$$p = \frac{\partial^2 \Phi(x;\boldsymbol{\alpha}_R)}{\partial \lambda_{12}^2}$$

$$= \lim_{\rho \to -1/(M-1)} \int_{-\infty}^{x} \cdots \int \left\{ \frac{\left[(1+\epsilon)x + \epsilon \sum_{3}^{M} \xi_i\right]^2}{\sigma_1^4} - \frac{\epsilon}{\sigma_1^2} \right\} G(x, x, \xi_3, \ldots, \xi_M; \mathbf{0}; \boldsymbol{\alpha}_\rho)\, d\xi_3 \cdots d\xi_M$$

$$= \lim_{\rho \to -1/(M-1)} G_2(x) \int_{-\infty}^{x} \cdots \int \left\{ \frac{\left[(1+\epsilon)x + \epsilon \sum_{3}^{M} \xi_i\right]^2}{\sigma_1^4} - \frac{\epsilon}{\sigma_1^2} \right\} G(\xi_3, \ldots, \xi_M | \xi_1 = x, \xi_2 = x)\, d\xi_3 \cdots d\xi_M \quad (15.82)$$

Note that

$$G(\xi_3, \ldots, \xi_M | \xi_1 = x, \xi_2 = x) = G(\xi_3, \ldots, \xi_M; \mathbf{m}_2; \mathbf{C}_2)$$

where

$$\mathbf{m}_2 \triangleq \begin{bmatrix} \dfrac{2\rho x}{1+\rho} \\ \cdot \\ \cdot \\ \cdot \\ \dfrac{2\rho x}{1+\rho} \end{bmatrix}$$

and the elements of $\mathbf{C}_2$ are

$$1 - \frac{2\rho^2}{1+\rho}$$

along the diagonal and

$$\rho - \frac{2\rho^2}{1+\rho}$$

off the diagonal. Let

$$\hat{\xi}_i = \xi_i - \frac{2\rho}{1+\rho} x \qquad i = 3, \ldots, M$$

Then

$$p = \lim_{\rho \to -1/(M-1)} G_2(x) \int_{-\infty}^{\gamma x} \cdots \int \Bigg(\frac{\{[(1+\epsilon) + 2\rho(M-2)\epsilon/(1+\rho)]x\}^2}{\sigma_1^4} + \frac{2[(1+\epsilon) + 2\rho(M-2)\epsilon/(1+\rho)]\,\epsilon x}{\sigma_1^4} \sum_{i=3}^{M} \hat{\xi}_i + \frac{\epsilon^2}{\sigma_1^4}\left(\sum_{i=3}^{M} \hat{\xi}_i\right)^2 - \frac{\epsilon}{\sigma_1^2}\Bigg) G(\hat{\xi}_3, \ldots, \hat{\xi}_M; \mathbf{0}; \mathbf{C}_2)\, d\hat{\xi}_3 \cdots d\hat{\xi}_M$$

where

$$\gamma = \frac{1-\rho}{1+\rho}$$

This can be reduced to

$$p = \left(\frac{M-1}{M-2}\right)^2 x^2 G_2(x) F_2(x) + \lim_{\rho \to -1/(M-1)} G_2(x) \underbrace{\int_{-\infty}^{\gamma x} \cdots \int}_{(M-2)\text{-fold}} \left[\frac{2\epsilon x}{(1+\rho)\sigma_1^2} \sum_{i=3}^{M} \hat{\xi}_i + \frac{\epsilon^2}{\sigma_1^4}\left(\sum_{i=3}^{M} \hat{\xi}_i\right)^2 - \frac{\epsilon}{\sigma_1^2}\right] G(\hat{\xi}_3, \ldots, \hat{\xi}_M; \mathbf{0}; \mathbf{C}_2)\, d\hat{\xi}_3 \cdots d\hat{\xi}_M \qquad (15.83)$$

Now we define

$$\hat{a} \triangleq G_2(x) \int_{-\infty}^{\gamma x} \cdots \int \hat{\xi}_i G(\hat{\xi}_3 \ldots, \hat{\xi}_M; \mathbf{0}; \mathbf{C}_2)\, d\hat{\xi}_3 \cdots d\hat{\xi}_M \qquad i = 3, \ldots, M \qquad (15.84)$$

As in the previous lemmas, $\hat{a}$ can be evaluated by noting that

$$G_3(x)F_3(x) = \frac{-2\epsilon x}{\sigma_1^2} G_2(x) F_2(x) - \frac{1 + (M-3)\epsilon}{\sigma_1^2} \underbrace{\int_{-\infty}^{x} \cdots \int}_{(M-2)\text{-fold}} \xi_i G(x, x, \xi_3, \ldots, \xi_M; \mathbf{0}; \boldsymbol{\alpha}_\rho)\, d\xi_3 \cdots d\xi_M \qquad (15.85)$$

and

$$\int_{-\infty}^{x} \cdots \int \xi_i G(x, x, \xi_3, \ldots, \xi_M; \mathbf{0}; \boldsymbol{\alpha}_\rho)\, d\xi_3 \cdots d\xi_M = G_2(x) \int_{-\infty}^{\gamma x} \cdots \int \left(\hat{\xi}_i + \frac{2\rho x}{1+\rho}\right) G(\hat{\xi}_3, \ldots, \hat{\xi}_M; \mathbf{0}; \mathbf{C}_2)\, d\hat{\xi}_3 \cdots d\hat{\xi}_M \qquad (15.86)$$

Substituting (15.86) into (15.85), we obtain

$$\hat{a} = \frac{-\sigma_1{}^2}{1 + (M - 3)\epsilon} G_3(x)F_3(x)$$

which, when substituted into (15.83), yields

$$p = \left(\frac{M-1}{M-2}\right)^2 x^2 G_2(x)F_2(x) - 2\frac{M-1}{M-2} xG_3(x)F_3(x) + \lim_{\rho \to -1/(M-1)} \left\{ \frac{\epsilon^2}{\sigma_1{}^4} [(M-2)\hat{C} + (M-2)(M-3)\hat{B}] - \frac{\epsilon}{\sigma_1{}^2} G_2(x)F_2(x) \right\} \quad (15.87)$$

where

$$\hat{C} = G_2(x) \int_{-\infty}^{\gamma x} \cdots \int \hat{\xi}_i{}^2 G(\hat{\xi}_3, \ldots, \hat{\xi}_M; \mathbf{0}; \mathbf{C}_2)\, d\hat{\xi}_3 \cdots d\hat{\xi}_M \qquad i = 3, \ldots, M \quad (15.88)$$

and

$$\hat{B} = G_2(x) \int_{-\infty}^{\gamma x} \cdots \int \hat{\xi}_i \hat{\xi}_j G(\xi_3, \ldots, \xi_M; \mathbf{0}; \mathbf{C}_2)\, d\hat{\xi}_3 \cdots d\hat{\xi}_M \qquad i \neq j;\ i, j \geqq 3 \quad (15.89)$$

Define

$$a_2 = \int_{-\infty}^{x} \cdots \int \xi_i G(x, x, \xi_3, \ldots, \xi_M; \mathbf{0}; \boldsymbol{\alpha}_\rho)\, d\xi_3 \cdots d\xi_M \quad (15.90)$$

which can be shown to be equal to

$$a_2 = -\frac{\sigma_1{}^2 G_3(x)F_3(x) + 2\epsilon x G_2(x)F_2(x)}{1 + (M-3)\epsilon} \quad (15.91)$$

Now,

$$G_4(x)F_4(x) = \lim_{\rho \to -1/(M-1)} \int_{-\infty}^{x} \cdots \int \frac{\partial^2}{\partial \xi_3\, \partial \xi_4} G(x, x, \xi_3, \ldots, \xi_M; \mathbf{0}; \boldsymbol{\alpha}_\rho)\, d\xi_3 \cdots d\xi_M$$

$$= \lim_{\rho \to -1/(M-1)} \left(\frac{4\epsilon^2 x^2}{\sigma_1{}^4} - \frac{\epsilon}{\sigma_1{}^2}\right) G_2(x)F_2(x) + \frac{4\epsilon x}{\sigma_1{}^4}[1 + (M-3)\epsilon]a_2 + \frac{1}{\sigma_1{}^4}([2\epsilon + (M-4)\epsilon^2]C' + \{1 + 2(M-4)\epsilon + [(M-4)(M-3) + 1]\epsilon^2\}B') \quad (15.92)$$

where

$$C'(x) = \int_{-\infty}^{x} \cdots \int \xi_i^2 G(x, x, \xi_3, \ldots, \xi_M; \mathbf{0}; \boldsymbol{\alpha}_\rho) \, d\xi_3 \cdots d\xi_M$$

$$i = 3, \cdots, M \quad (15.93)$$

and

$$B'(x) = \int_{-\infty}^{x} \cdots \int \xi_i \xi_j G(x, x, \xi_3, \ldots, \xi_M; \mathbf{0}; \boldsymbol{\alpha}_\rho) \, d\xi_3 \cdots d\xi_M$$

$$i \neq j; i, j \geqq 3 \quad (15.94)$$

By writing $C'(x)$ and $B'(x)$ in terms of $G_2(x)$ and substituting

$$\hat{\xi}_i = \xi_i - \frac{2\rho x}{1 + \rho} \qquad i = 3, \ldots, M$$

we obtain

$$B'(x) = \hat{B}(x) - \frac{4\rho\sigma_1^2 x}{[1 + (M-3)\epsilon](1+\rho)} G_3(x) F_3(x) + \frac{4\rho^2 x^2}{(1+\rho)^2} G_2(x) F_2(x) \quad (15.95)$$

and

$$C'(x) = \left(1 - \frac{2\rho^2}{1+\rho}\right) G_2(x) C(x) - \frac{4\rho\sigma_1^2 x}{[1 + (M-3)\epsilon](1+\rho)} G_3(x) F_3(x) + \frac{4\rho^2 x^2}{(1+\rho)^2} G_2(x) F_2(x) \quad (15.96)$$

Also,

$$\hat{C}(x) = \left(1 - \frac{2\rho^2}{1+\rho}\right) G_2(x) C(x) \quad (15.97)$$

Substituting (15.95) and (15.96) into (15.92), we obtain for $\hat{B}(x)$

$$\hat{B}(x) = \frac{\sigma_1^4}{1 + 2(M-4)\epsilon + [(M-4)(M-3) + 1]\epsilon^2} \left[G_4(x) F_4(x) + \frac{\epsilon G_2(x) F_2(x)}{\sigma_1^2} - \frac{2\epsilon + (M-4)\epsilon^2}{\sigma_1^4} \left(1 - \frac{2\rho^2}{1+\rho}\right) C(x) G_2(x) \right] \quad (15.98)$$

Finally, when (15.98) and (15.97) are substituted into (15.87) and we let $\rho \to -1/(M-1)$, we obtain the desired result.

With these three relationships we can now prove that the regular simplex is a local maximum in the tangent plane defined by (15.9). We define

$$b_i = \sum_{j=1}^{M} (\lambda'_{ij} - \lambda_{ij}{}^R) \tag{15.99}$$

and note that

$$\frac{\partial\Phi(x;\boldsymbol{\alpha}_R)}{\partial\lambda_{ij}} = G_2(x)F_2(x) \qquad i \neq j \tag{15.100}$$

Then

$$a_1(x) = \tfrac{1}{2}G_2(x)F_2(x) \sum_{i=1}^{M} b_i \tag{15.101}$$

LEMMA 15.8 *For any admissible* $\boldsymbol{\alpha}'$

$$a_2(x) = \left(\sum_{i>j}\sum \gamma_{ij}{}^2\right)(p - 2q + r) + \tfrac{1}{4}\left(\sum_{i=1}^{M} b_i\right)^2 r + \left(\sum_{i=1}^{M} b_i{}^2\right)(q - r) \tag{15.102}$$

where

$$\gamma_{ij} = \lambda'_{ij} - \lambda_{ij}{}^R \tag{15.103}$$

Proof: From (15.57),

$$a_2(x) = \sum_{i>j}\sum \sum_{k>l}\sum \gamma_{ij}\gamma_{kl}a_{ijkl}$$

which can be rewritten as

$$\begin{aligned} a_2(x) &= \left(\sum_{i>j}\sum \gamma_{ij}{}^2\right) p + \sum_{\substack{i>j \\ (i,j)\neq(k,l)}}\sum \sum_{k>l}\sum \gamma_{ij}\gamma_{kl}a_{ijkl} \\ &= \left(\sum_{i>j}\sum \gamma_{ij}{}^2\right) p + \tfrac{1}{4} \sum_{\substack{i>j \\ (i,j)\neq(k,l) \\ (i,j)\neq(l,k)}}\sum \sum_{k>l}\sum \gamma_{ij}\gamma_{kl}a_{ijkl} \\ &= \left(\sum_{i>j}\sum \gamma_{ij}{}^2\right) p + \sum_i \sum_j \gamma_{ij}\left[\frac{q}{2}\left(\sum_{\substack{k \\ k\neq j}} \gamma_{ik} + \sum_{\substack{k \\ k\neq i}} \gamma_{jk}\right)\right. \\ &\qquad \left. + \left(\sum_{\substack{k \\ k\neq i \\ k\neq j}} \sum_{\substack{l \\ l\neq i \\ l\neq j}} \gamma_{kl}\right)\frac{r}{4}\right] \end{aligned} \tag{15.104}$$

The coefficient of q is

$$\tfrac{1}{2}\sum_i \sum_j \gamma_{ij}[(b_i - \gamma_{ij}) + (b_j - \gamma_{ji})] = -2\sum_{i>j}\sum \gamma_{ij}{}^2 + \sum_i b_i{}^2$$

The coefficient of r is

$$\tfrac{1}{4}\sum_i \sum_j \gamma_{ij} \sum_{\substack{k \\ k\neq i \\ k\neq l}}\sum_{\substack{l \\ l\neq i \\ l\neq j}} \gamma_{kl} = \tfrac{1}{4}\sum_i \sum_j \gamma_{ij} \sum_{\substack{l \\ l\neq i \\ l\neq j}} (b_l - \gamma_{il} - \gamma_{jl})$$

$$= \sum_{i>j}\sum \gamma_{ij}^2 + \tfrac{1}{4}\Big(\sum_i b_i\Big)^2 - \sum_i b_i^2$$

Substituting these into (15.104) gives the desired result.

Now we restrict $\boldsymbol{\alpha}'$ to lie in the tangent plane given by (15.9). Therefore

$$\sum_i \sum_j \lambda'_{ij} = 0$$

which, from Lemma 14.5, implies

$$\sum_i \lambda'_{ij} = 0 \qquad j = 1, \ldots, M$$

Hence

$$b_i = 0 \qquad i = 1, \ldots, M$$

So when $\boldsymbol{\alpha}'$ is in this tangent plane we have

$$a_2(x) = \Big(\sum_{i>j}\sum \gamma_{ij}^2\Big)(p - 2q + r) \tag{15.105}$$

Now,

$$p - 2q + r = \left\{\left[\left(\frac{M-1}{M-2}\right)^2 x^2 - \frac{2(M-1)}{M(M-2)}\right] F_2(x) + \frac{M-1}{M(M-2)}\, C(x)\right\} G_2(x) + \frac{2(M-1)}{(M-2)(M-3)}\, xG_3(x)F_3(x) \tag{15.106}$$

From (14.43) we can write

$$\phi(\lambda;\boldsymbol{\alpha}') = \int_0^\infty \exp(\lambda x)\, \frac{d}{dx}[\Phi(x;\boldsymbol{\alpha}_R)]\, dx + \int_{-\infty}^\infty \exp(\lambda x)\, \frac{d}{dx}[\Phi(x;\boldsymbol{\alpha}') - \Phi(x;\boldsymbol{\alpha}_R)]\, dx$$

which, from (14.44), can be expressed as

$$\phi(\lambda;\boldsymbol{\alpha}') = \phi(\lambda;\boldsymbol{\alpha}_R) - \lambda \int_{-\infty}^\infty \exp(\lambda x)[\Phi(x;\boldsymbol{\alpha}') - \Phi(x;\boldsymbol{\alpha}_R)]\,dx \tag{15.107}$$

so that, employing (15.55), we have

$$\phi(\lambda;\boldsymbol{\alpha}') = \phi(\lambda;\boldsymbol{\alpha}_R) - \lambda \int_0^\infty \exp(\lambda x)\left[\sum_{k=1}^\infty a_k(x)\,\frac{\theta^k}{k!}\right] dx \tag{15.108}$$

But $a_1(x)$ vanishes when $\boldsymbol{\alpha}'$ is in the tangent plane; hence, in order to verify that $\boldsymbol{\alpha}_R$ is a local maximum in this tangent plane, we have only to prove the following.

THEOREM 15.5

$$\int_0^\infty \exp(\lambda x) a_2(x)\, dx > 0 \tag{15.109}$$

when $\boldsymbol{\alpha}'$ is in the tangent plane defined by

$$\sum_i \sum_j \lambda_{ij} = 0$$

Proof: Since in (15.105)

$$\sum_{i>j}\sum \gamma_{ij}^2 > 0$$

the proof reduces to showing that

$$\int_0^\infty \exp(\lambda x)(p - 2q + r)\, dx > 0 \tag{15.110}$$

where $p - 2q + r$ is given in (15.106). The integral over the last two terms of $p - 2q + r$ is clearly positive. Hence it is sufficient to verify that

$$\int_0^\infty \exp(\lambda x) \left(\frac{M-1}{M-2} x^2 - \frac{2}{M}\right) G_2(x) F_2(x)\, dx > 0 \tag{15.111}$$

To do this, we use the fact that

$$\frac{d}{dx} G_2(x) = -2 \frac{M-1}{M-2} x G_2(x)$$

to integrate

$$\int_0^\infty \exp(\lambda x) G_2(x) F_2(x)\, dx$$

by parts. This yields the relationship

$$\begin{aligned} \frac{M-1}{M-2} \int_0^\infty \exp(\lambda x) x^2 G_2(x) F_2(x)\, dx &= \tfrac{1}{2} \int_0^\infty \exp(\lambda x)\, G_2(x) F_2(x)(1 + x\lambda)\, dx \\ &\quad + \tfrac{1}{2} \int_0^\infty x \exp(\lambda x) G_2(x) \frac{dF_2(x)}{dx}\, dx \end{aligned} \tag{15.112}$$

which when substituted into (15.111) results in the inequality

$$\begin{aligned} \left(\frac{1}{2} - \frac{2}{M}\right) &\int_0^\infty \exp(\lambda x) G_2(x) F_2(x)\, dx \\ &+ \int_0^\infty x \exp(\lambda x) G_2(x) \left[\lambda F_2(x) + \frac{dF_2(x)}{dx}\right] dx > 0 \end{aligned} \tag{15.113}$$

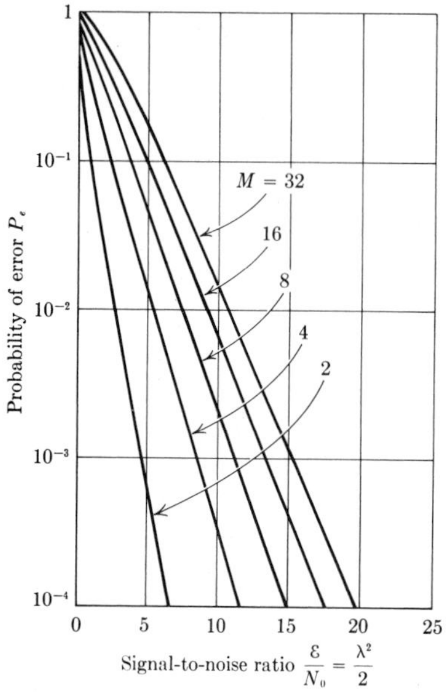

Fig. 15.2 Probability of error versus signal-to-noise ratio for the regular simplex signal structure.

Both these integrals are positive for $\lambda > 0$, and the coefficient of the first is positive for $M > 3$. Since the optimal solution for the $M = 3$ case has already been found, the proof is complete.

This completes the discussion of second-order conditions for the regular simplex, from which we can conclude that for any $\boldsymbol{\alpha}'$ in the tangent plane and in the neighborhood of $\boldsymbol{\alpha}_R$.

$$\phi(\lambda;\boldsymbol{\alpha}') < \phi(\lambda;\boldsymbol{\alpha}_R)$$

In summary, we have proved the following main results about the regular simplex signal structure:

1. In $\mathcal{A}$ the regular simplex is a local maximum at every signal-to-noise ratio and is the only signal structure that is a local maximum at all signal-to-noise ratios.
2. The regular simplex is the global optimum for sufficiently small signal-to-noise ratios and for signal-to-noise ratios sufficiently large.

Probability of error is plotted against signal-to-noise ratio for the regular simplex in Fig. 15.2.

15.5 MAXIMIZING THE MINIMUM DISTANCE

Because of its relative simplicity, the criterion of maximizing the minimum distance between the signal vectors has been a common one. In the class of all admissible signal sets we have the following.

LEMMA 15.9 *In the class $\mathcal{A}$ of all admissible $\boldsymbol{\alpha}$ the regular simplex is the only polytope which maximizes the minimum distance between the signal vectors.*

Remark: In this case of no dimensionality restriction and for the case when D is restricted to 2, the problem of maximizing the minimum distance has a unique solution. However, this criterion does not in general have a unique solution. In addition, different signal sets with the same minimum distance will be shown in Chap. 16 to have different probabilities of detection, making it a somewhat questionable criterion. All the signal-design solutions to date, however, do maximize the minimum distance in the subclass of $\mathcal{A}$ in which they are the optimum.

In early work, maximizing the minimum distance was the accepted criterion, so its relation to probability of detection should be known. It was most likely the first criterion used in signal design and appeared attractive because of its intuitive connection with maximum-likelihood decision rules and the divergence criterion.

Proof: Maximizing the minimum distance is synonymous to minimizing

$$\max_{i \neq j} \lambda_{ij} \tag{15.114}$$

The minimum value (15.114) we can attain is $-1/(M-1)$, since for any $\boldsymbol{\alpha}$

$$\sum_{i>j}\sum \lambda_{ij} \geq -\frac{M}{2}$$

Thus we must show that $\boldsymbol{\alpha}_R$ is the only polytope which attains this minimum value. Assume that $\boldsymbol{\alpha}$ represents another polytope with

$$\max_{i \neq j} \lambda_{ij} = \frac{-1}{M-1}$$

and reorder the λ_{ij} so that

$$\lambda_{12} = \max_{i \neq j} \lambda_{ij}$$

We know that $D(\boldsymbol{\alpha}) = 0$, since it has already been shown that

$$\sum_{i>j}\sum \lambda_{ij} > \frac{-M}{2}$$

whenever $D(\boldsymbol{\alpha}) > 0$, which implies

$$\max_{i \neq j} \lambda_{ij} > \frac{-1}{M-1}$$

Therefore for $\boldsymbol{\alpha}$

$$\sum_{i>j}\sum \lambda_{ij} = \frac{-M}{2}$$

from which (using Lemma 14.5)

$$\sum_{i=1}^{M} \lambda_{1i} = 0$$

or

$$\lambda_{13} + \cdots + \lambda_{1M} = \frac{-(M-2)}{M-1}$$

If any λ_{1i} for $i \geq 3$ is less than $-1/(M-1)$, then at least one of the others has to be greater than this term, which is a contradiction.

Then

$$\lambda_{1i} = \frac{-1}{M-1} \qquad i \geq 2$$

Using this and repeating the argument for

$$\sum_{j=1}^{M} \lambda_{ij} = 0 \qquad i \geq 2$$

we see that

$$\lambda_{ij} = \frac{-1}{M-1} \qquad \text{for all } i \neq j$$

The global optimality of the regular simplex signal structure has recently been shown over the sets of four signals which have 3 degrees of freedom. In their proof, Landau and Slepian [15.13] derive an upper bound for the probability of detection for an arbitrary M-ary signal set in $M - 1$ space using a sequence of intricate geometric inequalities. The conditions for equality are indicated and shown to be uniquely satisfied by the regular simplex when $M = 4$.

Since their inequality is satisfied for all signal-to-noise ratios, and the uniqueness of the regular simplex is independent of signal-to-noise ratio, the proof of the global optimality of $\boldsymbol{\alpha}_R$ is complete when $M = 4$. As indicated at the outset of this chapter, the advantage of the variational approach is its ability to provide additional solutions when constraints are imposed on the allowed dimensionality. These results are presented in the next three chapters.

A study of the properties of optimal signal sets which do not have equally likely a priori probabilities or do not have equal energies (but more generally have an average power constraint) is given by Dunbridge [15.14].

PROBLEM

15.1 Show that the regular simplex signal set is also the optimal choice when the model is the same as assumed here, with the exception that the amplitude of the transmitted waveform is any nonnegative random variable unknown to the receiver.

REFERENCES

15.1 Balakrishnan, A. V.: A Contribution to the Sphere Packing Problem of Communication Theory, *J. Math. Anal. Appl.*, vol. 3, December, 1961, pp. 485–506.

15.2 Ziv, J.: Generation of Optimal Codes by a "Pyramid-Packing" Argument, *IEEE Trans. Inform. Theory*, vol. IT-10, no. 3, July, 1964, pp. 253–255.

15.3 Balakrishnan, A. V.: "Advances in Communication," Academic Press, New York, 1965.

15.4 Stutt, C. A.: Regular Polyhedron Codes, *General Electric Res. Lab., Res. Rept. 59-R6-2202*, March, 1959.

15.5 Balakrishnan, A. V., and J. E. Taber: Error Rates in Coherent Communication Systems, *IRE Trans. Commun. Systems*, vol. CS-10, March, 1962, pp. 86–90.

15.6 Nuttall, A. H.: Error Probabilities for Equicorrelated M-ary Signals under Phase Coherent and Phase Incoherent Reception, *IRE Trans. Inform. Theory*, vol. IT-8, July, 1962, pp. 305–315.

15.7 Max, J.: Signals Sets With Uniform Correlation Properties, *J. SIAM*, vol. 10, March, 1962, pp. 113–118.

15.8 Stutt, C. A.: Information Rate in a Continuous Channel for Regular Simplex Codes, *IRE Trans. Inform. Theory*, vol. IT-6, December, 1960, pp. 516–522.

15.9 Chang, S. S. L., B. Harris, and J. J. Metzner: Optimum Message Transmission in a Finite Time, *IRE Trans. Inform. Theory*, vol. IT-8, September, 1962, pp. 215–224.

15.10 Golomb, S. W., L. D. Baumert, M. F. Easterling, J. J. Stiffler, and A. J. Viterbi: "Digital Communications," Prentice-Hall, Englewood Cliffs, N.J., 1964.

15.11 Slepian, D.: Permutation Modulation, *Proc. IEEE*, vol. 53, March, 1965, pp. 228–236.

15.12 Slepian, D.: Bounds on Communication, *Bell Systems Tech. J.*, vol. 42, May, 1963, pp. 681–707.

15.13 Landau, H. J., and D. Slepian: On the Optimality of the Regular Simplex Code, *Bell Systems Tech. J.*, vol. 45, no. 8, October, 1966.

15.14 Dunbridge, B.: Asymmetric Signal Design for the Coherent Gaussian Channel, *IEEE Trans. Inform. Theory*, vol. IT-13, no. 3, July, 1967.

16

Optimality for Coherent Systems When the Dimensionality Is Restricted to $D \leq M - 2$

The results derived in the previous chapter indicate that when dimensionality is unrestricted, the optimal signal set is the regular simplex which requires $M - 1$ degrees of freedom. We now extend Balakrishnan's variational theory of the signal-design problem to provide optimal signal sets for the case where the dimensionality is restricted to be less than $M - 1$. In particular, we now reduce by 1 the allowed degrees of freedom from that required for the regular simplex and look for optimal signal sets when $D \leq M - 2$, which corresponds to an equivalent reduction on the allowed bandwidth.

In the class of admissible $\boldsymbol{\alpha}$ for which $D \leq M - 2$ the signal sets which have inner-product matrices of the form

$$\boldsymbol{\alpha}_0 = \left[\begin{array}{ccc|ccc} 1 & & \frac{-1}{M_1-1} & & & \\ & \ddots & & & 0 & \\ \frac{-1}{M_1-1} & & 1 & & & \\ \hline & & & 1 & & \frac{-1}{M_2-1} \\ & 0 & & & \ddots & \\ & & & \frac{-1}{M_2-1} & & 1 \end{array}\right] \tag{16.1}$$

where M_1 can assume the values 2, 3, . . . , $M-2$, and $M_1 + M_2 = M$, are local extrema for all signal-to-noise ratios; that is, they satisfy necessary (first-order) conditions to be optimal. These local extrema consist of two regular simplices placed orthogonal to one another, with dimensionalities $M_1 - 1$ and $M_2 - 1$, thus satisfying the restriction of M signals in $M-2$ (or fewer) dimensions. These necessary conditions provide $M-3$ local extrema, which must be classified into local maxima, local minima, or saddle points.

We shall show that each of the above extrema satisfies sufficient (second-order) conditions to be a local maximum in $\mathcal{A}_2$, where for convenience we define $\mathcal{A}_2$ as that subclass of $\mathcal{A}$ for which $D \leq M-2$. Finally, we shall order these $M-3$ local maxima with respect to probability of error for large signal-to-noise ratios, with the conclusion that, of their local maxima, the preferred one has the signal-vector inner-product matrix

$$\boldsymbol{\alpha}_{0_{\text{opt}}} \triangleq \left[\begin{array}{c|c} \boldsymbol{\alpha}_{R_2} & \mathbf{0} \\ \hline \mathbf{0} & \boldsymbol{\alpha}_{R_{M-2}} \end{array}\right] \tag{16.2}$$

where $\boldsymbol{\alpha}_{R_k}$ is the regular simplex inner-product matrix for k vectors.

It should not be surprising that optimal signal sets are not equicorrelated for the cases when the dimensionality restrictions are less than $M-1$, since the only signal set which is equicorrelated and does not have dimensionality $D = M$ is $\boldsymbol{\alpha}_R$, for which $D = M - 1$. Hence these optimal sets necessarily cannot be equicorrelated.

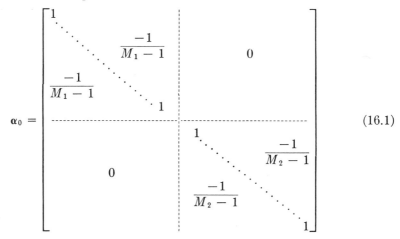

As a special case of interest, these results show that for five points in three dimensions the choice which corresponds to placing four points on the equator and one on the pole is not optimal, while the choice of three points on the equator and one on each pole is the optimum, even though both have the same minimum distance; this settles another heretofore unresolved point in this theory. In particular, we note that the mean width is larger in the latter case.

16.1 NECESSARY (FIRST-ORDER) CONDITIONS

We are looking for admissible $\boldsymbol{\alpha}$ which maximize the probability of detection subject to the restriction that $D \leq M - 2$. We shall obtain necessary conditions for a local extremum, and those $\boldsymbol{\alpha}_0$ in (16.1) will be shown to satisfy these conditions.

First, note that from (14.44) the probability of detection can be written as

$$P_d(\lambda;\boldsymbol{\alpha}) = \exp\left(-\tfrac{1}{2}\lambda^2\right) \int_0^\infty \exp(\lambda x) G(x) [\phi(x)]^{M-1}\,dx - \frac{\lambda}{M} \exp\left(-\tfrac{1}{2}\lambda^2\right) \int_0^\infty \exp(\lambda x) \{\Phi(x,\boldsymbol{\alpha}) - [\phi(x)]^M\}\,dx \tag{16.3}$$

where, as before, we integrate only over the region $(0, \infty)$, since we already know that the optimal solutions have convex hulls which contain the origin. Hence minimizing

$$J(\lambda;\boldsymbol{\alpha}) = \int_0^\infty \exp(\lambda x) \{\Phi(x;\boldsymbol{\alpha}) - [\phi(x)]^M\}\,dx \tag{16.4}$$

is equivalent to maximizing $P_d(\lambda;\boldsymbol{\alpha})$.

Second, note that the M^2 constraint equations

$$C_{ij} = 0 \qquad i, j = 1, \ldots, M \tag{16.5}$$

where, as before, C_{ij} is the cofactor, (including sign) of λ_{ij} in $\boldsymbol{\alpha}$, are sufficient to restrict $\boldsymbol{\alpha}$ to $M - 2$ degrees of freedom. Since the $\boldsymbol{\alpha}$ space consists of only symmetric matrices, the number of constraint equations can be reduced to

$$C_{ij} = 0 \qquad i \geq j; j = 1, \ldots, M \tag{16.6}$$

with no loss of generality.

Let Δ_2 denote the class of all $\boldsymbol{\alpha}$ whose elements satisfy (16.6). Observe that Δ_2 includes nonadmissible as well as admissible $\boldsymbol{\alpha}$; that is, there may be $\boldsymbol{\alpha}$ satisfying (16.6) that are not nonnegative-definite. But

$$\mathcal{A}_2 = \Delta_2 \cap \mathcal{A} \tag{16.7}$$

In addition, if extrema are found in Δ_2, and they are admissible (that is, they are also elements of $\mathcal{A}_2$), then they are also extrema of $\mathcal{A}_2$. This is the approach we adopt.

Now, consider some $\boldsymbol{\alpha}'$ in $\mathfrak{A}_2$; that is, $\boldsymbol{\alpha}'$ is a point on each of the surfaces in $\boldsymbol{\alpha}$ space defined by $C_{ij} = 0$ for $i \geq j$ and $j = 1, \ldots, M$. The vector normal to the surface $C_{ij} = 0$ at $\boldsymbol{\alpha}'$ is given by the $[M(M-1)/2]$-by-1 gradient vector

$$\nabla C_{ij}\Big|_{\boldsymbol{\alpha}'} = \begin{bmatrix} \left.\dfrac{\partial C_{ij}}{\partial \lambda_{12}}\right|_{\boldsymbol{\alpha}'} \\ \left.\dfrac{\partial C_{ij}}{\partial \lambda_{13}}\right|_{\boldsymbol{\alpha}'} \\ \cdot \\ \cdot \\ \cdot \\ \left.\dfrac{\partial C_{ij}}{\partial \lambda_{M-1\,M}}\right|_{\boldsymbol{\alpha}'} \end{bmatrix} \tag{16.8}$$

and the plane tangent to the surface at $\boldsymbol{\alpha}'$ consists of those vectors $\mathbf{t}$ satisfying

$$\nabla C_{ij}\Big|_{\boldsymbol{\alpha}'}^{T} \mathbf{t} = 0 \tag{16.9}$$

If $\boldsymbol{\alpha}'$ is to be a local minimum with respect to points in the neighborhood of $\boldsymbol{\alpha}'$ satisfying $C_{ij} = 0$ for all i and j, then

$$\nabla \phi(\lambda;\boldsymbol{\alpha}')^T \mathbf{t} = 0 \tag{16.10}$$

for all $\mathbf{t}$ satisfying (16.9), or, equivalently,

$$\nabla J(\lambda;\boldsymbol{\alpha}')^T \mathbf{t} = 0 \tag{16.11}$$

where

$$\nabla J(\lambda;\boldsymbol{\alpha}') \triangleq \begin{bmatrix} \dfrac{\partial J(\lambda;\boldsymbol{\alpha}')}{\partial \lambda_{12}} \\ \cdot \\ \cdot \\ \cdot \\ \dfrac{\partial J(\lambda;\boldsymbol{\alpha}')}{\partial \lambda_{M-1,M}} \end{bmatrix} \triangleq \frac{-1}{\lambda} \begin{bmatrix} \dfrac{\partial \phi(\lambda;\boldsymbol{\alpha}')}{\partial \lambda_{12}} \\ \cdot \\ \cdot \\ \cdot \\ \dfrac{\partial \phi(\lambda;\boldsymbol{\alpha}')}{\partial \lambda_{M-1,M}} \end{bmatrix} = \frac{-1}{\lambda} \nabla \phi(\lambda;\boldsymbol{\alpha}') \tag{16.12}$$

and where $\phi(\lambda;\boldsymbol{\alpha})$ is defined by (14.12).

If (16.11) were not satisfied, there would then exist a curve in $C_{ij} = 0$ through $\boldsymbol{\alpha}'$ on which the projection of $\nabla J(\lambda;\boldsymbol{\alpha}')$ would be negative, thus decreasing $J(\lambda;\boldsymbol{\alpha})$. The existence of such a curve is guaranteed, since C_{ij} is a polynomial in the λ_{kl} and is therefore differentiable. We have, however, $M(M+1)/2$ surfaces and, correspondingly, $M(M+1)/2$ tangent planes to which $\nabla J(\lambda;\boldsymbol{\alpha}')$ must be orthogonal if $\boldsymbol{\alpha}'$ is to be a local minimum of $J(\lambda;\boldsymbol{\alpha})$ in $\mathfrak{A}_2$. The flat which is tangent to all the surfaces at $\boldsymbol{\alpha}'$ consists of those vectors $\mathbf{t}$ such that

$$\nabla C_{ij}\Big|_{\boldsymbol{\alpha}'}^{T} \mathbf{t} = 0 \qquad i \geq j; j = 1, \ldots, M \tag{16.13}$$

Note that as the dimensionality of the manifold spanned by the set of normal vectors $\{\nabla C_{ij}\}$ increases, the dimensionality of the flat defined by (16.13) decreases. If $\boldsymbol{\alpha}'$ is to be a local minimum with respect to those $\boldsymbol{\alpha}$ which are in all $M(M+1)/2$ surfaces, then the relationship

$$\nabla J(\lambda;\boldsymbol{\alpha}')^T\mathbf{t} = 0 \tag{16.14}$$

must hold for all $\mathbf{t}$ in the flat defined in (16.13). Finally, since the flat defined in (16.13) is orthogonal to the manifold spanned by the normal vectors $\{\nabla C_{ij}\}$, and their union is the whole $\boldsymbol{\alpha}$ space, we can conclude that (16.14) will be satisfied if and only if $\nabla J(\lambda;\boldsymbol{\alpha}')$ can be expressed as a linear combination of the ∇C_{ij}; that is, if

$$\nabla J(\lambda;\boldsymbol{\alpha}') = \sum_{i}^{M}\sum_{j}^{M} \nu_{ij}\,\nabla C_{ij}\Big|_{\boldsymbol{\alpha}'} \tag{16.15}$$

Equation (16.15) is the necessary condition that must be satisfied by $\boldsymbol{\alpha}'$ for it to be a local extremum in $\mathcal{A}_2$.

THEOREM 16.1 *The $\boldsymbol{\alpha}_0$ given by*

$$\boldsymbol{\alpha}_0 = \left[\begin{array}{ccc|ccc} 1 & & \frac{-1}{M_1-1} & & & \\ & \ddots & & & 0 & \\ \frac{-1}{M_1-1} & & 1 & & & \\ \hline & & & 1 & & \frac{-1}{M_2-1} \\ & 0 & & & \ddots & \\ & & & \frac{-1}{M_2-1} & & 1 \end{array}\right] \tag{16.1}$$

where $M_1 = 2, \ldots, M-2$ and $M_1 + M_2 = M$, satisfy the necessary condition in (16.15) *and are therefore local extrema in $\mathcal{A}_2$.*

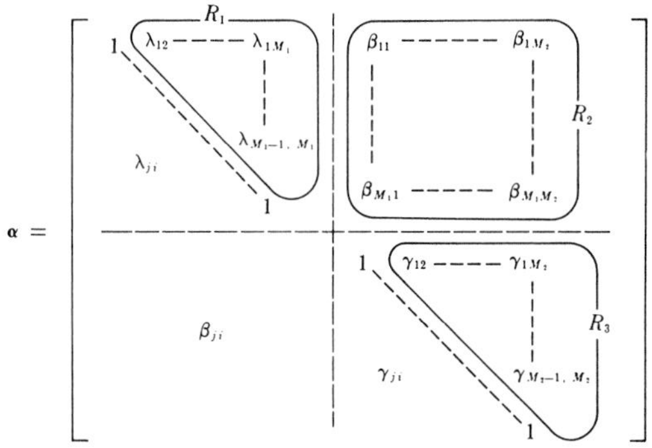

Fig. 16.1 Identification and indexing diagram for an arbitrary $\boldsymbol{\alpha}$ and fixed M_1 and M_2.

Proof: It is immediate that these $\boldsymbol{\alpha}_0$ are admissible and have dimensionality equal to $M - 2$.

Henceforth for M_1 and M_2 fixed such that $M_1 + M_2 = M$ we shall identify and index the elements of an arbitrary $\boldsymbol{\alpha}$ as indicated in Fig. 16.1. Identify the λ_{ij} elements as being in $\mathbf{R}_1$, the β_{ij} in $\mathbf{R}_2$, and the γ_{ij} in $\mathbf{R}_3$. Assume components of the gradient and normal vectors to be arranged in order over $\mathbf{R}_1$, then $\mathbf{R}_2$, and finally over $\mathbf{R}_3$.

Denote the M_1-by-M_1 regular simplex by

$$\boldsymbol{\alpha}_{R_{M_1}} \triangleq \begin{bmatrix} 1 & & \dfrac{-1}{M_1 - 1} \\ & \ddots & \\ \dfrac{-1}{M_1 - 1} & & 1 \end{bmatrix} \tag{16.16}$$

and, similarly, the M_2-by-M_2 regular simplex by

$$\boldsymbol{\alpha}_{R_{M_2}} \triangleq \begin{bmatrix} 1 & & \dfrac{-1}{M_2 - 1} \\ & \ddots & \\ \dfrac{-1}{M_2 - 1} & & 1 \end{bmatrix} \tag{16.17}$$

Hence

$$\boldsymbol{\alpha}_0 = \left[\begin{array}{c|c} \boldsymbol{\alpha}_{R_{M_1}} & \mathbf{0} \\ \hline \mathbf{0} & \boldsymbol{\alpha}_{R_{M_2}} \end{array}\right] \tag{16.18}$$

where $M_1 + M_2 = M$, and from (14.49),

$$J(\lambda;\boldsymbol{\alpha}_0) = \int_0^\infty \exp \lambda x \Bigg\{ \underbrace{\int \cdots \int_{-\infty}^{x}}_{M\text{-fold}} G(\boldsymbol{\xi};\mathbf{0};\boldsymbol{\alpha}_0)\, d\boldsymbol{\xi} - [\phi(x)]^M \Bigg\}\, dx$$

$$= \int_0^\infty \exp \lambda x \Bigg\{ \underbrace{\int \cdots \int_{-\infty}^{x}}_{M_1\text{-fold}} G(\boldsymbol{\xi};\mathbf{0};\boldsymbol{\alpha}_{R_{M_1}})\, d\boldsymbol{\xi}$$

$$\underbrace{\int \cdots \int_{-\infty}^{x}}_{M_2\text{-fold}} G(\mathbf{n};\mathbf{0};\boldsymbol{\alpha}_{R_{M_2}})\, d\mathbf{n} - [\phi(x)]^M \Bigg\}\, dx \tag{16.19}$$

Because of the symmetry in $\boldsymbol{\alpha}_{R_{M_1}}$ and $\boldsymbol{\alpha}_{R_{M_2}}$ (and introducing the symbols c_1, c_2, and c_3), we can write

$$\frac{\partial J(\lambda;\boldsymbol{\alpha}_0)}{\partial \lambda_{ij}} = \frac{\partial J(\lambda;\boldsymbol{\alpha}_0)}{\partial \lambda_{12}} = c_1 = \int_0^\infty dx \exp \lambda x$$

$$\underbrace{\int \cdots \int_{-\infty}^{x}}_{(M_1-2)\text{-fold}} G(x, x, \xi_3, \xi_4, \ldots, \xi_{M_1}; \mathbf{0}; \boldsymbol{\alpha}_{R_{M_1}})\, d\xi_3 \cdots d\xi_{M_1}$$

$$\underbrace{\int \cdots \int_{-\infty}^{x}}_{M_2\text{-fold}} G(\mathbf{n};\mathbf{0};\boldsymbol{\alpha}_{R_{M_2}})\, d\mathbf{n} \tag{16.20}$$

for all λ_{ij} in $\mathbf{R}_1$. Also,

$$\frac{\partial J(\lambda;\boldsymbol{\alpha}_0)}{\partial \gamma_{ij}} = \frac{\partial J(\lambda;\boldsymbol{\alpha}_0)}{\partial \gamma_{12}} = c_3 = \int_0^\infty dx \exp \lambda x$$

$$\underbrace{\int \cdots \int_{-\infty}^{x}}_{M_1\text{-fold}} G(\boldsymbol{\xi};\mathbf{0};\boldsymbol{\alpha}_{R_{M_1}})\, d\boldsymbol{\xi}$$

$$\underbrace{\int \cdots \int_{-\infty}^{x}}_{(M_2-2)\text{-fold}} G(x, x, \eta_3, \eta_4, \ldots, \eta_{M_2}; \mathbf{0}; \boldsymbol{\alpha}_{R_{M_2}})\, d\eta_3 \cdots d\eta_{M_2} \tag{16.21}$$

for all γ_{ij} in $\mathbf{R}_3$; and

$$\frac{\partial J(\lambda;\boldsymbol{\alpha}_0)}{\partial \beta_{ij}} = \frac{\partial J(\lambda;\boldsymbol{\alpha}_0)}{\partial \beta_{11}} = c_2 = \int_0^\infty dx \exp \lambda x \underbrace{\int_{-\infty}^{x} \cdots \int}_{(M_1-1)\text{-fold}} G(x, \xi_2, \ldots, \xi_{M_1}; \mathbf{0}; \boldsymbol{\alpha}_{R_{M_1}})\, d\xi_2 \cdots d\xi_{M_1} \underbrace{\int_{-\infty}^{x} \cdots \int}_{(M_2-1)\text{-fold}} G(x, \eta_2, \ldots, \eta_{M_2}; \mathbf{0}; \boldsymbol{\alpha}_{R_{M_2}})\, d\eta_2 \cdots d\eta_{M_2} \quad (16.22)$$

for all β_{ij} in $\mathbf{R}_2$. Therefore the gradient vector at $\boldsymbol{\alpha}_0$ is of the form

$$\nabla J(\lambda;\boldsymbol{\alpha}_0) = \begin{bmatrix} \dfrac{\partial J(\lambda;\boldsymbol{\alpha}_0)}{\partial \lambda_{12}} \\ \cdot \\ \cdot \\ \cdot \\ \dfrac{\partial J(\lambda;\boldsymbol{\alpha}_0)}{\partial \lambda_{M_1-1,M_1}} \\ \dfrac{\partial J(\lambda;\boldsymbol{\alpha}_0)}{\partial \beta_{11}} \\ \cdot \\ \cdot \\ \cdot \\ \dfrac{\partial J(\lambda;\boldsymbol{\alpha}_0)}{\partial \beta_{M_1M_2}} \\ \dfrac{\partial J(\lambda;\boldsymbol{\alpha}_0)}{\partial \gamma_{12}} \\ \cdot \\ \cdot \\ \cdot \\ \dfrac{\partial J(\lambda;\boldsymbol{\alpha}_0)}{\partial \gamma_{M_2-1,M_2}} \end{bmatrix} = \begin{bmatrix} c_1 \\ \cdot \\ \cdot \\ \cdot \\ c_1 \\ c_2 \\ \cdot \\ \cdot \\ \cdot \\ c_2 \\ c_3 \\ \cdot \\ \cdot \\ \cdot \\ c_3 \end{bmatrix} \quad (16.23)$$

Let us now look at the form of the $\{\nabla C_{ij}\}$ at $\boldsymbol{\alpha}_0$. We define C_{ij}^{kl} to be the determinant of $\boldsymbol{\alpha}$ after removing the ith and kth rows and the jth and lth columns and multiplying by $(-1)^{i+j+k+l}$. We also define $\overset{mn}{C_{ij}^{kl}}$ to be the determinant of $\boldsymbol{\alpha}$ after removing the ith, kth, and mth rows, and the jth, lth, and nth columns and multiplying by $(-1)^{i+j+k+l+m+n}$. From Lemma 14.6, the following relationships for any α can be readily verified:

$$\frac{\partial C_{ij}}{\partial \lambda_{ij}} = C_{ii}{}^{jj} \qquad i \neq j \tag{16.24}$$

$$\frac{\partial C_{jj}}{\partial \lambda_{ij}} = 0 \qquad i \neq j \tag{16.25}$$

$$\frac{\partial C_{ij}}{\partial \lambda_{jk}} = C_{ij}{}^{jk} = C_{jj}{}^{ik} \qquad i \neq j;\, j \neq k;\, k \neq i \tag{16.26}$$

$$\frac{\partial C_{jj}}{\partial \lambda_{kl}} = 2C_{jj}{}^{ki} \qquad k \neq l;\, l \neq j;\, j \neq k \tag{16.27}$$

and

$$\frac{\partial C_{ij}}{\partial \lambda_{kl}} = \sum_{\substack{m=1\\ m\neq i\\ m\neq k}}^{M} C_{ij}{}^{ml}_{kk} \text{ all indices different} \tag{16.28}$$

We shall use the fact that the N-by-N matrix

$$\alpha_\rho = \begin{bmatrix} 1 & & \rho \\ & \ddots & \\ \rho & & 1 \end{bmatrix} \tag{16.29}$$

has diagonal cofactors

$$C_{ii}(\rho) = [1 + (N-2)\rho](1-\rho)^{N-2} \qquad i = 1, \ldots, N \tag{16.30}$$

and off-diagonal cofactors

$$C_{ij}(\rho) = -\rho(1-\rho)^{N-2} \qquad i \neq j \tag{16.31}$$

On applying (16.24) to (16.31) to α_0 we can readily verify the following relationships, to which we assign the symbols $b_0, b_1, \ldots b_6$:

$$C_{ii}{}^{jj}\Big|_{\alpha_0} = \begin{cases} 0 & \text{if, } i, j \leq M_1 \text{ or } i, j > M_1 \\ C_{iiM_1}C_{jjM_2} = b_0 \neq 0 & \text{if } i \leq M_1;\, j > M_1 \end{cases} \tag{16.32}$$

where C_{iiM_1} is the diagonal cofactor of $\alpha_{R_{M_1}}$ and C_{iiM_2} is the diagonal cofactor of $\alpha_{R_{M_2}}$.

$$C_{jj}{}^{kl} = \begin{cases} C_{jjM_1}C_{klM_2} = b_1 & \text{if } jj \in R_1;\, kl \in R_2 \\ C_{jjM_2}C_{klM_1} = b_2 & \text{if } jj \in R_2;\, kl \in R_1 \\ 0 & \text{if } jj \in R_1 \text{ or } R_3;\, kl \in R_2 \end{cases} \tag{16.33}$$

if all indices are less than or equal to M_1 or greater than M_1, that is, if all indices are in $\mathbf{R}_1$ or all indices are in $\mathbf{R}_2$.

$$\frac{\partial C_{ij}}{\partial \lambda_{kl}} = \begin{cases} 2C_{ijM_2}C_{klM_1} = b_3 \neq 0 & \text{if } kl \in \mathbf{R}_1;\ ij \in \mathbf{R}_3 \\ C_{ijM_1}M_2C^{ml}_{kkM_2} = b_4 \neq 0 & \text{if } kl \in \mathbf{R}_3;\ ij \in \mathbf{R}_1 \\ C^{il}_{kkM_1}[C_{jjM_2} + (M_2 - 1)C_{mjM_2}] = b_5 & \\ & \text{if } kl \in \mathbf{R}_1;\ ij \in \mathbf{R}_2 \\ & \text{or } kl \in \mathbf{R}_2;\ ij \in \mathbf{R}_1 \\ C^{il}_{kkM_2}[C_{jjM_1} + (M_1 - 1)C_{mjM_1}] = b_6 \neq 0 & \\ & \text{if } kl \in \mathbf{R}_3;\ ij \in \mathbf{R}_2 \\ & \text{or } kl \in \mathbf{R}_2;\ ij \in \mathbf{R}_3 \end{cases} \tag{16.34}$$

In Fig. 16.2 enough of the normal vectors are described to form $\nabla J(\lambda;\boldsymbol{\alpha}_0)$. The $\mathbf{R}_1$ portion of $\nabla J(\lambda;\boldsymbol{\alpha}_0)$ comes from ∇C_{ii}, where ii is in $\mathbf{R}_3$; and the $\mathbf{R}_3$ part of $\nabla J(\lambda;\boldsymbol{\alpha}_0)$ can come from ∇C_{ii} for ii in $\mathbf{R}_1$.

	$\nabla\phi$	∇C_{ii} ii in $\mathbf{R}_1$	∇C_{ii} ii in $\mathbf{R}_3$	∇C_{ij} ij in $\mathbf{R}_2$	∇C_{ij} ij in $\mathbf{R}_1$
$\mathbf{R}_1$	c_1 ⋮ c_1	0 ⋮ 0	$2b_2$ ⋮ $2b_2$	b_5 ⋮ b_5	0 ⋮ 0
$\mathbf{R}_2$	c_2 ⋮ c_2	0 ⋮ 0	0 ⋮ 0	0 ⋮ 0	b_5 ⋮ b_5, 0 ⋮ 0 (ith row), b_5 ⋮ b_5
$\mathbf{R}_3$	c_3 ⋮ c_3	$2b_1$ ⋮ $2b_1$	0 ⋮ 0	b_6 ⋮ b_6	b_6 ⋮ b_6

Fig. 16.2 Normal vectors needed to form $\nabla J(\lambda;\boldsymbol{\alpha}_0)$.

Summing ∇C_{ij} for ij in $\mathbf{R}_1$ over one row in $\mathbf{R}_1$ results in

$$\sum_{i=1}^{M_1} \nabla C_{ij} \bigg|_{ij \in \mathbf{R}_1} = \begin{bmatrix} 0 \\ \cdot \\ \cdot \\ \cdot \\ 0 \\ \hdashline (M_1 - 1)b_5 \\ \cdot \\ \cdot \\ \cdot \\ (M_1 - 1)b_5 \\ \hdashline M_1 b_6 \\ \cdot \\ \cdot \\ \cdot \\ M_1 b_6 \end{bmatrix} \tag{16.35}$$

By subtracting from this a scalar times ∇C_{ii} (ii in $\mathbf{R}_1$), we have a vector of the form

$$\begin{bmatrix} 0 \\ \cdot \\ \cdot \\ \cdot \\ 0 \\ \hdashline b \\ \cdot \\ \cdot \\ \cdot \\ b \\ \hdashline 0 \\ \cdot \\ \cdot \\ \cdot \\ 0 \end{bmatrix} \tag{16.36}$$

from which the $\mathbf{R}_2$ part of $\nabla J(\lambda;\boldsymbol{\alpha}_0)$ can be written. This shows that $\nabla J(\lambda;\boldsymbol{\alpha}_0)$ can be expressed as a linear combination of the normal vectors at $\boldsymbol{\alpha}_0$, and therefore proves that these $\boldsymbol{\alpha}_0$ do satisfy the necessary first-order conditions to be local minima of $J(\lambda;\boldsymbol{\alpha})$ in $\mathcal{A}_2$.

We have established that these $\boldsymbol{\alpha}_0$ are local extrema in the class of admissible $\boldsymbol{\alpha}$ for which $D \leq M - 2$. Next we must consider sufficient (that is, second-order) conditions in order to conclude that these local extrema are indeed local minima of $J(\lambda;\boldsymbol{\alpha})$.

16.2 SUFFICIENT (SECOND-ORDER) CONDITIONS

We now use the results obtained from the second-order variations about the regular simplex to determine second-order variations for the $\boldsymbol{\alpha}_0$ of Sec. 16.1.

Note that $\boldsymbol{\alpha}_0$ is in the plane defined by

$$\sum_{i=1}^{M} \sum_{j=1}^{M} \lambda_{ij} = 0 \tag{16.37}$$

and further, that at $\boldsymbol{\alpha}_0$ the only admissible directions are in this tangent plane and in the boundary of $\mathcal{A}$, where by "admissible" we now mean directions or paths which restrict $\boldsymbol{\alpha}$ to $\mathcal{A}_2$. This is the case, since directions toward the interior of $\mathcal{A}$ from $\boldsymbol{\alpha}_0$ have dimensionality $D = M$.

Let $\boldsymbol{\alpha}'$ be any other admissible signal set in the neighborhood of $\boldsymbol{\alpha}_0$, for which we write

$$J(\lambda;\boldsymbol{\alpha}') - J(\lambda;\boldsymbol{\alpha}_0) = \sum_{\mathbf{R}_1} (\lambda'_{ij} - \lambda^{\circ}_{ij})c_1 + \sum_{\mathbf{R}_2} (\beta'_{ij} - \beta^{\circ}_{ij})c_2 + \sum_{\mathbf{R}_3} (\gamma'_{ij} - \gamma^{\circ}_{ij})c_3 + \cdots \tag{16.38}$$

where c_1, c_2, and c_3 are given by (16.20), (16.21), and (16.22) and are nonnegative.

For $\boldsymbol{\alpha}_0$ we have that

$$\sum_{\mathbf{R}_1} \lambda^{\circ}_{ij} + \sum_{\mathbf{R}_2} \beta^{\circ}_{ij} + \sum_{\mathbf{R}_3} \gamma^{\circ}_{ij} = -\frac{M}{2} \tag{16.39}$$

$$\sum_{\mathbf{R}_1} \lambda^{\circ}_{ij} = -\frac{M_1}{2} \tag{16.40}$$

$$\sum_{\mathbf{R}_2} \beta^{\circ}_{ij} = 0 \tag{16.41}$$

$$\sum_{\mathbf{R}_3} \gamma^{\circ}_{ij} = -\frac{M_2}{2} \tag{16.42}$$

For $\boldsymbol{\alpha}'$, from Theorem 15.1 we can say

$$\sum_{\mathbf{R}_1} \lambda'_{ij} + \sum_{\mathbf{R}_2} \beta'_{ij} + \sum_{\mathbf{R}_3} \gamma'_{ij} \geq -\frac{M}{2} \tag{16.43}$$

$$\sum_{\mathbf{R}_1} \lambda'_{ij} \geq -\frac{M_1}{2} \tag{16.44}$$

$$\sum_{\mathbf{R}_3} \gamma'_{ij} \geq -\frac{M_2}{2} \tag{16.45}$$

If equality exists in (16.44) and (16.45), then

$$\sum_{\mathbf{R}_2} \beta'_{ij} \geq 0 \tag{16.46}$$

from which we have the first-order variations in (16.38) are nonnegative.

It is immediate that there are cases for which there is strict inequality in (16.46). Hence $\boldsymbol{\alpha}_0$ cannot be a local minimum of probability of detection in $\mathcal{A}_2$ and is therefore a local maximum, or possibly a saddle point. To conclude that $\boldsymbol{\alpha}_0$ is indeed a local maximum in $\mathcal{A}_2$ [local minimum of $J(\lambda;\boldsymbol{\alpha}_0)$], we examine the second-order variations when the first-order variations vanish. It is clear that the first-order variations in (16.38) vanish when there is equality in (16.44), (16.45), and (16.46), for which we have the following.

THEOREM 16.2 *The second-order variations in the neighborhood of $\boldsymbol{\alpha}_0$ of $P_d(\lambda;\boldsymbol{\alpha}_0)$ are negative for all signal-to-noise ratios in all admissible directions in $\mathcal{A}_2$ for which the first-order variations vanish and*

$$\sum_{\mathbf{R}_1} \lambda'_{ij} = -\frac{M_1}{2} \qquad \sum_{\mathbf{R}_2} \beta'_{ij} = 0 \qquad \sum_{\mathbf{R}_3} \gamma'_{ij} = -\frac{M_2}{2} \tag{16.47}$$

Proof: As before, we look at the second-order variations of $J(\lambda;\boldsymbol{\alpha}_0)$ and prove that they are positive. Employing Lemma 14.5 and (16.47), and noting that (16.47) implies equality in (16.43), we have

$$\begin{aligned} \sum_i \beta'_{ij} &= 0 \qquad j = 1, \ldots, M_2 \\ \sum_j \beta'_{ij} &= 0 \qquad i = 1, \ldots, M_1 \end{aligned} \tag{16.48}$$

We now make the following definitions for

$$\begin{aligned} \Phi(x;\boldsymbol{\alpha}_0) &= \underbrace{\int_{-\infty}^{x} \cdots \int}_{M_1\text{-fold}} G(\boldsymbol{\xi};\mathbf{0};\boldsymbol{\alpha}_{R_{M_1}})\, d\boldsymbol{\xi} \underbrace{\int_{-\infty}^{x} \cdots \int}_{M_2\text{-fold}} G(\mathbf{n};\mathbf{0};\boldsymbol{\alpha}_{R_{M_2}})\, d\mathbf{n} \\ &= \Phi(x;\boldsymbol{\alpha}_{R_{M_1}})\Phi(x;\boldsymbol{\alpha}_{R_{M_2}}) \end{aligned} \tag{16.49}$$

We define

$$p_1 \triangleq \frac{\partial^2\Phi(x;\boldsymbol{\alpha}_0)}{\partial\lambda_{ij}^2} = \frac{\partial^2\Phi(x;\boldsymbol{\alpha}_0)}{\partial\lambda_{12}^2}$$
$$= \underbrace{\int_{-\infty}^{x}\cdots\int}_{M_1\text{-fold}} \frac{\partial^4}{\partial\xi_1^2\,\partial\xi_2^2} G(\boldsymbol{\xi};\mathbf{0};\boldsymbol{\alpha}_{R_{M_1}})\,d\boldsymbol{\xi} \underbrace{\int_{-\infty}^{x}\cdots\int}_{M_2\text{-fold}} G(\mathbf{n};\mathbf{0};\boldsymbol{\alpha}_{R_{M_2}})\,d\mathbf{n} \tag{16.50}$$

$$p_2 \triangleq \frac{\partial^2\Phi(x;\boldsymbol{\alpha}_0)}{\partial\gamma_{ij}^2} = \frac{\partial^2\Phi(x;\boldsymbol{\alpha}_0)}{\partial\gamma_{12}^2}$$
$$= \underbrace{\int_{-\infty}^{x}\cdots\int}_{M_1\text{-fold}} G(\boldsymbol{\xi};\mathbf{0};\boldsymbol{\alpha}_{R_{M_1}})\,d\boldsymbol{\xi} \underbrace{\int_{-\infty}^{x}\cdots\int}_{M_2\text{-fold}} \frac{\partial^4}{\partial\eta_1^2\,\partial\eta_2^2} G(\mathbf{n};\mathbf{0};\boldsymbol{\alpha}_{R_{M_2}})\,d\mathbf{n} \tag{16.51}$$

$$q_1 \triangleq \frac{\partial^2\Phi(x;\boldsymbol{\alpha}_0)}{\partial\lambda_{12}\,\partial\lambda_{13}} \qquad r_1 \triangleq \frac{\partial^2\Phi(x;\boldsymbol{\alpha}_0)}{\partial\lambda_{12}\,\partial\lambda_{34}} \tag{16.52}$$

$$q_2 \triangleq \frac{\partial^2\Phi(x;\boldsymbol{\alpha}_0)}{\partial\gamma_{12}\,\partial\gamma_{13}} \qquad r_2 \triangleq \frac{\partial^2\Phi(x;\boldsymbol{\alpha}_0)}{\partial\gamma_{12}\,\partial\gamma_{34}} \tag{16.53}$$

$$s_1 \triangleq \frac{\partial^2\Phi(x;\boldsymbol{\alpha}_0)}{\partial\lambda_{ij}\,\partial\beta_{ii}} = \frac{\partial^2\Phi(x;\boldsymbol{\alpha}_0)}{\partial\lambda_{12}\,\partial\beta_{11}}$$
$$= \underbrace{\int_{-\infty}^{x}\cdots\int}_{M_1\text{-fold}} \frac{\partial^3}{\partial\xi_1^2\,\partial\xi_2} G(\boldsymbol{\xi};\mathbf{0};\boldsymbol{\alpha}_{R_{M_1}})\,d\boldsymbol{\xi}$$
$$\underbrace{\int_{-\infty}^{x}\cdots\int}_{(M_2-1)\text{-fold}} G(x,\eta_2,\ldots,\eta_{M_2};\mathbf{0};\boldsymbol{\alpha}_{R_{M_2}})\,d\eta_2\cdots d\eta_M \tag{16.54}$$

$$s_2 \triangleq \frac{\partial^2\Phi(x;\boldsymbol{\alpha}_0)}{\partial\gamma_{12}\,\partial\beta_{11}} \tag{16.55}$$

$$t_1 \triangleq \frac{\partial^2\Phi(x;\boldsymbol{\alpha}_0)}{\partial\lambda_{12}\,\partial\beta_{31}} \qquad t_2 \triangleq \frac{\partial^2\Phi(x;\boldsymbol{\alpha}_0)}{\partial\gamma_{12}\,\partial\beta_{13}} \tag{16.56}$$

$$u \triangleq \frac{\partial^2\Phi(x;\boldsymbol{\alpha}_0)}{\partial\lambda_{12}\,\partial\gamma_{12}} \qquad v \triangleq \frac{\partial^2\Phi(x;\boldsymbol{\alpha}_0)}{\partial\beta_{ij}^2} \tag{16.57}$$

and finally,

$$w_1 \triangleq \frac{\partial^2\Phi(x;\boldsymbol{\alpha}_0)}{\partial\beta_{11}\,\partial\beta_{21}} \qquad w_2 = \frac{\partial^2\Phi(x;\boldsymbol{\alpha}_0)}{\partial\beta_{11}\,\partial\beta_{12}} \tag{16.58}$$

With a_{ijkl} defined as in (15.58), where $\boldsymbol{\alpha}_R$ is replaced by $\boldsymbol{\alpha}_0$, the second-order variation of $J(\lambda;\boldsymbol{\alpha}') - J(\lambda;\boldsymbol{\alpha}_0)$ can be written as

$$\Big[\sum_{\mathbf{R}_1} (\lambda'_{ij} - \lambda^\circ_{ij}) + \sum_{\mathbf{R}_2} (\beta'_{ij} - \beta^\circ_{ij}) + \sum_{\mathbf{R}_3} (\gamma'_{ij} - \gamma^\circ_{ij})\Big] \Big[\sum_{\mathbf{R}_1} (\lambda'_{kl} - \lambda^\circ_{kl}) + \sum_{\mathbf{R}_2} (\beta'_{kl} - \beta^\circ_{kl}) + \sum_{\mathbf{R}_3} (\gamma'_{kl} - \gamma^\circ_{kl})\Big] a_{ijkl} = S_1 + S_2 + S_3 + 2\,(S_4 + S_5 + S_6) \qquad (16.59)$$

where

$$S_1 = \Big[\sum_{\mathbf{R}_1} (\lambda'_{ij} - \lambda^\circ_{ij})\Big] \Big[\sum_{\mathbf{R}_1} (\lambda'_{kl} - \lambda^\circ_{kl})\Big] a_{ijkl} \qquad (16.60)$$

$$S_2 = \Big[\sum_{\mathbf{R}_2} (\beta'_{ij} - \beta^\circ_{ij})\Big] \Big[\sum_{\mathbf{R}_2} (\beta'_{kl} - \beta^\circ_{kl})\Big] a_{ijkl} \qquad (16.61)$$

$$S_3 = \Big[\sum_{\mathbf{R}_3} (\gamma'_{ij} - \gamma^\circ_{ij})\Big] \Big[\sum_{\mathbf{R}_3} (\gamma'_{kl} - \gamma^\circ_{kl})\Big] a_{ijkl} \qquad (16.62)$$

$$S_4 = \Big[\sum_{\mathbf{R}_1} (\lambda'_{ij} - \lambda^\circ_{ij})\Big] \Big[\sum_{\mathbf{R}_2} (\beta'_{kl} - \beta^\circ_{kl})\Big] a_{ijkl} \qquad (16.63)$$

$$S_5 = \Big[\sum_{\mathbf{R}_1} (\lambda'_{ij} - \lambda^\circ_{ij})\Big] \Big[\sum_{\mathbf{R}_3} (\gamma'_{kl} - \gamma^\circ_{kl})\Big] a_{ijkl} \qquad (16.64)$$

$$S_6 = \Big[\sum_{\mathbf{R}_2} (\beta'_{ij} - \beta^\circ_{ij})\Big] \Big[\sum_{\mathbf{R}_3} (\gamma'_{kl} - \gamma^\circ_{kl})\Big] a_{ijkl} \qquad (16.65)$$

Now

$$S_5 = u \sum_{\mathbf{R}_1} (\lambda'_{ij} - \lambda^\circ_{ij}) \sum_{\mathbf{R}_3} (\gamma'_{kl} - \gamma^\circ_{kl}) \qquad (16.66)$$

and if we apply (16.40), (16.42), and (16.47), it is immediate that $S_5 = 0$. For S_4 we have

$$S_4 = \sum_{\mathbf{R}_1} (\lambda'_{ij} - \lambda^\circ_{ij}) \Big\{ s_1 \sum_{l=1}^{M_2} [(\beta'_{il} - \beta^\circ_{il}) + (\beta'_{jl} - \beta^\circ_{jl})] + t_1 \sum_{\substack{k=1 \\ k\neq i \\ k\neq j}}^{M_1} \sum_{l=1}^{M_2} (\beta'_{kl} - \beta^\circ_{kl}) \Big\} \qquad (16.67)$$

which, when (16.48) is applied, is seen to vanish. Similarly, S_6 is also shown to vanish.

From the results of the previous chapter and the hypotheses of this theorem, we have

$$S_1 = \Big[\sum_{\mathbf{R}_1} (\lambda_{ij} - \lambda_{ij})^2\Big] (p_1 - 2q_1 + r_1) \qquad (16.68)$$

From (15.106),

$$\int_0^\infty \exp(\lambda x) S_1\, dx = \sum_{\mathbf{R}_1} (\lambda'_{ij} - \lambda^\circ_{ij})^2 \int_0^\infty \exp(\lambda x)\Phi(x;\boldsymbol{\alpha}_{R_{M_2}})$$
$$\left(\left\{\left[\left(\frac{M_1-1}{M_1-2}\right)^2 x^2 - \frac{2(M_1-1)}{M_1(M_1-2)}\right] F_2(x) + \frac{M_1-1}{M_1(M_1-2)} C(x)\right\}\right.$$
$$\left. G_2(x) + \frac{2(M_1-1)}{(M_1-2)(M_1-3)} x G_3(x) F_3(x)\right) dx \quad (16.69)$$

which we must show is positive. The integration over the last two terms is clearly positive, so it is sufficient to verify that

$$\int_0^\infty \exp(\lambda x)\Phi(x;\boldsymbol{\alpha}_{R_{M_2}}) \left(\frac{M_1-1}{M_1-2} x^2 - \frac{2}{M_1}\right) G_2(x) F_2(x)\, dx > 0 \quad (16.70)$$

As in Theorem 15.5, we integrate

$$\int_0^\infty \exp(\lambda x)\Phi(x;\boldsymbol{\alpha}_{R_{M_2}}) G_2(x) F_2(x)\, dx$$

by parts; when substituted into (16.70), results in

$$\int_0^\infty \exp(\lambda x)\Phi(x;\boldsymbol{\alpha}_{R_{M_2}}) \left(\frac{M_1-1}{M_1-2} x^2 - \frac{2}{M_1}\right) G_2(x) F_2(x)\, dx$$
$$= \left(\frac{1}{2} - \frac{2}{M_1}\right) \int_0^\infty \exp(\lambda x)\Phi(x;\boldsymbol{\alpha}_{R_{M_2}}) G_2(x) F_2(x)\, dx$$
$$+ \frac{1}{2}\int_0^\infty \exp(\lambda x) G_2(x) \left[\lambda x \Phi(x;\boldsymbol{\alpha}_{R_{M_2}}) F_2(x) + p_{M_2}(x;\boldsymbol{\alpha}_{R_{M_2}}) F_2(x)\right.$$
$$\left. + \Phi(x;\boldsymbol{\alpha}_{R_{M_2}}) \frac{dF_2(x)}{dx}\right] dx \quad (16.71)$$

which is seen to be positive for $M_1 > 3$.

Similarly, the integral over S_3 is shown to be positive.

Finally, by again making use of (16.48), we have

$$S_2 = \sum_{\mathbf{R}_2} (\beta'_{ij} - \beta^\circ_{ij})^2 (v - w_1 - w_2 + u) \quad (16.72)$$

Now

$$u = \underbrace{\int_{-\infty}^{x} \cdots \int}_{(M_1-2)\text{-fold}} G(x, x, \xi_3, \ldots, \xi_{M_1}; \mathbf{0}; \boldsymbol{\alpha}_{R_{M_1}})\, d\xi_3 \cdots d\xi_{M_1}$$
$$\underbrace{\int_{-\infty}^{x} \cdots \int}_{(M_2-2)\text{-fold}} G(x, x, \eta_3, \ldots, \eta_{M_2}; \mathbf{0}; \boldsymbol{\alpha}_{R_{M_2}})\, d\eta_3 \cdots d\eta_{M_2}$$
$$= G_2(x;\boldsymbol{\alpha}_{R_{M_1}}) F_2(x;\boldsymbol{\alpha}_{R_{M_1}}) G_2(x;\boldsymbol{\alpha}_{R_{M_1}}) F_2(x;\boldsymbol{\alpha}_{R_{M_1}}) \quad (16.73)$$

where the notation introduced here is apparent, and

$$v = \underbrace{\int_{-\infty}^{x} \cdots \int}_{M_1\text{-fold}} \frac{\partial^2}{\partial \xi_1{}^2} G(\boldsymbol{\xi};\mathbf{0};\boldsymbol{\alpha}_{R_{M_1}})\, d\boldsymbol{\xi} \underbrace{\int_{-\infty}^{x} \cdots \int}_{M_2\text{-fold}} \frac{\partial^2}{\partial \eta_1{}^2} G(\mathbf{n};\mathbf{0};\boldsymbol{\alpha}_{R_{M_2}})\, d\mathbf{n} \tag{16.74}$$

Following the same technique used in partially evaluating q in Lemma 15.6, we introduce

$$\boldsymbol{\alpha}_{\rho_1} \triangleq \begin{bmatrix} 1 & & \rho_1 \\ & \ddots & \\ \rho_1 & & 1 \end{bmatrix} \quad \text{and} \quad \boldsymbol{\alpha}_{\rho_2} \triangleq \begin{bmatrix} 1 & & \rho_2 \\ & \ddots & \\ \rho_2 & & 1 \end{bmatrix}$$

as M_1-by-M_1 and M_2-by-M_2 matrices, respectively. Then we perform the necessary algebraic manipulations and finally take limits as $\rho_1 \rightarrow -1/(M_1 - 1)$ and $\rho_2 \rightarrow -1/(M_2 - 1)$. In this way v can be shown to be (after limits are taken)

$$\begin{aligned} v = {} & x^2 G_1(x;\boldsymbol{\alpha}_{R_{M_1}})F_1(x;\boldsymbol{\alpha}_{R_{M_1}})G_1(x;\boldsymbol{\alpha}_{R_{M_2}})F_1(x;\boldsymbol{\alpha}_{R_{M_2}}) \\ & - x[G_2(x;\boldsymbol{\alpha}_{R_{M_1}})F_2(x;\boldsymbol{\alpha}_{R_{M_1}})G_1(x;\boldsymbol{\alpha}_{R_{M_2}})F_1(x;\boldsymbol{\alpha}_{R_{M_2}}) \\ & + G_2(x;\boldsymbol{\alpha}_{R_{M_2}})F_2(x;\boldsymbol{\alpha}_{R_{M_2}})G_1(x;\boldsymbol{\alpha}_{R_{M_1}})F_1(x;\boldsymbol{\alpha}_{R_{M_1}})] \\ & + G_2(x;\boldsymbol{\alpha}_{R_{M_1}})F_2(x;\boldsymbol{\alpha}_{R_{M_1}})G_2(x;\boldsymbol{\alpha}_{R_{M_2}})F_2(x;\boldsymbol{\alpha}_{R_{M_2}}) \end{aligned} \tag{16.75}$$

Then w_1 and w_2 can be similarly evaluated and shown to be (again after limits are taken)

$$\begin{aligned} w_1 = {} & -xG_2(x;\boldsymbol{\alpha}_{R_{M_1}})F_2(x;\boldsymbol{\alpha}_{R_{M_1}})G_1(x;\boldsymbol{\alpha}_{R_{M_2}})F_1(x;\boldsymbol{\alpha}_{R_{M_2}}) \\ & + G_2(x;\boldsymbol{\alpha}_{R_{M_1}})F_2(x;\boldsymbol{\alpha}_{R_{M_1}})G_2(x;\boldsymbol{\alpha}_{R_{M_2}})F_2(x;\boldsymbol{\alpha}_{R_{M_2}}) \end{aligned}$$

and

$$\begin{aligned} w_2 = {} & -xG_2(x;\boldsymbol{\alpha}_{R_{M_2}})F_2(x;\boldsymbol{\alpha}_{R_{M_2}})G_1(x;\boldsymbol{\alpha}_{R_{M_1}})F_1(x;\boldsymbol{\alpha}_{R_{M_1}}) \\ & + G_2(x;\boldsymbol{\alpha}_{R_{M_1}})F_2(x;\boldsymbol{\alpha}_{R_{M_1}})G_2(x;\boldsymbol{\alpha}_{R_{M_2}})F_2(x;\boldsymbol{\alpha}_{R_{M_2}}) \end{aligned}$$

After substitution it is immediate that

$$v - w_1 - w_2 + u = x^2 G_1(x;\boldsymbol{\alpha}_{R_{M_1}})F_1(x;\boldsymbol{\alpha}_{R_{M_1}})G_1(x;\boldsymbol{\alpha}_{R_{M_2}})F_1(x;\boldsymbol{\alpha}_{R_{M_2}})$$

and finally that

$$\int_0^\infty \exp(\lambda x)(v - w_1 - w_2 + u)\, dx > 0 \tag{16.76}$$

Hence $J(\lambda;\boldsymbol{\alpha}_0)$ is a local minimum in $\mathcal{A}_2$, or, equivalently, $P_d(\lambda;\boldsymbol{\alpha}_0)$ is a local maximum in $\mathcal{A}_2$.

16.3 CHOOSING THE LARGEST OF SEVERAL LOCAL MAXIMA

Several signal sets have been shown to be local maxima in the class of admissible $\boldsymbol{\alpha}$ for which $D \leq M - 2$ (that is, $\mathfrak{A}_2$). For large signal-to-noise ratios at any rate, the largest of these can be determined.

THEOREM 16.3 *Of the local maxima in $\mathfrak{A}_2$, at large signal-to-noise ratios the one with the largest probability of detection is the one for which $M_1 = 2$.*

Remark: Since the signal design is being performed on the assumption of an equilikely a priori distribution, M_1 and M_2 are symmetric variables; that is, M_1 and M_2 can be interchanged with no change in the probability of error of the system. Thus it is sufficient to order those local maxima for which $M_1 \leq M_2$. This will then order all the local maxima (of probability of detection) in $\mathfrak{A}_2$ at large signal-to-noise ratios.

In fact, for all the signal design discussed thus far we have assumed an equilikely a priori distribution. In the case of the regular simplex, however, the resultant probability of detection is independent of the a priori distribution. That is, even though the optimization was carried out for an equilikely a priori distribution, the probability of detection for the regular simplex does not depend on the a priori distribution. This is not the case, however, when two regular simplices which do not have the same number of signals in each are placed orthogonal to one another.

The independence of a priori–distribution property occurs when the signal structure is uniform, that is, when the array of signals as viewed from a given vector is identical for all signals. This is clearly the case for the regular simplex. For two regular simplices placed orthogonal to one another, a signal chosen from one of them is then uniform with respect to the other vectors in that simplex. However, it will not be uniform with respect to the signal vectors of the other regular simplex unless both simplices have the same number of signals. Equivalently stated, the region in which a signal is decided upon when the received vector falls into it has the same shape for all signals.

When the uniform property is present, then,

$$
\begin{aligned}
P_d &= \sum_{i=1}^{M} \Pr(\mathbf{S}_i \text{ was transmitted}) \Pr(\mathbf{S}_i \text{ was decided } | \mathbf{S}_i \text{ was transmitted}) \\
&= \Pr(\mathbf{S}_i \text{ was decided } | \mathbf{S}_i \text{ was transmitted}) \sum_{i=1}^{M} \Pr(\mathbf{S}_i \text{ was transmitted}) \\
&= \Pr(\mathbf{S}_i \text{ was decided } | \mathbf{S}_i \text{ was transmitted})
\end{aligned}
$$

and is independent of i.

Proof: From Theorem 14.6, for any $\boldsymbol{\alpha}$ and large λ

$$P_e(\lambda;\boldsymbol{\alpha}) \approx \frac{1}{2M} \sum_{i>j}\sum \frac{1}{\sqrt{2\pi}\,\gamma_{ij}\lambda} \exp -\tfrac{1}{2}\gamma_{ij}^2\lambda^2 \tag{14.79}$$

where

$$\gamma_{ij} = \sqrt{\frac{1-\lambda_{ij}}{1+\lambda_{ij}}} \tag{14.80}$$

Applying this result to the $\boldsymbol{\alpha}_0$ in this chapter, we obtain

$$\begin{aligned} P_e(\lambda,\boldsymbol{\alpha}_0,M_1) \approx \frac{1}{\sqrt{2\pi}\,2M\lambda} \Bigg[& M_1(M-M_1) \exp -\tfrac{1}{2}\lambda^2 \\ & + \frac{(M_1-1)\sqrt{M_1(M_1-2)}}{2} \exp\left(-\tfrac{1}{2}\lambda^2 \frac{M_1}{M_1-2}\right) \\ & + \frac{(M-M_1-1)\sqrt{(M-M_1)(M-M_1-2)}}{2} \\ & \exp\left(-\tfrac{1}{2}\lambda^2 \frac{M-M_1}{M-M_1-2}\right)\Bigg] \end{aligned} \tag{16.77}$$

where M_2 has been replaced by $M - M_1$ and the notation of P_e is altered to include the parameter M_1, thus completely specifying a given $\boldsymbol{\alpha}_0$. We shall prove that as M_1 decreases the probability of error in (16.77) decreases. That is, for $2 < M_1 \leq M/2$

$$P_e(\lambda;\boldsymbol{\alpha}_0;M_1) - P_e(\lambda;\boldsymbol{\alpha}_0;M_1' = M_1 - 1) > 0 \tag{16.78}$$

for large λ.

To see this we note that

$$\begin{aligned} P_e(\lambda;\boldsymbol{\alpha}_0;M_1) \approx & \frac{1}{\sqrt{2\pi}\,2M\lambda} \exp(-\tfrac{1}{2}\lambda^2) \Bigg[M_1(M-M_1) \\ & + \frac{(M_1-1)\sqrt{M_1(M_1-2)}}{2} \exp \frac{-\lambda^2}{M_1-2} \\ & + \frac{(M-M_1-1)\sqrt{(M-M_1)(M-M_1-2)}}{2} \exp \frac{-\lambda^2}{M-M_1-2}\Bigg] \end{aligned} \tag{16.79}$$

and

$$\begin{aligned} P_e(\lambda;\boldsymbol{\alpha}_0;M_1' = M_1 - 1) \approx & \frac{1}{\sqrt{2\pi}\,2M\lambda} \exp(-\tfrac{1}{2}\lambda^2) \Bigg[(M_1-1) \\ & (M-M_1+1) + \frac{(M_1-2)\sqrt{(M_1-1)(M_1-3)}}{2} \exp \frac{-\lambda^2}{M_1-3} \\ & + \frac{(M-M_1)\sqrt{(M-M_1+1)(M-M_1-1)}}{2} \exp \frac{-\lambda^2}{M-M_1-1}\Bigg] \end{aligned} \tag{16.80}$$

For large λ the second and third terms in (16.79) and (16.80) become negligible in comparison with the constant first term. The coefficients are identical and

$$M_1(M - M_1) - (M_1 - 1)(M - M_1 + 1) = M + 1 - 2M_1 > 0 \qquad M_1 \leq \frac{M}{2} \qquad (16.81)$$

which proves (16.78) and hence the theorem.

Therefore choosing M_1 (or M_2) to be its smallest value, namely, 2, minimizes the probability of error for large signal-to-noise ratios. Hence for a given M the optimal signal-vector inner-product matrix is

$$\boldsymbol{\alpha}_{0_{2,M-2}} = \left[\begin{array}{c|c} \boldsymbol{\alpha}_{R_2} & \mathbf{0} \\ \hline \mathbf{0} & \boldsymbol{\alpha}_{R_{M-2}} \end{array}\right] \qquad (16.82)$$

The criterion for choosing the best of the several local maxima could be equivalently stated as choosing the one whose inner-product matrix has the least number of zeros.

16.4 FIVE SIGNAL VECTORS IN THREE DIMENSIONS

As an application of the results of this chapter we shall consider the case of five points in three dimensions. Figure 16.3 is a pictorial diagram of the two systems to be compared; system B, with three points on the equator and one on each pole, is a special case of $\boldsymbol{\alpha}_0$ when $M = 5$, $M_1 = 3$, and $M_2 = 2$ and is therefore a local maximum of $P_d(\lambda;\boldsymbol{\alpha})$.

Consider a small arbitrary perturbation of system A and approximate it by

$$J(\lambda;\boldsymbol{\alpha}(\theta)) \approx J(\lambda;\boldsymbol{\alpha}_A) + \left[\sum_{i>j}\sum \frac{\partial J(\lambda;\boldsymbol{\alpha}_A)}{\partial \lambda_{ij}} \frac{\partial \tilde{\lambda}_{ij}(\theta)}{\partial \theta}\bigg|_{\theta=0}\right]\theta \qquad (16.83)$$

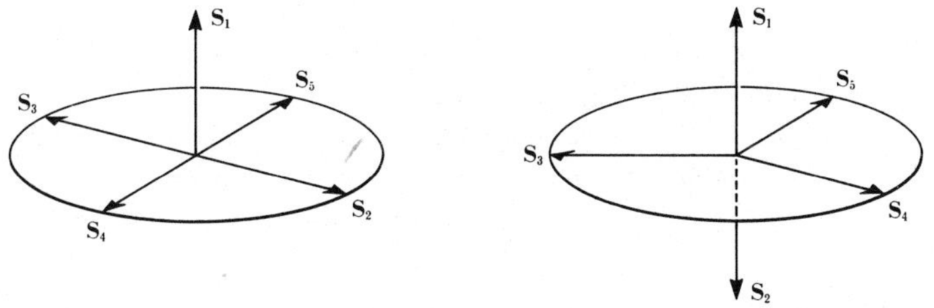

System $A = \alpha_A$ System $B = \alpha_B$

Fig. 16.3 Two systems of five signals in three dimensions.

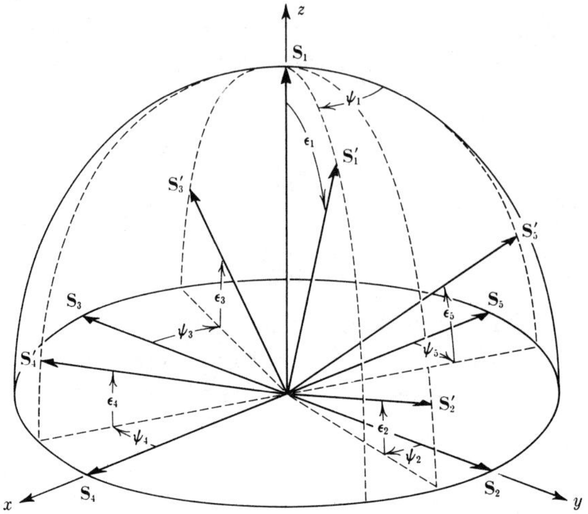

Fig. 16.4 System A: perturbed.

for small θ, where

$$\tilde{\lambda}_{ij}(\theta) = \mathbf{S}_i'(\theta)^T\mathbf{S}_j'(\theta) \tag{16.84}$$

The coordinates of system A, are from Fig. 16.4,

$$\begin{aligned} \mathbf{S}_1 &= (0,0,1) \\ \mathbf{S}_2 &= (0,1,0) \\ \mathbf{S}_3 &= (0,-1,0) \\ \mathbf{S}_4 &= (1,0,0) \\ \mathbf{S}_5 &= (-1,0,0) \end{aligned} \tag{16.85}$$

The coordinates of $\mathbf{S}_i'(\theta)$ can be expressed in terms of trigonometric functions from Fig. 16.4 at $\theta = 1$, and are as follows:

$$\begin{aligned} \mathbf{S}_1'(\theta) &= (\sin\theta\epsilon_1 \sin\theta\psi_1,\ \sin\theta\epsilon_1\cos\theta\psi_1,\ \cos\theta\epsilon_1) \\ \mathbf{S}_2'(\theta) &= (\cos\theta\epsilon_2\sin\theta\psi_2,\ \cos\theta\epsilon_2\cos\theta\psi_2,\ \sin\theta\epsilon_2) \\ \mathbf{S}_3'(\theta) &= (-\cos\theta\epsilon_3\sin\theta\psi_3,\ -\cos\theta\epsilon_3\cos\theta\psi_3,\ \sin\theta\epsilon_3) \\ \mathbf{S}_4'(\theta) &= (\cos\theta\epsilon_4\cos\theta\psi_4,\ -\cos\theta\epsilon_4\sin\theta\psi_4,\ \sin\theta\epsilon_4) \\ \mathbf{S}_5'(\theta) &= (-\cos\theta\epsilon_5\cos\theta\psi_5,\ \cos\theta\epsilon_5\sin\theta\psi_5,\ \sin\theta\epsilon_5) \end{aligned} \tag{16.86}$$

where the ϵ_i and the ψ_i are arbitrary perturbation angles as indicated in Fig. 16.4. By constructing the pertubation in this manner, the restriction of $D \leq 3$ is automatically satisfied.

Now,

$$J(\lambda;\boldsymbol{\alpha}_A) = \int_0^\infty dx \exp(\lambda x) \left\{ \int_{-\infty}^x G(\xi_1)\, d\xi_1 \int \underset{-\infty}{\overset{x}{\cdots}} \int G(\xi_2,\xi_3;\mathbf{0};\boldsymbol{\alpha}_{R_2}) \right.$$
$$\left. d\xi_2\, d\xi_3 \int \underset{-\infty}{\overset{x}{\cdots}} \int G(\xi_4,\xi_5;\mathbf{0};\boldsymbol{\alpha}_{R_2})\, d\xi_4\, d\xi_5 - [\phi(x)]^5 \right\}$$

$$\frac{\partial J(\lambda;\boldsymbol{\alpha}_A)}{\partial \lambda_{12}} = \frac{\partial J(\lambda;\boldsymbol{\alpha}_A)}{\partial \lambda_{13}} = \frac{\partial J(\lambda;\boldsymbol{\alpha}_A)}{\partial \lambda_{14}} = \frac{\partial J(\lambda;\boldsymbol{\alpha}_A)}{\partial \lambda_{15}} = a$$
$$\frac{\partial J(\lambda;\boldsymbol{\alpha}_A)}{\partial \lambda_{24}} = \frac{\partial J(\lambda;\boldsymbol{\alpha}_A)}{\partial \lambda_{25}} = \frac{\partial J(\lambda;\boldsymbol{\alpha}_A)}{\partial \lambda_{34}} = \frac{\partial J(\lambda;\boldsymbol{\alpha}_A)}{\partial \lambda_{35}} = b$$
$$\frac{\partial J(\lambda;\boldsymbol{\alpha}_A)}{\partial \lambda_{23}} = \frac{\partial J(\lambda;\boldsymbol{\alpha}_A)}{\partial \lambda_{45}} = 0$$

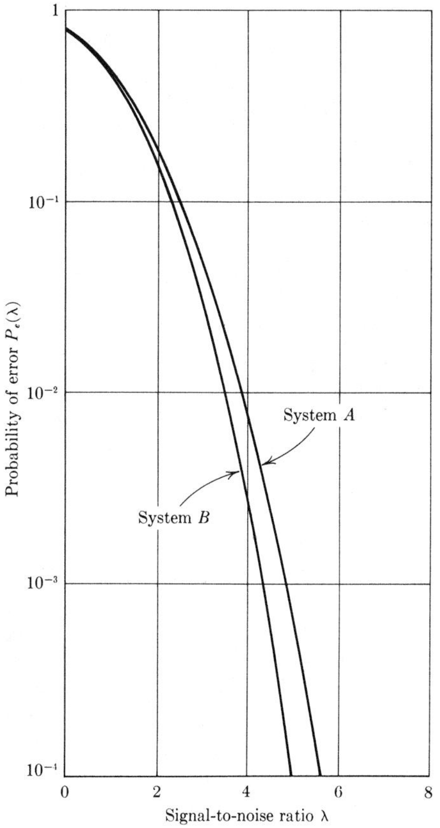

Fig. 16.5 Performance curves for system A and system B.

Substitution yields

$$J(\lambda;\boldsymbol{\alpha}(\theta)) \approx J(\lambda;\boldsymbol{\alpha}_A) + \theta a \sum_{j=2}^{5} \frac{\partial \tilde{\lambda}_{1j}(\theta)}{\partial \theta}\bigg|_{\theta=0}$$

$$+ b\theta \left[\frac{\partial \tilde{\lambda}_{24}(\theta)}{\partial \theta}\bigg|_{\theta=0} + \frac{\partial \tilde{\lambda}_{25}(\theta)}{\partial \theta}\bigg|_{\theta=0} + \frac{\partial \tilde{\lambda}_{34}(\theta)}{\partial \theta}\bigg|_{\theta=0} + \frac{\partial \tilde{\lambda}_{25}(\theta)}{\partial \theta}\bigg|_{\theta=0} \right]$$

which can be shown to be

$$J(\lambda;\boldsymbol{\alpha}(\theta)) \approx J(\lambda;\boldsymbol{\alpha}_A) + a(\epsilon_2 + \epsilon_3 + \epsilon_4 + \epsilon_5)\theta$$

Since the ϵ_i are arbitrary, they can be chosen to either increase or decrease $J(\lambda;\boldsymbol{\alpha}(\theta))$ in the neighborhood of $\theta = 0$. Therefore, not only is system A (consisting of four points on the equator and one on a pole) not a local maximum, but it is also not even an extremum. Figure 16.5 is a plot of the probability of error for these two systems, indicating the preference of system B to system A at all signal-to-noise ratios. Further, it should be pointed out that the minimum distance between the different vectors is the same for both these systems, thus proving that the relationship between maximizing the probability of detection and maximizing the minimum distance which exists when there is no dimensionality restriction does not exist when there are restrictions on the dimensionality more severe than $D \leq M - 1$.

REFERENCES

16.1 Weber, C. L.: "New Results in Optimal Signal Selection for Coherent Channels," doctoral dissertation, University of California, Los Angeles, Department of Engineering, June, 1964.

16.2 Weber, C. L.: Optimal Signal Selection for Coherent Channels, *Intern. Conf. Microwaves, Circuit Theory, Inform. Theory,* Tokyo, September, 1964.

16.3 Weber, C. L.: New Solutions to the Signal Design Problem for Coherent Channels, *IEEE Trans. Inform. Theory,* vol. IT-12, no. 2, April, 1966.

17

Optimality for Coherent Systems When the Dimensionality Is Restricted to $D \leq M - K$, Where $K \leq M/2$

We have obtained solutions to the signal-design problem for the following cases: (1) where the allowed bandwidth, and therefore the allowed dimensionality, is unrestricted, in which case the solution is the regular simplex signal structure (requiring $M - 1$ dimensions); (2) where the dimensionality is restricted to be such that it is less than that required for the regular simplex, $D \leq M - 2$, in which case the solution consists of two lower-dimensional regular simplices placed orthogonal to one another, one simplex containing two signals and the other $M - 2$; and (3) where the dimensionality is $D = 2$. In this chapter we look for optimal signal structures in the class of admissible signal sets whose dimensionality is restricted to $D \leq M - K$, where K is any given positive integer which is at most $M/2$. We shall prove that the concept of placing lower-dimensional regular simplices mutually orthogonal to one another can be extended and is indeed the solution for these cases. That is, for a given K the solution to the signal-design problem is shown to consist of K lower-dimensional regular simplices placed mutually orthogonal to one another, where the number of signals in each regular simplex is at least 2. More precisely, in the class of all admissible signal structures whose dimensionality is $D \leq M - K$, such signal structures are proved to be local extrema for all signal-to-noise ratios. For these extrema, if we define M_j as the number of signals in the jth regular simplex, we require

$$\sum_{j=1}^{K} M_j = M \tag{17.1}$$

and hence the number of dimensions that the signal structure occupies is

$$\sum_{j=1}^{K} (M_j - 1) = M - K \tag{17.2}$$

thus satisfying the requirement of M signals in at most $M - K$ dimensions. The signal-vector inner-product matrix for such signal structures is given by

$$\alpha_0 \triangleq \left[\begin{array}{c|c|c|c}
\begin{matrix} 1 & & \dfrac{-1}{M_1 - 1} \\ & \ddots & \\ \dfrac{-1}{M_1 - 1} & & 1 \end{matrix} & 0 & 0 & \\
\hline
0 & \begin{matrix} 1 & & \dfrac{-1}{M_2 - 1} \\ & \ddots & \\ \dfrac{-1}{M_2 - 1} & & 1 \end{matrix} & & \\
\hline
0 & 0 & \ddots & 0 \\
\hline
 & & 0 & \begin{matrix} 1 & & \dfrac{-1}{M_K - 1} \\ & \ddots & \\ \dfrac{-1}{M_K - 1} & & 1 \end{matrix}
\end{array}\right] \tag{17.3}$$

Finally, it will be shown that at least for large signal-to-noise ratios, of these local extrema, the one with the smallest probability of error is the one for which

$$\begin{aligned} M_i &= 2 \qquad i = 1, \ldots, K - 1 \\ M_K &= M - 2(K - 1) \end{aligned} \tag{17.4}$$

Since only one of these local extrema satisfies (17.4), the preferred one is in this case unique.

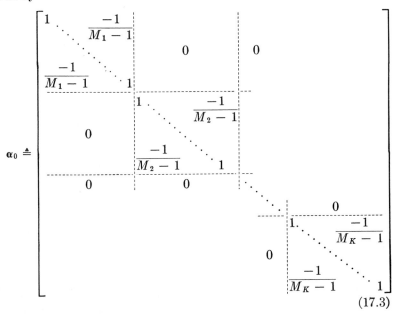

A special case of practical importance in this class of solutions is that where M is even and $D = M/2$, the optimal choice being $M/2$ regular simplices, all mutually orthogonal to one another, and each consisting of two signals in one dimension. This is the well-known *biorthogonal* signal structure. Thus the long-standing conjecture that the biorthogonal signal structure is indeed optimal in the class of M signals restricted to $M/2$ degrees of freedom has rigorous evidence substantiating it.

17.1 NECESSARY (FIRST-ORDER) CONDITIONS

We are looking for admissible $\boldsymbol{\alpha}$ which maximize the probability of detection subject to the restriction that $D \leq M - K$, where $K \leq M/2$. From Chap. 16 we have that minimizing $J(\lambda;\boldsymbol{\alpha})$, defined by (16.4), is equivalent to maximizing $P_d(\lambda;\boldsymbol{\alpha})$. Further, the type of constraint equations that were used in the previous chapter to restrict the dimensionality of $\boldsymbol{\alpha}$ to $D \leq M - 2$,

$$C_{ij} = 0 \qquad i, j = 1, \ . \ . \ . \ , M$$

where C_{ij} is the cofactor of λ_{ij} in $\boldsymbol{\alpha}$, can be extended to restricting the dimensionality to $D \leq M - 3$ by the constraint equations

$$C_{\mathbf{ij}} = 0 \tag{17.5}$$

where $C_{\mathbf{ij}}$ is the cofactor of the cofactor of $\boldsymbol{\alpha}$, that is, the determinant of the matrix which results from removing the i_1th and i_2th rows and the j_1th and j_2th columns of $\boldsymbol{\alpha}$, where we adopt the vector notation

$$\mathbf{i} = \begin{bmatrix} i_1 \\ i_2 \end{bmatrix} \qquad \mathbf{j} = \begin{bmatrix} j_1 \\ j_2 \end{bmatrix} \tag{17.6}$$

to represent the rows and columns, respectively, which have been removed from $\boldsymbol{\alpha}$. To restrict $\boldsymbol{\alpha}$ to be at most of dimensionality $M - 3$, (17.5) must be satisfied for all possible indices $i_1 < i_2$ and $j_1 < j_2$ for i_1 and $j_1 = 1, \ . \ . \ . \ , M - 1$. In general, therefore, in order to restrict the dimensionality (or rank) of $\boldsymbol{\alpha}$ to be at most $M - K$, the determinants of all square submatrices of order $M - (K - 1)$ must vanish; that is, the determinants of all matrices which result from removing $K - 1$ rows and $K - 1$ columns from $\boldsymbol{\alpha}$ must vanish. If we let $\mathbf{i}$ now represent the $(K - 1)$-by-1 vector whose elements (all different) are the numbers of the $K - 1$ rows that have been removed and let $\mathbf{j}$ represent the $(K - 1)$-by-1 vector whose elements (again all different) are the numbers of the $K - 1$ columns that have been removed, then the constraint equations are

$$C_{\mathbf{ij}} = 0 \tag{17.7}$$

for all possible values of $\mathbf{i}$ and $\mathbf{j}$. Using the same method of deriving the necessary conditions for a matrix $\boldsymbol{\alpha}'$ to be a local extremum in $\mathcal{A}_2$ (the class of admissible $\boldsymbol{\alpha}$ for which $D \leq M - 2$), the necessary conditions for an admissible $\boldsymbol{\alpha}'$ to be a local extremum in $\mathcal{A}_K$ (the class of admissible $\boldsymbol{\alpha}$ for which $D \leq M - K$) is

$$\nabla J(\lambda;\boldsymbol{\alpha}') = \sum_{\mathrm{i}} \sum_{\mathrm{j}} \nu_{\mathrm{ij}} \nabla C_{\mathrm{ij}} \bigg|_{\boldsymbol{\alpha}'} \tag{17.8}$$

where the gradient operator ∇ is as indicated in (16.12). This condition can also be attained by using the Lagrange variational technique and forming the lagrangian functional

$$L(\lambda;\boldsymbol{\alpha}) = J(\lambda;\boldsymbol{\alpha}) - \sum_{\mathrm{i}} \sum_{\mathrm{j}} \nu_{\mathrm{ij}} C_{\mathrm{ij}} \tag{17.9}$$

where ν_{ij} are the langragian multipliers. If $\boldsymbol{\alpha}'$ is indeed a local extremum in the class of $\boldsymbol{\alpha}$ satisfying the constraint equations in (17.7), then the gradient operator ∇ operating on the lagrangian in (17.9) must vanish at $\boldsymbol{\alpha}'$. This condition is the same as that expressed in (17.8).

THEOREM 17.1 *For a given K which is at most $M/2$ the $\boldsymbol{\alpha}_0$ described by* (17.3) *satisfy the necessary (first-order) conditions given by* (17.8) *to be local extrema for all signal-to-noise ratios in the class $\mathcal{A}_K$ of admissible signal sets whose dimensionality is $D \leq M - K$.*

Proof: Because of the similarity between this proof and that given for Theorem 16.1, and since there is considerably more algebraic detail involved in this proof, we present here only an outline of the essential points. We denote the M_k-by-M_k regular simplex by

$$\boldsymbol{\alpha}_{R_{M_k}} = \begin{bmatrix} 1 & & \dfrac{-1}{M_k - 1} \\ & \ddots & \\ \dfrac{-1}{M_k - 1} & & 1 \end{bmatrix} \tag{17.10}$$

Then the signal sets $\boldsymbol{\alpha}_0$, which we will show are local extrema in $\mathcal{A}_K$, can be written in partitioned form as

$$\boldsymbol{\alpha}_0 = \begin{bmatrix} \boldsymbol{\alpha}_{R_{M_1}} & \mathbf{0} & & \\ \mathbf{0} & \boldsymbol{\alpha}_{R_{M_2}} & & \\ & & \ddots & \\ & & & \boldsymbol{\alpha}_{R_{M_K}} \end{bmatrix} \tag{17.11}$$

where

$$\sum_{k=1}^{K} M_k = M \tag{17.12}$$

Also, $J(\lambda,\boldsymbol{\alpha}_0)$ can be expressed as

$$J(\lambda;\boldsymbol{\alpha}_0) = \int_0^\infty dx \exp(\lambda x)\left[\prod_{k=1}^{K} \underbrace{\int_{-\infty}^{x} \cdots \int}_{M_k\text{-fold}} G(\boldsymbol{\xi};\mathbf{0};\boldsymbol{\alpha}_{R_{M_k}})\, d\boldsymbol{\xi} - [\phi(x)]^M\right] \tag{17.13}$$

We partition an arbitrary $\boldsymbol{\alpha}$ in $\mathcal{Q}_K$ as we did $\boldsymbol{\alpha}_0$ and adopt the following notation in order to identify the signal-vector inner-product matrix:

$$\left[\begin{array}{c|c|c|c} \begin{matrix} 1 & & \lambda_{ij}{}^{11} \\ & \ddots & \\ \lambda_{ji}{}^{11} & & 1 \end{matrix} & \lambda_{ij}{}^{12} & \cdots & \lambda_{ij}{}^{1K} \\ \hline \lambda_{ji}{}^{12} & \begin{matrix} 1 & & \lambda_{ij}{}^{22} \\ & \ddots & \\ \lambda_{ji}{}^{22} & & 1 \end{matrix} & & \vdots \\ \hline & & \ddots & \\ \hline & & & \begin{matrix} 1 & & \lambda_{ij}{}^{KK} \\ & \ddots & \\ \lambda_{ji}{}^{KK} & & 1 \end{matrix} \end{array}\right] \tag{17.14}$$

where the indices i and j in $\lambda_{ij}{}^{kl}$ are from $i = 1, \ldots, M_k$ and $j = 1, \ldots, M_l$. Now, because of the symmetry in $\boldsymbol{\alpha}_0$, and introducing the symbols d_{kl}, we can write

$$\begin{aligned} \frac{\partial J(\lambda;\boldsymbol{\alpha}_0)}{\partial \lambda_{ij}{}^{kl}} &= \frac{\partial J(\lambda;\boldsymbol{\alpha}_0)}{\partial \lambda_{11}{}^{kl}} = d_{kl} \\ &= \int_0^\infty dx \exp \lambda x \\ &\quad \underbrace{\int_{-\infty}^{x} \cdots \int}_{(M_k-1)\text{-fold}} G(x, \xi_2, \ldots, \xi_{M_k}; \mathbf{0}; \boldsymbol{\alpha}_{R_{M_k}})\, d\xi_2 \cdots d\xi_{M_k} \\ &\quad \underbrace{\int_{-\infty}^{x} \cdots \int}_{(M_l-1)\text{-fold}} G(x, \eta_2, \ldots, \eta_{M_l}; \mathbf{0}; \boldsymbol{\alpha}_{R_{M_l}})\, d\eta_2 \cdots d\eta_{M_l} \\ &\quad \prod_{\substack{p=1 \\ p\neq k \\ p \neq l}}^{K} \underbrace{\int_{-\infty}^{x} \cdots \int}_{M_p\text{-fold}} G(\zeta_1, \ldots, \zeta_p; \mathbf{0}; \boldsymbol{\alpha}_{R_{M_p}})\, d\zeta_1 \cdots d\zeta_p \end{aligned} \tag{17.15}$$

for $k \neq l$, $i = 1, \ldots, M_k$, and $j = 1, \ldots, M_l$. Similarily, for

$l = k$ we have that

$$\frac{\partial J(\lambda;\boldsymbol{\alpha}_0)}{\partial \lambda_{ij}{}^{kk}} = \frac{\partial J(\lambda;\boldsymbol{\alpha}_0)}{\partial \lambda_{12}{}^{kk}} = d_{kk}$$

$$= \int_0^\infty dx \exp \lambda x \underbrace{\int_{-\infty}^{x} \cdots \int}_{(M_k-2)\text{-fold}} G(x, x, \xi_3, \ldots, \xi_{M_k}; \mathbf{0}; \boldsymbol{\alpha}_{R_{M_k}})\, d\xi_3 \cdots d\xi_{M_k}$$

$$\prod_{\substack{p=1\\p\neq k}}^{K} \underbrace{\int_{-\infty}^{x} \cdots \int}_{M_p\text{-fold}} G(\eta_1, \ldots, \eta_{M_p}; \mathbf{0}; \boldsymbol{\alpha}_{R_{M_p}})\, d\eta_1 \cdots d\eta_{M_p} \tag{17.16}$$

for $i > j$ and, i and $j = 1, \ldots, M_k$. If we now define the vector $\mathbf{d}_{kl}$, consisting of $M_k M_l$ elements if $k \neq l$ and $M_k(M_k - 1)/2$ elements if $k = l$, to be

$$\mathbf{d}_{kl} = \begin{bmatrix} d_{kl} \\ \cdot \\ \cdot \\ \cdot \\ d_{kl} \end{bmatrix} \tag{17.17}$$

the gradient vector at $\boldsymbol{\alpha}_0$ is then of the form

$$\nabla J(\lambda;\boldsymbol{\alpha}_0) = \begin{bmatrix} \mathbf{d}_{11} \\ \mathbf{d}_{12} \\ \cdot \\ \cdot \\ \cdot \\ \mathbf{d}_{1K} \\ \mathbf{d}_{22} \\ \mathbf{d}_{23} \\ \cdot \\ \cdot \\ \cdot \\ \mathbf{d}_{2K} \\ \mathbf{d}_{33} \\ \cdot \\ \cdot \\ \cdot \\ \mathbf{d}_{K-1,K} \\ \mathbf{d}_{KK} \end{bmatrix} \tag{17.18}$$

To prove the theorem we must demonstrate that there exists a linear combination of ∇C_{ij} which can be added to equal the above gradient vector. That this is indeed the case follows from the same kind of combination of terms as used in the proof for the case $D \leq M - 2$. To form $\mathbf{d}_{11}$, for example, set $\mathbf{i}$ and $\mathbf{j}$ equal to

$$\mathbf{i} = \mathbf{j} = \begin{bmatrix} M_1 + 1 \\ M_1 + M_2 + 1 \\ \cdot \\ \cdot \\ \cdot \\ M - M_K + 1 \end{bmatrix} \triangleq \mathbf{i}_0 \tag{17.19}$$

That is, one row and one column (the same row and column) are removed from $\boldsymbol{\alpha}_{R_{M_2}}, \boldsymbol{\alpha}_{R_{M_3}}, \ldots, \boldsymbol{\alpha}_{R_{M_K}}$. Then

$$\nabla C_{\mathbf{i}_0\mathbf{i}_0} = \begin{bmatrix} \mathbf{a} \\ 0 \\ 0 \\ \cdot \\ \cdot \\ \cdot \\ 0 \end{bmatrix} \tag{17.20}$$

where $\mathbf{a}$ is of the same order as $\mathbf{d}_{11}$. Thus $\mathbf{d}_{11}$ can be obtained. $\mathbf{d}_{kk}$ can be similarly formed. To form $\mathbf{d}_{kl}$, where $k \neq l$, we again apply the same technique used in the proof of Theorem 16.1. For example, to attain $\mathbf{d}_{12}$, choose $\mathbf{i}$ and $\mathbf{j}$ such that they differ only in the first element and set the other elements equal to the corresponding elements of $\mathbf{i}_0$. That is, remove the same row and column from $\boldsymbol{\alpha}_{R_{M_3}}, \ldots, \boldsymbol{\alpha}_{R_{M_K}}$ and for the present assume that these choices are fixed. Now, for the matrix

$$\left[\begin{array}{c|c} \boldsymbol{\alpha}_{R_{M_1}} & \mathbf{0} \\ \hline \mathbf{0} & \boldsymbol{\alpha}_{R_{M_2}} \end{array}\right]$$

proceed, as in the proof of Theorem 16.1, to remove one row and column from $\boldsymbol{\alpha}_{R_{M_1}}$ and/or $\boldsymbol{\alpha}_{R_{M_2}}$, and specify it by the first element of $\mathbf{i}$ and $\mathbf{j}$, respectively. The result is a linear combination of these particular $\nabla C_{\mathbf{ij}}$ which can be added to give

$$\begin{bmatrix} 0 \\ \mathbf{d}_{12} \\ 0 \\ \cdot \\ \cdot \\ \cdot \\ 0 \end{bmatrix}$$

The other $\mathbf{d}_{ij}$ are similarly formed, and a linear combination of the ∇C_{ij} can be found which result in the gradient vector in (17.18), thus completing the proof.

17.2 CHOOSING THE LARGEST OF SEVERAL LOCAL EXTREMA

The class of admissible signal sets whose dimensionality is $D \leq M - K$ has been shown to have many local extrema. At large signal-to-noise ratios the one with the largest probability of detection can be determined. Since the probability of detection is independent of any permutation of the $\{M_i;\ i = 1, \ldots, K\}$, we shall assume that the M_i are arranged in ascending order.

THEOREM 17.2 *For a given M and $K \leq M/2$, of the local extrema given by (17.3), at high signal-to-noise ratios the one with the smallest probability of error is the one in which $M_i = 2$ for $i = 1, \ldots, K - 1$ and $M_K = M - 2(K - 1)$, with the corresponding signal-vector inner-product matrix*

$$\alpha_{0_{\text{opt}}} = \begin{bmatrix} \alpha_{R_2} & 0 & & & \\ 0 & \alpha_{R_2} & & & \\ & & \ddots & & \\ & & & \alpha_{R_2} & 0 \\ & & & 0 & \alpha_{R_{M-2(K-1)}} \end{bmatrix} \tag{17.21}$$

Proof: For a given set of $\{M_i; i = 1, \ldots, K\}$ at large signal-to-noise ratios we have the approximation

$$P_e(\lambda;\alpha_0;\{M_i\}) = \frac{1}{2M\sqrt{2\pi}\,\lambda}$$
$$\left\{\sum_{i=1}^{K} \frac{(M_i - 1)\sqrt{M_i(M_i - 2)}}{2} \exp\left(-\tfrac{1}{2}\frac{M_i}{M_i - 2}\lambda^2\right)\right.$$
$$\left. + \left[\frac{M(M - 1)}{2} - \sum_{i=1}^{K} \frac{M_i(M_i - 2)}{2}\right] \exp -\tfrac{1}{2}\lambda^2\right\} \tag{17.22}$$

Assume that the $\{M_i\}$ are arranged in ascending order and that $M_1 > 2$. Then decreasing M_1 by 1 and increasing M_K by 1 decreases the probability of error at high signal-to-noise ratios, the other M_i

remaining fixed. This is so because

$$
\begin{aligned}
&P_e(\lambda; \boldsymbol{\alpha}_0; M_1, \ldots, M_K) - P_e(\lambda; \boldsymbol{\alpha}_0; M_1' \\
&= M_1 - 1, M_2' = M_2, \ldots, M_{K-1}' = M_{K-1}, M_K' = M_K + 1) \\
&\quad \approx \frac{\exp(-\frac{1}{2}\lambda^2)}{2M\sqrt{2\pi}\,\lambda} \left\{ \frac{\sqrt{M_1(M_1-2)}\,(M_1-1)}{2} \exp\frac{-\lambda^2}{M_1-2} \right. \\
&\qquad + \frac{\sqrt{M_K(M_K-2)}\,(M_K-1)}{2} \exp\frac{-\lambda^2}{M_K-2} \\
&\qquad\qquad - \frac{M_1(M_1-1)}{2} - \frac{M_K(M_K-1)}{2} \\
&\qquad - \left[\frac{\sqrt{(M_1-1)(M_1-3)}\,(M_1-2)}{2} \exp\frac{-\lambda^2}{M_1-3} \right. \\
&\qquad + \frac{\sqrt{(M_K+1)(M_K-1)}\,M_K}{2} \exp\frac{-\lambda^2}{M_K-1} \\
&\qquad\qquad \left.\left. - \frac{(M_1-1)(M_1-2)}{2} - \frac{(M_K+1)M_K}{2} \right] \right\}
\end{aligned}
\tag{17.23}
$$

for large λ. The predominant terms are the constants, which become

$$
\begin{aligned}
&\tfrac{1}{2}[(M_1-1)(M_1-2) + (M_K+1)M_K \\
&\qquad - M_1(M_1-1) - M_K(M_K-1)] = M_K + 1 - M_1
\end{aligned}
$$

and are strictly positive, since the M_i have been arranged in ascending order. If M_1' is not 2, the above procedure is iterated until $M_1' = 2$. Then, if $M_2 > 2$, we set $M_2' = M_2 - 1$ and iterate again. It is clear that regardless of the original choice of the $\{M_i\}$, this iterative procedure will always arrive at the signal structure $\boldsymbol{\alpha}_{0\text{opt}}$ in (17.21), with each successive iteration reducing the probability of error. Therefore at large signal-to-noise ratios $\boldsymbol{\alpha}_{0\text{opt}}$ is the most preferred of the many local extrema in $\mathcal{A}_K$.

When M is even and $K = M/2$, we note that there is only one local extremum of the probability of detection of this form, the biorthogonal signal structure, which consists of $M/2$ one-dimensional regular simplices mutually orthogonal to one another. The biorthogonal signal structure is analyzed in Refs. [17.2, 17.3, 17.4].

17.3 THE EFFECT OF DIMENSIONALITY ON SYSTEM PERFORMANCE

The optimal signal design obtained for the various cases thus far permits us to determine to some extent the effect on probability of error of a reduction in the allowed dimensionality of the signal set. In Fig. 17.1 probability of error is plotted against signal-to-noise ratio for $M = 8$ for the optimal signal set when $D = 7$ (regular simplex), $D = 4$ (biorthogonal), and $D = 2$ (equally spaced unit circle). A similar plot is given in

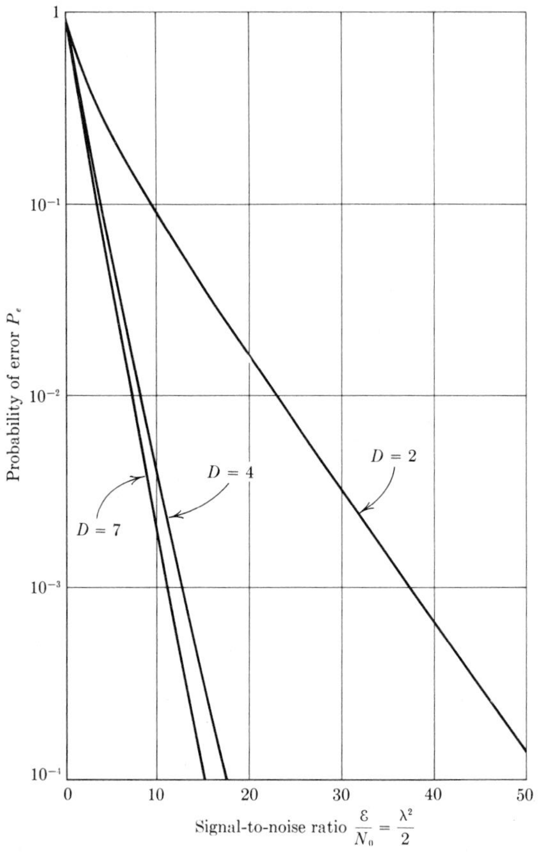

Fig. 17.1 Effect of dimensionality on optimal system performance when $M = 8$.

Fig. 17.2 for $M = 16$. Since the biorthogonal and regular simplex curves are very close to each other, the cost in probability of error due to reduction in dimensionality is small when $M/2 \leq D \leq M - 1$. Although the optimal choice is still open when the allowed dimensionality is reduced below that required for the biorthogonal signal set, with the exception $D = 2$, it is clear from Figs. 17.1 and 17.2 that as the allowed dimensionality is successively reduced, the probability of error significantly increases.

PROBLEM

17.1 Show that all the lower-dimensional solutions to the coherent-signal-design problem are also solutions when the amplitude of the transmitted waveform is assumed to be any nonnegative random variable unknown to the receiver.

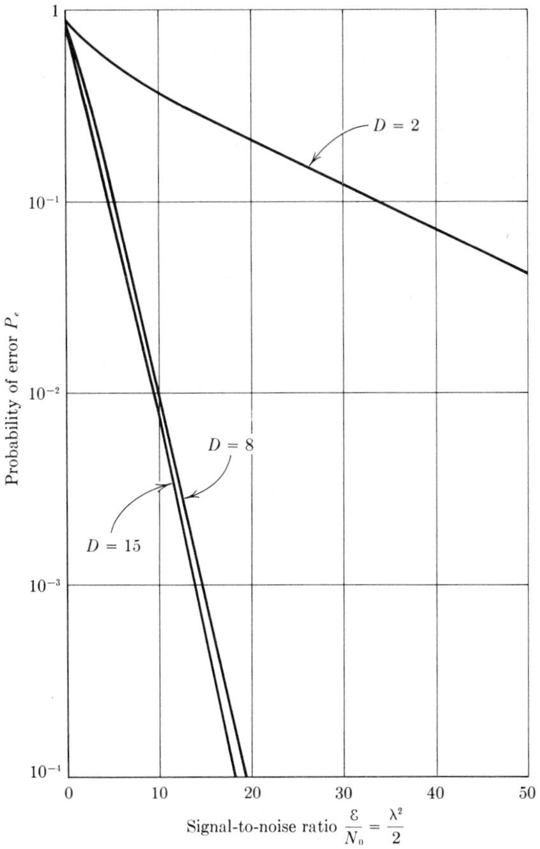

Fig. 17.2 Effect of dimensionality on optimal system performance when $M = 16$.

REFERENCES

17.1 Weber, C. L.: New Solutions to the Signal Design Problem for Coherent Channels, *IEEE Trans. Inform. Theory,* vol. IT-12, no. 2, April, 1966.

17.2 Sanders, R. W.: Digilock Communication System, *IRE WESCON Conv. Record,* part 5, September, 1960, pp. 125–131.

17.3 Balakrishnan, A. V., and J. E. Taber: Error Rates in Coherent Communication Systems, *IRE Trans. Commun. Systems,* vol. CS-10, March, 1962, pp. 86–90.

17.4 Golomb, S. W., L. D. Baumert, M. F. Easterling, J. J. Stiffler, and A. J. Viterbi: "Digital Communications," Prentice-Hall, Englewood Cliffs, N.J., 1964.

18
Additional Solutions for Three-Dimensional Signal Structures

With probability of error as the criterion, when $D = 3$ the signal-design problem has been solved for $M = 2$ through $M = 6$. For $M > 6$, however, the solution is still open with the exceptions $M = 8$ and $M = 12$. We can arrive at some insight into these solutions by examining the somewhat easier criterion of maximizing the minimum distance between the signal vectors (constrained to the unit sphere). Unfortunately, this is also a very difficult problem with few solutions. Solutions that have been found, other than those for which $M \leq 6$, are for the cases $M = 7$, 8, 9, 12, and 24.

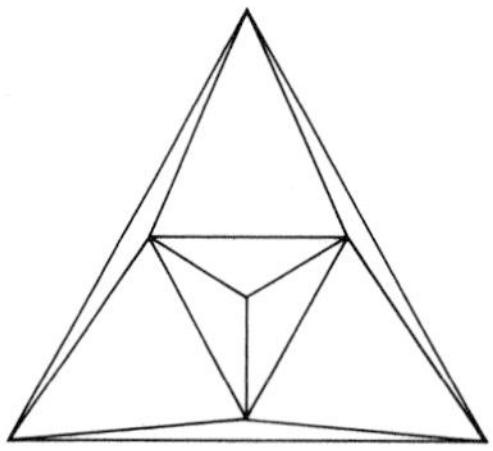

Fig. 18.1 Maximizing the minimum distance, $M = 7$, $D = 3$.

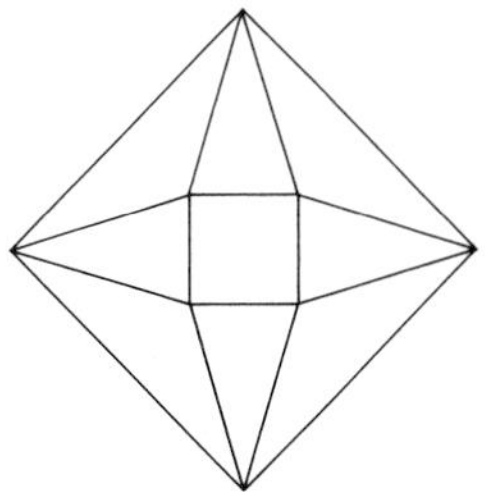

Fig. 18.2 Maximizing the minimum distance, $M = 8$, $D = 3$.

Since this criterion is also independent of orthogonal transformations, specification of the inner-product matrix is sufficient to identify the solutions. It is more convenient in specifying these solutions, however, to describe the vertices and faces of the convex hull that the signal structure generates.

The solution for $M = 7$ is diagrammed in Fig. 18.1. It consists of one vertex at a pole and three equally spaced vertices on each of two circles, one above and one below the equator and parallel to it.

For $M = 8$ the solution (see Fig. 18.2) is the *cuboctahedron* (or twisted cube), which has one square and three equilateral triangles meeting at each vertex.

Fig. 18.3 diagrams the solution for $M = 9$. This solution consists of three vertices on each of three mutually parallel circles on the sphere. The solution for $M = 12$ is similar, consisting of three points on each of four mutually parallel circles. Detailed accounts of these solutions are given by Schutte and van der Waerden [18.1, 18.2].

The $M = 8$, the *cuboctahedron*, and $M = 12$, the dodecahedron, solutions fall into the class of symmetric polyhedra, which are defined by Landau and Slepian [18.4] as signal sets which are such that the conditions for equality in their upper bound for the probability of detection described in Chap. 15 are met. Therefore the global solution has been obtained for $M = 8$ and $M = 12$ when $D = 3$.

The case for $M = 24$ is given by Robinson [18.3] and is diagrammed in Fig. 18.4. He has shown that the answer is the polyhedron with four triangles and one square at each vertex.

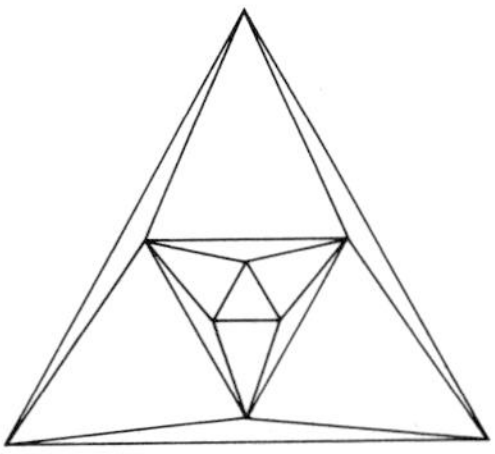

Fig. 18.3 Maximizing the minimum distance, $M = 9$, $D = 3$.

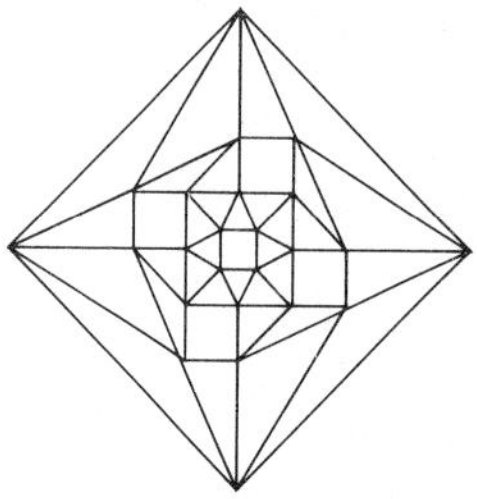

Fig. 18.4 Maximizing the minimum distance, $M = 24$, $D = 3$.

It is clear that regular polyhedra do not yield solutions to the problem of maximization of the minimum distance. The solutions for $M = 8$ and $M = 24$ are *semiregular polyhedra*, by which we mean polyhedra that have the same number and kind of regular faces meeting at each vertex. Regular or semiregular polyhedra have been found for several other M and thus are conjectured solutions, but none has yet been shown to maximize the minimum distance or the probability detection.

In conclusion, it might be said that for a given M, as the allowed dimensionality is continually decreased, the problems of maximizing the probability of detection and maximizing the minimum distance become progressively more difficult. With the exception $D = 2$, when $D < M/2$, not only is the solution in general still open, but in most cases conjectured solutions do not even exist.

REFERENCES

18.1 Schutte, K., and B. L. van der Waerden: Auf welcher Kugel haben 5, 6, 7, 8, oder 9 Punkte mit Mindestabstand Eins Platz?, *Math. Ann.*, vol. 123, 1951, pp. 96–124.

18.2 van der Waerden, B. L., Punkte auf der Kugel, Drei Zusatze, *Math. Ann.*, vol. 125, 1952, pp. 213–222.

18.3 Robinson, R. M.: Arrangement of 24 Points on a Sphere, *Math. Ann.*, vol. 144, 1961, pp. 17–49.

18.4 Landau, H. J., and D. Slepian: On the Optimality of the Regular Simplex Code, *Bell Sys. Tech. J.*, vol. 45, no. 8, October, 1966.

19
Signal-Design Concepts for Noncoherent Channels

All our results thus far apply exclusively to coherent channels. We now consider the more difficult problem of determining optimal signal waveforms for telemetry systems which are still synchronous, that is, where the receiver knows the time interval during which the signal is to arrive but not the carrier-phase angle of the arriving signal. The waveform emitted at the transmitter during the interval $[0,T]$ is still assumed to be one of M equally powered equally likely signals, but after transmission through the channel, coherent phase information of the carrier is assumed to be lost; in addition, we have the previous assumption of corruption of the signal by additive white gaussian noise. The received signal $y(t)$ is then of the form

$$y(t) = Vs_j(t;\phi) + n(t) \tag{19.1}$$

where, as before, $V^2/2$ is the average received signal power and $n(t)$ is the additive white gaussian noise (independent of the transmitted waveforms). For the noncoherent receiver, however, the received signal is of the form

$$s_j(t;\phi) = A_j(t) \cos [\omega_c t + \theta_j(t) + \phi] \tag{19.2}$$

where $A_j(t)$ is the envelope of the signal with

$$\frac{1}{T} \int_0^T [A_j(t)]^2 \, dt = 1 \qquad j = 1, \ldots, M \tag{19.3}$$

ϕ is a random variable (unknown to the receiver) which is uniformly distributed over $(0,2\pi)$, and $A_j(t)$ and $\theta_j(t)$ are assumed to be waveforms which are narrowband with respect to the carrier frequency ω_c. For the present, no further bandwidth restrictions will be considered.

As before, our goal is to design the $\{A_j(t)\}$ and $\{\theta_j(t)\}$ to maximize the probability of detection at each signal-to-noise ratio, assuming that each given signal set has associated with it an optimal noncoherent receiver. It should be noted that the form of the $\{\theta_j(t)\}$ is now limited to comply to the restriction that the receiver does not have knowledge of the absolute phase of the carrier. The receiver, however, can still detect phase differences, as long as we assume that ϕ does not change during the time interval $[0,T]$. Such telemetry links are termed *differential phase systems*.

For a given signal set $\{s_j(t;\phi)\}$ the optimal receiver (optimal in the sense of maximizing the probability of detection) derived in Chap. 10 is the one which forms

$$F_i' \triangleq \frac{1}{2\pi} \int_0^{2\pi} d\psi \, \exp \left[\frac{2}{N_0} \int_0^T y(t) s_i(t;\psi) \, dt \right] \qquad i = 1, \ldots, M \tag{19.4}$$

where N_0 is the one-sided spectral density of the additive white gaussian noise and decides that the jth signal has been transmitted if

$$F_j' = \max_i F_i' \tag{19.5}$$

For the particular form that has been chosen for the transmittable signals, the integration over ϕ in (19.4) can be carried out, and the optimal receiver can be equivalently stated (see Chap. 7) as the one which forms (with complex notation)

$$F_i \triangleq \left| \int_0^T y(t) A_i(t) \exp [-j(\omega_c t + \theta_i(t))] \, dt \right| \qquad i = 1, \ldots, M \tag{19.6}$$

and again applies the decision rule in (19.5). With trigonometric functions

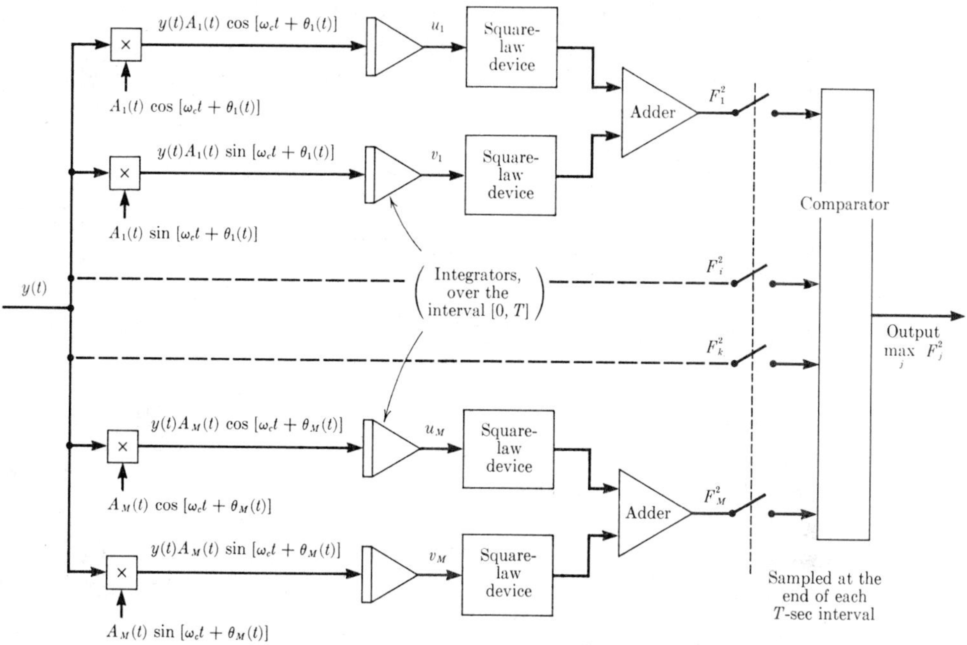

Fig. 19.1 Form of the optimal noncoherent receiver.

(19.6) can be expressed equivalently as

$$F_i^2 = \left\{\int_0^T y(t)A_i(t)\cos[\omega_c t + \theta_i(t)]\,dt\right\}^2 + \left\{\int_0^T y(t)A_i(t)\sin[\omega_c t + \theta_i(t)]\,dt\right\}^2 \qquad i = 1, \ldots, M \tag{19.7}$$

Thus the optimal receiver matches the received waveform with the M possible signals, and with these same M possibilities with the carrier phase angle shifted by $\pi/2$, it integrates, squares, and adds, as diagrammed in Fig. 19.1. At time T the output of the ith adder is F_i^2. The comparator then chooses the largest of the F_i^2 (which is equivalent to choosing the largest of the F_i, since they are nonnegative).

An examination of various representations of the probability of error for noncoherent systems is given by Balakrishnan and Abrams [19.4] and for noncoherent systems with equicorrelated signal sets by Nuttall [19.5].

The incoming signal is of unknown phase, but this lack of information has already been taken into account in the design of the optimal receiver. It can be demonstrated without difficulty that the F_i^2 in (19.7) do not depend on the carrier phase angle of the incoming signal. Therefore, to simplify the analysis, we assume $\phi = 0$ for the arriving signal in (19.1).

Using the narrowband assumption on $A_i(t)$ and $\theta_i(t)$, we can reduce F_i^2 to

$$F_i^2 = \left\{\frac{V}{2}\int_0^T A_j(t)A_i(t)\cos[\theta_j(t) - \theta_i(t)]\,dt + \int_0^T n(t)A_i(t)\cos[\omega_c t + \theta_i(t)]\,dt\right\}^2 + \left\{\frac{V}{2}\int_0^T A_j(t)A_i(t)\sin[\theta_i(t) - \theta_j(t)]\,dt + \int_0^T n(t)A_i(t)\sin[\omega_c t + \theta_i(t)]\,dt\right\}^2 \qquad i = 1, \ldots, M \tag{19.8}$$

Proceeding in a manner somewhat similar to that used for the coherent systems, we define

$$\lambda_{ij} \triangleq \frac{1}{T}\int_0^T A_i(t)A_j(t)\cos[\theta_i(t) - \theta_j(t)]\,dt \tag{19.9}$$

and

$$\beta_{ij} \triangleq \frac{1}{T}\int_0^T A_i(t)A_j(t)\sin[\theta_i(t) - \theta_j(t)]\,dt \tag{19.10}$$

Clearly, $\lambda_{ii} = 1$ and $\beta_{ii} = 0$ for all i. Also, $\lambda_{ij} = \lambda_{ji}$ and $\beta_{ij} = -\beta_{ji}$. Further, we define

$$\xi_i \triangleq \sqrt{\frac{4}{TN_0}}\int_0^T n(t)A_i(t)\cos[\omega_c t + \theta_i(t)]\,dt \qquad i = 1, \ldots, M \tag{19.11}$$

and

$$\eta_i \triangleq \sqrt{\frac{4}{TN_0}}\int_0^T n(t)A_i(t)\sin[\omega_c t + \theta_i(t)]\,dt \qquad i = 1, \ldots, M \tag{19.12}$$

Then the $\{\xi_i\}$ and $\{\eta_i\}$ are gaussian random variables with

$$E(\xi_i) = E(\eta_i) = 0 \qquad \text{for all } i$$

and

$$E(\xi_i^2) = E(\eta_i^2) = 1 \qquad \text{for all } i \tag{19.13}$$

As in the coherent case, multiplication of the output of each of the matched filters by a scalar does not alter the decision rule. So, again as in the coherent case, we multiply the output by $\sqrt{4/TN_0}$ and, as usual, define the signal-to-noise ratio λ^2 as

$$\lambda^2 \triangleq \frac{V^2T}{N_0} = \frac{2\mathcal{E}}{N_0} \tag{19.14}$$

where, as before, $\mathcal{E} = V^2T/2$ is total signal energy in $[0,T]$ and N_0 is the one-sided noise spectral density. Then, including the scale factor, the F_i^2 can be expressed (provided the jth signal has been transmitted) as

$$F_i^2 = (\lambda\lambda_{ij} + \xi_i)^2 + (\lambda\beta_{ij} + \eta_i)^2 \qquad i = 1, \ldots, M \tag{19.15}$$

where

$$\begin{aligned} E(\xi_i\xi_j) &= \lambda_{ij} \\ E(\eta_i\eta_j) &= \lambda_{ij} \\ E(\eta_i\xi_j) &= \beta_{ij} \end{aligned} \tag{19.16}$$

Equivalently, we write

$$F_i^2 = u_i^2 + v_i^2$$

where

$$u_i \triangleq \lambda\lambda_{ij} + \xi_i$$

and

$$v_i \triangleq \lambda\beta_{ij} + \eta_i \qquad i = 1, \ldots, M$$

If we now define the nonnegative symmetric matrix of the $\{\lambda_{ij}\}$ as

$$\boldsymbol{\alpha} \triangleq \begin{bmatrix} 1 & & \lambda_{ij} \\ & \ddots & \\ \lambda_{ji} & & 1 \end{bmatrix} \qquad \lambda_{ij} = \lambda_{ji} \tag{19.17}$$

and the matrix of $\{\beta_{ij}\}$ as

$$\boldsymbol{\beta} \triangleq \begin{bmatrix} 0 & & \beta_{ij} \\ & \ddots & \\ -\beta_{ji} & & 0 \end{bmatrix} \qquad \beta_{ij} = -\beta_{ji} \tag{19.18}$$

and further let

$$\boldsymbol{\xi} \triangleq \begin{bmatrix} \xi_1 \\ \cdot \\ \cdot \\ \cdot \\ \xi_M \end{bmatrix} \tag{19.19}$$

and

$$\mathbf{n} \triangleq \begin{bmatrix} \eta_1 \\ \cdot \\ \cdot \\ \cdot \\ \eta_M \end{bmatrix} \tag{19.20}$$

then the covariance matrix of $\begin{bmatrix} \boldsymbol{\xi} \\ \mathbf{n} \end{bmatrix}$ is

$$\boldsymbol{\Gamma} \triangleq \begin{bmatrix} \boldsymbol{\alpha} & \boldsymbol{\beta} \\ \boldsymbol{\beta}^T & \boldsymbol{\alpha} \end{bmatrix} \tag{19.21}$$

It is clear that for noncoherent systems the class of admissible $\boldsymbol{\Gamma}$ consists of those symmetric nonnegative definite $2M$-by-$2M$ matrices with 1s along the main diagonal and of the form indicated by (19.21). Note that the $\{\lambda_{ij}\}$ and $\{\beta_{ij}\}$ in the noncoherent case represent quantities different from the $\{\lambda_{ij}\}$ in the coherent systems.

Finally, we can write the probability of detection as a function only of λ and $\boldsymbol{\Gamma}$ in the following way:

$$P_d(\lambda;\boldsymbol{\Gamma}) = \sum_{k=1}^{M} P(s_k) \iint_{-\infty}^{\infty} du_k\, dv_k \iint_{C_k} du_1\, dv_1 \cdots \iint_{C_k} du_{k-1}\, dv_{k-1}$$
$$\iint_{C_k} du_{k+1}\, dv_{k+1} \cdots \iint_{C_k} du_M\, dv_M\, G\left(\begin{bmatrix} \mathbf{u} \\ \mathbf{v} \end{bmatrix}; \lambda \begin{bmatrix} \boldsymbol{\lambda}_k \\ \boldsymbol{\beta}_k \end{bmatrix}; \boldsymbol{\Gamma}\right) \tag{19.22}$$

where

$$\mathbf{u} \triangleq \begin{bmatrix} u_1 \\ \cdot \\ \cdot \\ \cdot \\ u_M \end{bmatrix} \qquad \mathbf{v} \triangleq \begin{bmatrix} v_1 \\ \cdot \\ \cdot \\ \cdot \\ v_M \end{bmatrix} \tag{19.23}$$

where $P(s_k)$ is the a priori probability that s_k was transmitted, C_k is the region inside the two-dimensional circle centered at the origin with radius given by $\sqrt{u_k^2 + v_k^2}$, and $\boldsymbol{\lambda}_k$ and $\boldsymbol{\beta}_k$ are the kth column of $\boldsymbol{\alpha}$ and $\boldsymbol{\beta}$, respectively. The $2M$-fold integral in (19.22) is the probability that s_k was decided upon at the receiver, given that s_k was transmitted; we denote this as $P(s_k|s_k)$. Equivalently stated, given that s_k was sent, the integration is over that region in which $F_k = \max_i F_i$.

A simplified form for the probability of detection is derived in the following lemma.

LEMMA 19.1[1] *The probability of detection as expressed in Eq. (19.22) can be written equivalently as*

$$P_d(\lambda;\mathbf{\Gamma}) = \frac{1}{M} \exp\left(-\tfrac{1}{2}\lambda^2\right) \int \cdots \int_{E^{2M}} I_0\left(\lambda \max_k \sqrt{u_k^2 + v_k^2}\right) G(\mathbf{w};\mathbf{0};\mathbf{\Gamma})\, d\mathbf{w}$$

$$= \frac{1}{M} \exp\left(-\tfrac{1}{2}\lambda^2\right) E\left(I_0\left(\lambda \max_k \sqrt{u_k^2 + v_k^2}\right)\right) \qquad (19.24)$$

where E^{2M} is the real euclidean space of $2M$ dimensions,

$$\mathbf{w} = \begin{bmatrix} \mathbf{u} \\ \mathbf{v} \end{bmatrix} \qquad (19.25)$$

and I_0 is the modified Bessel function of the first kind.

Proof: Assume for the present that $\mathbf{\Gamma}$ is a positive-definite covariance matrix. Then, assuming equilikely a priori probabilities, (19.22) can be explicitly expressed as

$$P_d(\lambda;\mathbf{\Gamma}) = \frac{1}{M} \sum_{k=1}^{M} \int \cdots \int_{\Lambda_k} \frac{1}{(2\pi)^M \sqrt{|\mathbf{\Gamma}|}} \exp\left[-\tfrac{1}{2}(\mathbf{w} - \lambda\mathbf{\Gamma}_k)^T \mathbf{\Gamma}^{-1} (\mathbf{w} - \lambda\mathbf{\Gamma}_k)\right] d\mathbf{w}$$

where $\mathbf{\Gamma}_k$ is the kth column of $\mathbf{\Gamma}$ and Λ_k is the region where

$$u_k^2 + v_k^2 = \max_j (u_j^2 + v_j^2)$$

Since

$$\mathbf{\Gamma}_k^T \mathbf{\Gamma}^{-1} \mathbf{\Gamma}_k = 1 \qquad k = 1, \ldots, M$$

and

$$\tfrac{1}{2}(\mathbf{\Gamma}_k^T \mathbf{\Gamma}^{-1} \mathbf{w} + \mathbf{w}^T \mathbf{\Gamma}^{-1} \mathbf{\Gamma}_k) = u_k \qquad k = 1, \ldots, M$$

we have

$$P_d(\lambda;\mathbf{\Gamma}) = \frac{1}{M} \exp\left(-\tfrac{1}{2}\lambda^2\right) \sum_{k=1}^{M} \int \cdots \int_{\Lambda_k} \frac{1}{(2\pi)^M \sqrt{|\mathbf{\Gamma}|}} \exp\left(\lambda u_k - \tfrac{1}{2}\mathbf{w}^T \mathbf{\Gamma}^{-1} \mathbf{w}\right) d\mathbf{w} \qquad (19.26)$$

In (19.26) introduce the transformation

$$u_j' \triangleq u_j \cos\psi - v_j \sin\psi$$
$$v_j' \triangleq u_j \sin\psi + v_j \cos\psi \qquad j = 1, \ldots, M$$

[1] Due to L. Welch, private communication.

This is simply a rotation of coordinates by the angle ψ and will certainly not change the value of the expression. Note that

$$u_k'^2 + v_k'^2 = u_k^2 + v_k^2 \qquad k = 1, \ldots, M$$

and hence the regions of integration remain unchanged. Set

$$\mathbf{u}' \triangleq \begin{bmatrix} u_1' \\ \cdot \\ \cdot \\ \cdot \\ u_M' \end{bmatrix} \qquad \mathbf{v}' \triangleq \begin{bmatrix} v_1' \\ \cdot \\ \cdot \\ \cdot \\ v_M' \end{bmatrix} \qquad \mathbf{w}' \triangleq \begin{bmatrix} \mathbf{u}' \\ \mathbf{v}' \end{bmatrix}$$

The covariance matrix of $\mathbf{w}'$ is also $\mathbf{\Gamma}$, so that (19.26) can be written as

$$P_d(\lambda;\mathbf{\Gamma}) = \frac{1}{M} \exp -\tfrac{1}{2}\lambda^2 \sum_{k=1}^{M} \int \cdots \int_{\Lambda_k} \frac{\exp\left[-\tfrac{1}{2}\mathbf{w}'^T\mathbf{\Gamma}^{-1}\mathbf{w}' + \lambda(u_k' \cos\psi + v_k' \sin\psi)\right]}{(2\pi)^M \sqrt{|\mathbf{\Gamma}|}}\, d\mathbf{w}' \tag{19.27}$$

There is nothing significant about the particular choice of ψ. In fact, it has been previously indicated that the decision at the receiver is independent of the relative phase angle between the arriving signal waveform and the signal waveforms stored in the receiver. Therefore, since $P_d(\lambda;\mathbf{\Gamma})$ in (19.27) is independent of ψ, we can average ψ uniformly over $(0,2\pi)$ without altering the value of P_d; this gives us

$$\begin{aligned} P_d(\lambda;\mathbf{\Gamma}) &= \frac{1}{M} \exp -\tfrac{1}{2}\lambda^2 \sum_{k=1}^{M} \int \cdots \int_{\Lambda_k} G(\mathbf{w}';\mathbf{0};\mathbf{\Gamma}) I_0\left(\lambda \sqrt{u_k'^2 + v_k'^2}\right) d\mathbf{w}' \\ &= \frac{1}{M} \exp -\tfrac{1}{2}\lambda^2 \sum_{k=1}^{M} \int \cdots \int_{\Lambda_k} G(\mathbf{w}';\mathbf{0};\mathbf{\Gamma}) I_0\left(\lambda \max_j \sqrt{u_j'^2 + v_j'^2}\right) d\mathbf{w}' \end{aligned}$$

In this form the integrand is independent of k, and since

$$\bigcup_{k=1}^{M} \Lambda_k = E^{2M}$$

we have finally

$$P_d(\lambda;\mathbf{\Gamma}) = \frac{1}{M} \exp -\tfrac{1}{2}\lambda^2 \int \cdots \int_{E^{2M}} I_0\left(\lambda \max_j \sqrt{u_j^2 + v_j^2}\right) G(\mathbf{w};\mathbf{0};\mathbf{\Gamma})\, d\mathbf{w}$$

If $\mathbf{\Gamma}$ is nonnegative-definite but not positive-definite, consider a sequence of positive-definite covariance matrices which converge to $\mathbf{\Gamma}$ and apply this proof to each matrix in the sequence. Hence the proof is valid for all admissible $\mathbf{\Gamma}$.

The form for the probability of detection for noncoherent systems sufficiently resembles that derived for coherent systems that the methods employed in demonstrating that certain signal structures are optimum in different cases for coherent systems are also applicable in the noncoherent case. The significant reasons for this are that, as in the coherent case, it is possible to express the probability of detection in terms of only the signal-to-noise ratio λ^2 and the elements of the symmetric nonnegative-definite matrix $\mathbf{\Gamma}$, thus making it a variational problem, and that the expression for $P_d(\lambda;\mathbf{\Gamma})$ could be expressed as an integration over the whole $2M$-dimensional euclidean space. Hence the signal-design problem for noncoherent systems has been reduced to finding that admissible $\mathbf{\Gamma}$ which maximizes $P_d(\lambda;\mathbf{\Gamma})$ in (19.24) and, as before, determining the dependence of the optimal signal structure on the signal-to-noise ratio.

19.1 NECESSARY (FIRST-ORDER) CONDITIONS FOR NONCOHERENT OPTIMALITY

The results that follow do not apply when the restriction on the allowed bandwidth is severe enough to require the use of admissible $\mathbf{\Gamma}$ which have degrees of freedom less than $2M$. Equivalently, we assume that $\mathbf{\Gamma}$ can have full rank. These results also apply when there is no bandwidth restriction.

For the orthogonal signal structure $\mathbf{\Gamma}_0$, that is, the one for which

$$\boldsymbol{\beta} = \mathbf{0}$$

and (19.28)

$$\boldsymbol{\alpha} = \mathbf{I}$$

(the M-by-M identity matrix), we have the following main result.

THEOREM 19.1 *In the class of all admissible signal sets for noncoherent systems the orthogonal signal structure satisfies necessary (first-order) conditions to be a local extremum in all directions at every signal-to-noise ratio.*

Proof: Since the orthogonal signal structure is in the interior of the admissible $\mathbf{\Gamma}$ space, the admissible directions consist of all possible directions. Therefore there are no constraint conditions to be satisfied.

For any fixed λ write the probability of detection for any signal structure $\mathbf{\Gamma}'$ in the neighborhood of the orthogonal signal structure in the form of a Taylor's series expansion about the orthogonal signal set as

$$P_d(\lambda;\mathbf{\Gamma}') = P_d(\lambda;\mathbf{\Gamma}_0) + \sum_{i>j}\sum \lambda'_{ij}\,\frac{\partial P_d(\lambda;\mathbf{\Gamma}_0)}{\partial \lambda_{ij}} + \sum_{i>1}\sum \beta'_{ij}\,\frac{\partial P_d(\lambda;\mathbf{\Gamma}_0)}{\partial \beta_{ij}} + \cdots \tag{19.29}$$

To prove the theorem it is sufficient to prove

$$\frac{\partial P_d(\lambda;\mathbf{\Gamma}_0)}{\partial \lambda_{ij}} = 0 \qquad \text{for all } i > j \tag{19.30}$$

and

$$\frac{\partial P_d(\lambda;\mathbf{\Gamma}_0)}{\partial \beta_{ij}} = 0 \qquad \text{for all } i > j \tag{19.31}$$

Consider (19.30) first. For this we have

$$\left.\frac{\partial P_d(\lambda;\mathbf{\Gamma})}{\partial \lambda_{ij}}\right|_{\mathbf{\Gamma}=\mathbf{\Gamma}_0} = \frac{1}{M}\exp -\tfrac{1}{2}\lambda^2 \int_{E^{2M}}\cdots\int I_0(\lambda \max_k \sqrt{u_k^2+v_k^2})\,\frac{\partial}{\partial \lambda_{ij}} G(\mathbf{w};\mathbf{0};\mathbf{\Gamma})\,d\mathbf{w}\,\bigg|_{\mathbf{\Gamma}=\mathbf{\Gamma}_0}$$

This differentiation can be best carried out by expressing $G(\mathbf{w};\mathbf{0};\mathbf{\Gamma})$ in terms of its characteristic function $C(\mathbf{t},\mathbf{\Gamma})$, as we did for coherent systems. Thus

$$\frac{\partial}{\partial \lambda_{ij}} G(\mathbf{w};\mathbf{0};\mathbf{\Gamma}) = \frac{\partial}{\partial \lambda_{ij}}\frac{1}{(2\pi)^M}\int_{E^{2M}}\cdots\int \exp(-i\mathbf{t}^T\mathbf{w})\exp(-\tfrac{1}{2}\mathbf{t}^T\mathbf{\Gamma}\mathbf{t})\,d\mathbf{t}$$

where $\mathbf{t}$ is a $2M$-dimensional vector of integration, which we shall express as

$$\mathbf{t} \triangleq \begin{bmatrix}\mathbf{r}\\ \mathbf{s}\end{bmatrix}$$

where $\mathbf{r}$ and $\mathbf{s}$ are each M-dimensional vectors. Then

$$\begin{aligned}
\frac{\partial}{\partial \lambda_{ij}} G(\mathbf{w};\mathbf{0};\mathbf{\Gamma})
&= \frac{1}{(2\pi)^M}\int_{E^{2M}}\cdots\int \exp(-i\mathbf{t}^T\mathbf{w})\,\frac{\partial}{\partial \lambda_{ij}}(\exp -\tfrac{1}{2}\mathbf{t}^T\mathbf{\Gamma}\mathbf{t})\,d\mathbf{t}\\
&= \frac{1}{(2\pi)^M}\int_{E^{2M}}\cdots\int \exp(-i\mathbf{t}^T\mathbf{w})(-1)(r_ir_j + s_is_j)\exp(-\tfrac{1}{2}\mathbf{t}^T\mathbf{\Gamma}\mathbf{t})\,d\mathbf{t}\\
&= \frac{1}{(2\pi)^M}\int_{E^{2M}}\cdots\int \left(\frac{\partial^2}{\partial u_i\,\partial u_j} + \frac{\partial^2}{\partial v_i\,\partial v_j}\right)\exp(-i\mathbf{t}^T\mathbf{w})\exp(-\tfrac{1}{2}\mathbf{t}^T\mathbf{\Gamma}\mathbf{t})\,d\mathbf{t}
\end{aligned}$$

Thus

$$\frac{\partial}{\partial \lambda_{ij}} G(\mathbf{w};\mathbf{0};\mathbf{\Gamma}) = \left(\frac{\partial^2}{\partial u_i\, \partial u_j} + \frac{\partial^2}{\partial v_i\, \partial v_j}\right) G(\mathbf{w};\mathbf{0};\mathbf{\Gamma}) \tag{19.32}$$

Substitution gives

$$\frac{\partial P_d(\lambda;\mathbf{\Gamma})}{\partial \lambda_{ij}} = \frac{1}{M} \exp\left(-\tfrac{1}{2}\lambda^2\right) \int_{E^{2M}} \cdots \int I_0\left(\lambda \max_k \sqrt{u_k^2 + v_k^2}\right) \left(\frac{\partial^2}{\partial u_i\, \partial u_j} + \frac{\partial^2}{\partial v_i\, \partial v_j}\right) G(\mathbf{w};\mathbf{0};\mathbf{\Gamma})\, d\mathbf{w} \tag{19.33}$$

At $\mathbf{\Gamma} = \mathbf{\Gamma}_0$ we have

$$\left.\frac{\partial P_d(\lambda;\mathbf{\Gamma})}{\partial \lambda_{ij}}\right|_{\mathbf{\Gamma}=\mathbf{\Gamma}_0} = \frac{1}{M} \exp -\tfrac{1}{2}\lambda^2 \int_{E^{2M}} \cdots \int I_0\left(\lambda \max_k \sqrt{u_k^2 + v_k^2}\right) (u_i u_j + v_i v_j) G(\mathbf{w};\mathbf{0};\mathbf{\Gamma}_0)\, d\mathbf{w}$$

which is the integral of an odd function over symmetric limits and therefore vanishes. Similarly, (19.31) can be written as

$$\left.\frac{\partial P_d(\lambda;\mathbf{\Gamma})}{\partial \beta_{ij}}\right|_{\mathbf{\Gamma}=\mathbf{\Gamma}_0} = \frac{1}{M} \exp -\tfrac{1}{2}\lambda^2 \int_{E^{2M}} \cdots \int I_0\left(\lambda \max_k \sqrt{u_k^2 + v_k^2}\right) (u_j v_i - u_i v_j) G(\mathbf{w};\mathbf{0};\mathbf{\Gamma}_0)\, d\mathbf{w} \tag{19.34}$$

which also is the integral of an odd function over symmetric limits and thus vanishes.

Therefore the orthogonal signal structure is a local extremum in the class of admissible signal sets at every signal-to-noise ratio.

19.2 EVALUATION OF PROBABILITY OF ERROR FOR THE ORTHOGONAL NONCOHERENT SIGNAL STRUCTURE

It is immediate that $P_d(\lambda;\mathbf{\Gamma}_0)$ is independent of the a priori probability distribution and can be expressed as

$$P_d(\lambda;\mathbf{\Gamma}_0) = \iint_{-\infty}^{\infty} dx\, dy\, \frac{\exp -\frac{1}{2}[(x-\lambda)^2 + y^2]}{2\pi} \left[\iint_C du\, dv\, \frac{\exp -\frac{1}{2}(u^2+v^2)}{2\pi}\right]^{M-1}$$

where C is the set of points inside the circle with radius $\sqrt{x^2 + y^2}$ centered at the origin. The integration over u and v can be carried out by transforming to polar coordinates, yielding

$$P_d(\lambda;\mathbf{\Gamma}_0) = \iint\limits_{-\infty}^{\infty} dx\,dy\,\frac{\exp -\frac{1}{2}[(x-\lambda)^2 + y^2]}{2\pi} \left[1 - \exp -\tfrac{1}{2}(x^2+y^2)\right]^{M-1} \tag{19.35}$$

If we expand the bracketed factor in a binomial-series expansion, the integrations over x and y can be separated and carried out, resulting in the following expression, which is well suited for numerical calculation at small λ:

$$P_d(\lambda;\mathbf{\Gamma}_0) = \exp\left(-\tfrac{1}{2}\lambda^2\right) \sum_{k=0}^{M-1} (-1)^k \binom{M-1}{k} \frac{1}{k+1} \exp\left(\tfrac{1}{2}\lambda^2 \frac{1}{k+1}\right) \tag{19.36}$$

From (19.35) we have the approximation for large λ in the following lemma.

LEMMA 19.2 *For large λ the probability of error for the noncoherent orthogonal signal structure has the approximation*

$$P_e(\lambda;\mathbf{\Gamma}_0) \approx \frac{M-1}{2} \exp -\tfrac{1}{4}\lambda^2$$

Proof: Let $z = x - \lambda$ in (19.35), and then approximate the binomial-series expansion of the resulting bracketed factor by the first two terms. This results in

$$P_e(\lambda;\mathbf{\Gamma}_0) \approx \frac{M-1}{2\pi} \iint\limits_{-\infty}^{\infty} \exp\left[-\tfrac{1}{2}(y^2+z^2)\right] \exp\left\{-\tfrac{1}{2}[(z+\lambda)^2 + y^2]\right\} dy\,dz$$

which, after direct integration, yields the desired result.

From (19.36) the probability of error can then be written explicitly as

$$P_e(\lambda;\mathbf{\Gamma}_0) = 1 - P_d(\lambda;\mathbf{\Gamma}_0) = \frac{1}{M} \exp\left(-\tfrac{1}{2}\lambda^2\right) \sum_{n=2}^{M} (-1)^n \binom{M}{n} \exp \frac{\lambda^2}{2n}$$

$P_e(\lambda;\mathbf{\Gamma}_0)$ is plotted in Fig. 19.2 for various M. This form of probability of error has been examined by Rieger [19.6] and Turin [19.7]. In Fig. 19.3, for $M = 2$, 4, and 16, probability of error is plotted against signal-to-noise ratio for the optimal coherent system (regular simplex) and optimal noncoherent system (orthogonal) when there is no restriction on the allowed bandwidth, thus indicating the additional average transmitter power needed to attain a specified probability of error for the optimal noncoher-

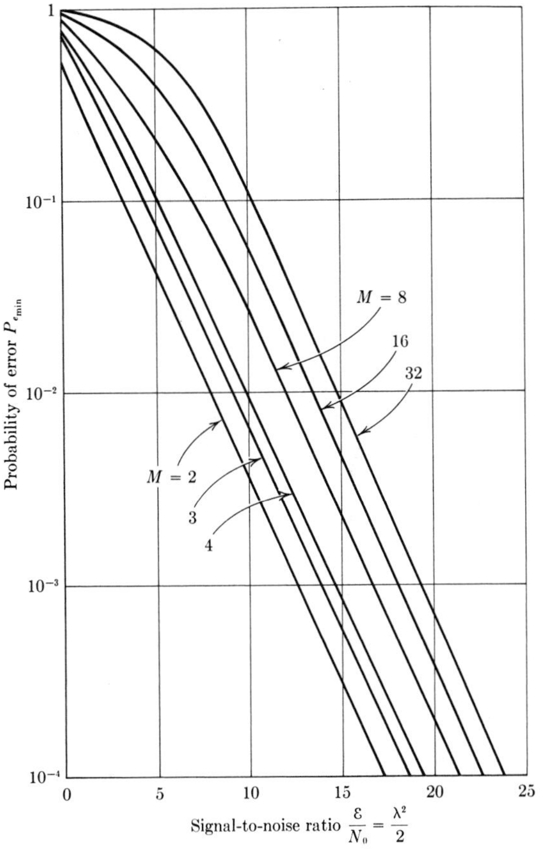

Fig. 19.2 Probability of error versus signal-to-noise ratio for the orthogonal noncoherent signal structure.

ent system over that for the optimal coherent system. One system, already in extensive use, which is a particular implementation of the noncoherent orthogonal signal set is the M-ary noncoherent frequency-shift-keying system.

19.3 SUFFICIENT (SECOND-ORDER) CONDITIONS FOR NONCOHERENT OPTIMALITY

In Sec. 19.1 the orthogonal signal structure was shown to be a local extremum in the class of all admissible signal sets for all signal-to-noise ratios. In this section we classify this extremum as a local maximum by proving that the orthogonal signal structure does indeed satisfy sufficient

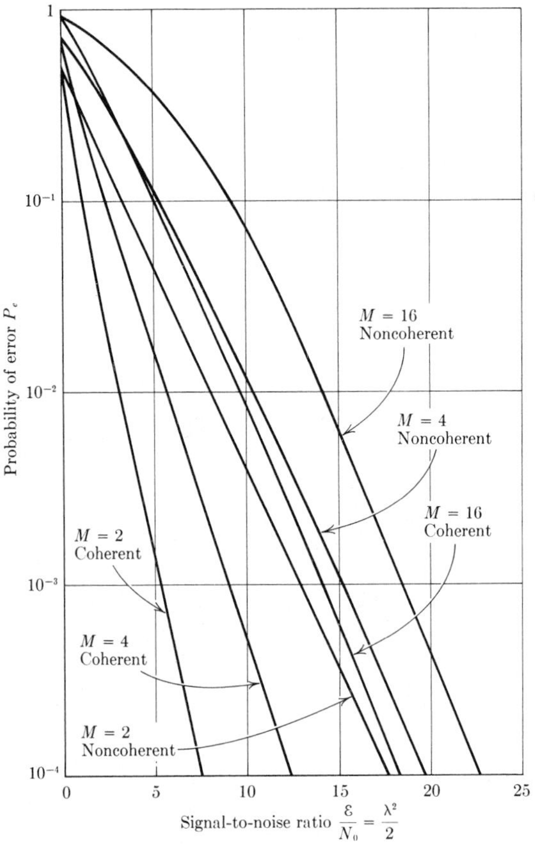

Fig. 19.3 Probability of error versus signal-to-noise ratio for the optimal coherent (regular simplex) and optimal noncoherent (orthogonal) signal structure.

conditions to locally maximize the probability of detection for all signal-to-noise ratios.

LEMMA 19.3 *For any admissible* $\mathbf{\Gamma}$

$$\left.\frac{\partial P_d(\lambda;\mathbf{\Gamma})}{\partial\lambda}\right|_{\lambda=0} = 0 \tag{19.37}$$

Equivalently, at small signal-to-noise ratios and for arbitrary $\mathbf{\Gamma}$ *there is no first-order variation of the probability of detection with respect to the signal-to-noise ratio; that is,* $P_d(\lambda;\mathbf{\Gamma})$ *behaves as* λ^2 *for small* λ.

Proof: The lemma follows immediately, since the expression as given by (19.24) is a function of λ^2.

We now proceed toward the main objective of this section. To start with, we rewrite the Taylor's series expansion in (19.29) for an arbitrary signal structure $\mathbf{\Gamma}'$ in the neighborhood of the orthogonal signal set and add the second-order variations. That is, for any fixed λ

$$P_d(\lambda;\mathbf{\Gamma}') = P_d(\lambda;\mathbf{\Gamma}_0) + \sum_{i>j}\sum \lambda'_{ij} \frac{\partial P_d(\lambda;\mathbf{\Gamma}_0)}{\partial \lambda_{ij}} + \sum_{i>j}\sum \beta'_{ij} \frac{\partial P_d(\lambda;\mathbf{\Gamma}_0)}{\partial \beta_{ij}}$$
$$+ \tfrac{1}{2}\Bigg[\sum_{i>j}\sum\sum_{k>l}\sum \lambda'_{ij}\lambda'_{kl} \frac{\partial^2 P_d(\lambda;\mathbf{\Gamma}_0)}{\partial \lambda_{ij}\,\partial \lambda_{kl}} + 2\sum_{i>j}\sum\sum_{k>l}\sum \lambda'_{ij}\beta'_{kl} \frac{\partial^2 P_d(\lambda;\mathbf{\Gamma}_0)}{\partial \lambda_{ij}\,\partial \beta_{kl}}$$
$$+ \sum_{i>j}\sum\sum_{k>l}\sum \beta'_{ij}\beta'_{kl} \frac{\partial^2 P_d(\lambda;\mathbf{\Gamma}_0)}{\partial \beta_{ij}\,\partial \beta_{kl}}\Bigg] + \cdots \tag{19.38}$$

We already know that the first-order variations in (19.36) are identically zero for arbitrary $\mathbf{\Gamma}'$. Our aim is to prove that the bracketed second-order variations are strictly negative for arbitrary $\mathbf{\Gamma}'$ in the neighborhood of $\mathbf{\Gamma}_0$. It is sufficient, therefore, to prove that the second-order variations in (19.38) are strictly negative for any admissible $\mathbf{\Gamma}'$.

THEOREM 19.2[1] *The second-order variations in* (19.38) *are strictly negative for any $\mathbf{\Gamma}'$ different from $\mathbf{\Gamma}_0$ for all $\lambda > 0$, and therefore in the class of all admissible $\mathbf{\Gamma}$ the orthogonal signal structure is a local maximum for all signal-to-noise ratios.*

Proof: The proof is given in parts.

Part 1

$$\frac{\partial^2 P_d(\lambda;\mathbf{\Gamma}_0)}{\partial \lambda_{ij}\,\partial \beta_{kl}} = 0 \qquad \text{for all } i > j,\ k > l \tag{19.39}$$

We have that

$$\frac{\partial^2 P_d(\lambda;\mathbf{\Gamma}_0)}{\partial \lambda_{ij}\,\partial \beta_{kl}} = \frac{1}{M} \exp -\tfrac{1}{2}\lambda^2$$
$$\int \cdots \int_{E^{2M}} I_0(\lambda \max_n \sqrt{u_n^2 + v_n^2}) \frac{\partial^2}{\partial \lambda_{ij}\,\partial \beta_{kl}} [G(\mathbf{w};\mathbf{0};\mathbf{\Gamma})]\, d\mathbf{w}$$

which, with the machinery from Theorem 19.1, can be written as

$$\frac{\partial^2 P_d(\lambda;\mathbf{\Gamma}_0)}{\partial \lambda_{ij}\,\partial \beta_{kl}} = \frac{1}{M} \exp(-\tfrac{1}{2}\lambda^2) \int \cdots \int_{E^{2M}} I_0(\lambda \max_n \sqrt{u_n^2 + v_n^2})$$
$$\left(\frac{\partial^2}{\partial u_k\,\partial v_l} - \frac{\partial^2}{\partial u_l\,\partial v_k}\right)\left(\frac{\partial^2}{\partial u_i\,\partial u_j} + \frac{\partial^2}{\partial v_i\,\partial v_j}\right) G(\mathbf{w};\mathbf{0};\mathbf{\Gamma}_0)\, d\mathbf{w}$$
$$= \frac{1}{M} \exp(-\tfrac{1}{2}\lambda^2) \int \cdots \int_{E^{2M}} I_0(\lambda \max_n \sqrt{u_n^2 + v_n^2})$$
$$(u_k v_l - u_l v_k)(u_i u_j + v_i v_j) G(\mathbf{w};\mathbf{0};\mathbf{\Gamma}_0)\, d\mathbf{w}$$

[1] This result was attained jointly with R. A. Scholtz [19.9].

After multiplication each of the four terms has at least one variable which makes the overall integrand an odd function; this again, with the symmetric limits, is sufficient for the term to vanish.

Part 2

$$\frac{\partial^2 P_d(\lambda;\boldsymbol{\Gamma}_0)}{\partial\lambda_{ij}\,\partial\lambda_{kl}} = 0 \qquad \text{for all } (i, j) \neq (k, l) \tag{19.40}$$

and

$$\frac{\partial^2 P_d(\lambda;\boldsymbol{\Gamma}_0)}{\partial\beta_{ij}\,\partial\beta_{kl}} = 0 \qquad \text{for all } (i, j) \neq (k, l) \tag{19.41}$$

As in Part 1, we use the fact that the resultant integrand is an odd function in at least one of the variables, and that the integration is over even limits.

Therefore the proof reduces to showing that

$$\sum_{i>j}\sum \lambda_{ij}'^2 \frac{\partial^2 P_d(\lambda;\boldsymbol{\Gamma}_0)}{\partial\lambda_{ij}^2} + \sum_{i>j}\sum \beta_{ij}'^2 \frac{\partial^2 P_d(\lambda;\boldsymbol{\Gamma}_0)}{\partial\beta_{ij}^2} \tag{19.42}$$

is strictly negative. Since $\lambda_{ij}'^2$ and $\beta_{ij}'^2$ are nonnegative and arbitrary, we must be able to show that

$$\frac{\partial^2 P_d(\lambda;\boldsymbol{\Gamma}_0)}{\partial\lambda_{ij}^2} < 0 \qquad \text{for every } i > j;\ \lambda > 0 \tag{19.43}$$

and

$$\frac{\partial^2 P_d(\lambda;\boldsymbol{\Gamma}_0)}{\partial\beta_{ij}^2} < 0 \qquad \text{for every } i > j;\ \lambda > 0 \tag{19.44}$$

Because of the symmetry of $\boldsymbol{\Gamma}_0$, we have

$$\frac{\partial^2 P_d(\lambda;\boldsymbol{\Gamma}_0)}{\partial\lambda_{ij}^2} = \frac{\partial^2 P_d(\lambda;\boldsymbol{\Gamma}_0)}{\partial\lambda_{12}^2} \qquad \text{for all } i > j \tag{19.45}$$

and

$$\frac{\partial^2 P_d(\lambda;\boldsymbol{\Gamma}_0)}{\partial\beta_{ij}^2} = \frac{\partial^2 P_d(\lambda;\boldsymbol{\Gamma}_0)}{\partial\beta_{12}^2} \qquad \text{for all } i > j \tag{19.46}$$

Thus the proof has been reduced to showing that

$$\frac{\partial^2 P_d(\lambda;\boldsymbol{\Gamma}_0)}{\partial\lambda_{12}^2} < 0 \qquad \text{for all } \lambda > 0 \tag{19.47}$$

and

$$\frac{\partial^2 P_d(\lambda;\boldsymbol{\Gamma}_0)}{\partial\beta_{12}^2} < 0 \qquad \text{for all } \lambda > 0 \tag{19.48}$$

Our aim now is to simplify (19.47) and (19.48) to the point where the desired inequalities become apparent. Consider (19.47) first.

Part 3

$$\frac{\partial^2 P_d(\lambda;\mathbf{\Gamma}_0)}{\partial\lambda_{12}{}^2} = \frac{2}{M}\exp\left(-\tfrac{1}{2}\lambda^2\right)\int_0^\infty r\exp\left(-\tfrac{1}{2}r^2\right)I_0(\lambda r)$$
$$F_2(r)[F_1(r)]^{M-3}[(r^2-2)-\tfrac{1}{2}(Mr^2-4)\exp -\tfrac{1}{2}r^2]\,dr \quad (19.49)$$

where

$$F_1(r) = 1-\exp -\tfrac{1}{2}r^2 \quad (19.50)$$

$$F_2(r) = -\tfrac{1}{2}r^2\exp -\tfrac{1}{2}r^2 \quad (19.51)$$

From (19.33) we have that

$$\frac{\partial^2 P_d(\lambda;\mathbf{\Gamma}_0)}{\partial\lambda_{12}{}^2} = \frac{1}{M}\exp -\tfrac{1}{2}\lambda^2$$
$$\int_{E^{2M}}\cdots\int I_0(\lambda\max_k\sqrt{u_k{}^2+v_k{}^2})\left(\frac{\partial^2}{\partial u_1\,\partial u_2}+\frac{\partial^2}{\partial v_1\,\partial v_2}\right)^2 G(\mathbf{w};\mathbf{0};\mathbf{\Gamma}_0)\,d\mathbf{w}$$

which reduces to

$$\frac{\partial^2 P_d(\lambda;\mathbf{\Gamma}_0)}{\partial\lambda_{12}{}^2} = \frac{2}{M}\exp -\tfrac{1}{2}\lambda^2$$
$$\int_{E^{2M}}\cdots\int I_0(\lambda\max_k\sqrt{u_k{}^2+v_k{}^2})\frac{\partial^4}{\partial u_1{}^2\,\partial u_2{}^2}G(\mathbf{w};\mathbf{0};\mathbf{\Gamma}_0)\,d\mathbf{w}$$

Performing the indicated differentiations, we have

$$\frac{\partial^2 P_d(\lambda;\mathbf{\Gamma}_0)}{\partial\lambda_{12}{}^2} = \frac{2}{M}\exp -\tfrac{1}{2}\lambda^2$$
$$\int_{E^{2M}}\cdots\int I_0(\lambda\max_k\sqrt{u_k{}^2+v_k{}^2})\,H_2(u_1)H_2(u_2)G(\mathbf{w};\mathbf{0};\mathbf{\Gamma}_0)\,d\mathbf{w}$$

where $H_2(u) = u^2-1$ is the second-order Hermite polynomial. This can be expressed as

$$\frac{\partial^2 P_d(\lambda;\mathbf{\Gamma}_0)}{\partial\lambda_{12}{}^2} = \frac{2}{M}\exp\left(-\tfrac{1}{2}\lambda^2\right)\sum_{j=1}^{M}\int_{\Lambda_j}\cdots\int I_0(\lambda\sqrt{u_j{}^2+v_j{}^2})$$
$$H_2(u_1)H_2(u_2)G(\mathbf{w};\mathbf{0};\mathbf{\Gamma}_0)\,d\mathbf{w}$$
$$= \frac{2}{M}\exp\left(-\tfrac{1}{2}\lambda^2\right)\Big[2\int_{\Lambda_1}\cdots\int I_0(\lambda\sqrt{u_1{}^2+v_1{}^2})$$
$$H_2(u_1)H_2(u_2)G(\mathbf{w};\mathbf{0};\mathbf{\Gamma}_0)\,d\mathbf{w}$$
$$+(M-2)\int_{\Lambda_M}\cdots\int I_0(\lambda\sqrt{u_M{}^2+v_M{}^2})$$
$$H_2(u_1)H_2(u_2)G(\mathbf{w};\mathbf{0};\mathbf{\Gamma}_0)\,d\mathbf{w}\Big]$$

where Λ_j is the region where $u_j{}^2+v_j{}^2 = \max_i\,(u_i{}^2+v_i{}^2)$.

We have used the fact that the M integrals above are of two types, one of which occurs twice and the second of which occurs in the remaining $M - 2$ terms. Rearranging terms, we have

$$\frac{\partial^2 P_d(\lambda;\mathbf{\Gamma}_0)}{\partial \lambda_{12}^2} = \frac{2}{M} \exp -\tfrac{1}{2}\lambda^2$$

$$\Bigg(2 \iint_{-\infty}^{\infty} du_1\, dv_1\, I_0(\lambda \sqrt{u_1^2 + v_1^2})\, H_2(u_1) \frac{\exp -\frac{1}{2}(u_1^2 + v_1^2)}{2\pi}$$

$$\iint_{C_{1_r}} du_2\, dv_2\, H_2(u_2) \frac{\exp -\frac{1}{2}(u_2^2 + v_2^2)}{2\pi}$$

$$\left\{ \iint_{C_{1_r}} \frac{\exp [-\frac{1}{2}(u_M^2 + v_M^2)]}{2\pi}\, du_M\, dv_M \right\}^{M-2}$$

$$+ (M-2) \iint_{-\infty}^{\infty} du_M\, dv_M\, I_0(\lambda \sqrt{u_M^2 + v_M^2}) \frac{\exp -\frac{1}{2}(u_M^2 + v_M^2)}{2\pi}$$

$$\left[\iint_{C_{M_r}} H_2(u_1)\, du_1\, dv_1 \frac{\exp -\frac{1}{2}(u_1^2 + v_1^2)}{2\pi} \right]^2$$

$$\left[\iint_{C_{M_r}} \frac{\exp -\frac{1}{2}(u_2^2 + v_2^2)}{2\pi}\, du_2\, dv_2 \right]^{M-3} \Bigg) \qquad (19.52)$$

where C_{j_r} is the region inside the circle centered at the origin with radius $\sqrt{u_j^2 + v_j^2} = r$. Note that

$$\iint_{C_r} \frac{\exp -\frac{1}{2}(u^2 + v^2)}{2\pi}\, du\, dv = 1 - \exp -\tfrac{1}{2}r^2 = F_1(r)$$

and

$$\iint_{C_r} H_2(u) \frac{\exp -\frac{1}{2}(u^2 + v^2)}{2\pi}\, du\, dv = -\tfrac{1}{2}r^2 \exp -\tfrac{1}{2}r^2 = F_2(r)$$

which when substituted into (19.52) results (dropping all subscripts since they are now superfluous) in

$$\frac{\partial^2 P_d(\lambda;\mathbf{\Gamma}_0)}{\partial \lambda_{12}^2} = \frac{2}{M} \exp -\tfrac{1}{2}\lambda^2$$

$$\Bigg\{ 2 \iint_{-\infty}^{\infty} I_0(\lambda \sqrt{u^2 + v^2})\, H_2(u) \frac{\exp -\frac{1}{2}(u^2 + v^2)}{2\pi} F_2(r)[F_1(r)]^{M-2}\, du\, dv$$

$$+ (M-2) \iint_{-\infty}^{\infty} I_0(\lambda \sqrt{u^2 + v^2}) \frac{\exp -\frac{1}{2}(u^2 + v^2)}{2\pi}$$

$$[F_2(r)]^2 [F_1(r)]^{M-3}\, du\, dv \Bigg\}$$

where $u^2 + v^2 = r^2$. Combining and transforming the (u,v) integration into polar coordinates, we obtain finally

$$\frac{\partial^2 P_d(\lambda;\mathbf{\Gamma}_0)}{\partial\lambda_{12}{}^2} = \frac{2}{M}\exp\left(-\tfrac{1}{2}\lambda^2\right)\left\{\int_0^\infty r\exp\left(-\tfrac{1}{2}r^2\right)I_0(\lambda r)F_2(r)[F_1(r)]^{M-3}\right.$$
$$\left.[(r^2-2)-\tfrac{1}{2}(Mr^2-4)\exp -\tfrac{1}{2}r^2]\,dr\right\}$$

The integrals in (19.49), which represent the even moments of the *rician distribution* [19.8], result in *Laguerre polynomials* times an exponential factor. Designating these integrals by $K_m(\lambda,\gamma)$, we have the identity

$$\begin{aligned} K_m(\lambda;\gamma) &= \int_0^\infty I_0(\lambda r)r^{2m+1}\exp\left(\frac{-\gamma r^2}{2}\right)dr \\ &= \frac{2^m m!}{\gamma^{m+1}}L_m\left(\frac{-\lambda^2}{2\gamma}\right)\exp\frac{\lambda^2}{2\gamma} \end{aligned} \tag{19.52}$$

where L_m is the mth-order Laguerre polynomial. In particular,

$$L_1\left(\frac{-\lambda^2}{2\gamma}\right) = 1 + \frac{\lambda^2}{2\gamma}$$

and

$$L_2\left(\frac{-\lambda^2}{2\gamma}\right) = 1 + 2\left(\frac{\lambda^2}{2\gamma}\right) + \frac{1}{2}\left(\frac{\lambda^2}{2\gamma}\right)^2$$

We expand $[F_1(r)]^{M-3}$ in a binomial series, and $\partial^2 P_d(\lambda;\mathbf{\Gamma}_0)/\partial\lambda_{12}{}^2$ as expressed in (19.49) becomes a series of integrals, each of which is in the form indicated by (19.52). Carrying out the indicated integrations and expressing the result in terms of $K_m(\lambda,\gamma)$, we have

$$\begin{aligned} \frac{\partial^2 P_d(\lambda;\mathbf{\Gamma}_0)}{\partial\lambda_{12}{}^2} = \frac{-1}{M}\exp\left(-\tfrac{1}{2}\lambda^2\right)\sum_{k=0}^{M-3}(-1)^k\binom{M-3}{k} \\ \left[-\frac{M}{2}K_2(\lambda, k+3) + 2K_1(\lambda, k+3)\right. \\ \left. + K_2(\lambda, k+2) - 2K_1(\lambda, k+2)\right] \end{aligned} \tag{19.53}$$

Part 4: From (19.53) we can simplify $\partial^2 P_d(\lambda;\mathbf{\Gamma}_0)/\partial\lambda_{12}{}^2$ algebraically and express it equivalently as

$$\begin{aligned} \frac{\partial^2 P_d(\lambda;\mathbf{\Gamma}_0)}{\partial\lambda_{12}{}^2} = \frac{-\lambda^2}{M}\exp -\tfrac{1}{2}\lambda^2 \\ \sum_{k=0}^{M-2}\frac{(-1)^k}{(k+2)^3}\binom{M-2}{k}\left[2+\frac{\lambda^2}{2(k+2)}\right]\exp\frac{\lambda^2}{2(k+2)} \end{aligned} \tag{19.54}$$

Directly from (19.53) we have

$$\frac{\partial^2 P_d(\lambda;\mathbf{\Gamma}_0)}{\partial \lambda_{12}{}^2} = \frac{-1}{M} \exp -\tfrac{1}{2}\lambda^2$$

$$\left\{ \sum_{k=1}^{M-2} (-1)^{k-1} \binom{M-3}{k-1} \left[-\frac{M}{2} K_2(\lambda, k+2) + 2K_1(\lambda, k+2) \right] \right.$$

$$\left. + \sum_{k=0}^{M-3} (-1)^k \binom{M-3}{k} [K_2(\lambda, k+2) - 2K_1(\lambda, k+2)] \right\}$$

The $k = M - 2$ term in the first sum is

$$(-1)^{M-2} \frac{\lambda^2}{M^3} \left(2 + \frac{\lambda^2}{2M}\right) \exp \frac{\lambda^2}{2M}$$

The $k = 0$ term in the second term is

$$\frac{\lambda^2}{8} \left(2 + \frac{\lambda^2}{4}\right) \exp \frac{\lambda^2}{4}$$

For $k \neq 0$ and $k \neq M - 2$ the kth term of each summation can be combined and written as

$$(-1)^k \frac{\lambda^2}{(k+2)^3} \binom{M-2}{k} \left[2 + \frac{\lambda^2}{2(k+2)}\right] \exp \frac{\lambda^2}{2(k+2)}$$

Combining these into one summation proves the claim.

The proof has now been reduced to showing that the sum over k in (19.54) is strictly positive for every $\lambda > 0$.

Part 5: Proving that the sum in (19.54) is strictly positive is the same as showing

$$\sum_{k=0}^{M-2} \frac{(-1)^k}{(k+2)^n} \binom{M-2}{k} > 0 \qquad \text{for } n = 3, 4, \ldots \tag{19.55}$$

Our goal is to prove that

$$\sum_{k=0}^{M-2} \frac{(-1)^k}{(k+2)^3} \binom{M-2}{k} \left[2 + \frac{\lambda^2}{2(k+2)}\right] \exp \frac{\lambda^2}{2(k+2)} \tag{19.56}$$

is strictly positive for all $\lambda > 0$. To show that this is equivalent to proving the inequalities in (19.55), we make the following observations. The sum in (19.56) is an alternating series with the first, third, etc., terms always positive and the second, fourth, etc., terms always negative. For each fixed λ the terms

$$\exp \frac{\lambda^2}{2(k+2)} \qquad \frac{1}{(k+2)^3} \qquad 2 + \frac{\lambda^2}{2(k+2)}$$

each decrease as k increases. Therefore, except for the binomial coefficient $\binom{M-2}{k}$, which increases for $k < (M-2)/2$, we could conclude that the $k = 0$ term is numerically greater than the $k = 1$ term, the $k = 2$ term is numerically greater than the $k = 3$ term, etc., and the proof would be complete. To take into account the binomial coefficients, we consider the following.

Let $\lambda^2/2 = \gamma$ and rewrite (19.56) as

$$h(\gamma) = \sum_{k=0}^{M-2} \frac{(-1)^k}{(k+2)^3} \binom{M-2}{k} \left(2 + \frac{\gamma}{k+2}\right) \exp \frac{\gamma}{k+2}$$

Taking the differential with respect to γ, we have

$$\frac{dh(\gamma)}{d\gamma} = \sum_{k=0}^{M-2} \frac{(-1)^k}{(k+2)^4} \binom{M-2}{k} \left(3 + \frac{\gamma}{k+2}\right) \exp \frac{\gamma}{k+2}$$

or

$$\frac{d^n h(\gamma)}{d\gamma^n} = \sum_{k=0}^{M-2} \frac{(-1)^k}{(k+2)^{n+3}} \binom{M-2}{k} \left(n + 2 + \frac{\gamma}{k+2}\right) \exp \frac{\gamma}{k+2}$$

The binomial coefficient $\binom{M-2}{k}$ remains unchanged with continued differentiation, while the power of $1/(k+2)$ increases. We see, then, that there exists an N such that each positive term in the sum dominates the succeeding negative term. This is clear since the rate of decresae of $1/(k+2)^{3+n}$ increases with continued differentiation, while the rate of increase of the binomial coefficients remains unchanged. Hence there exists an N such that

$$\frac{d^N h(\gamma)}{d\gamma^N} = \sum_{k=0}^{M-2} \frac{(-1)^k}{(k+2)^{N+3}} \binom{M-2}{k} \left(N + 2 + \frac{\gamma}{k+2}\right) \exp \frac{\gamma}{k+2}$$
$$> 0 \qquad \text{for all } \lambda > 0 \quad (19.57)$$

We can now reconstruct $h(\gamma)$ by successive integration of (19.57):

$$\int_0^{\gamma_1} \frac{d^N h(\gamma)}{d\gamma^N} \, d\gamma = \frac{d^{N-1} h(\gamma_1)}{d\gamma_1^{N-1}} - \left.\frac{d^{N-1} h(\gamma)}{d\gamma^{N-1}}\right|_{\gamma=0}$$

or

$$\frac{d^{N-1} h(\gamma_1)}{d\gamma_1^{N-1}} = \int_0^{\gamma_1} \frac{d^N h(\gamma)}{d\gamma^N} \, d\gamma + \sum_{k=0}^{M-2} \frac{(-1)^k}{(k+2)^{N+2}} \binom{M-2}{k} (N+1)$$

The integral is positive, since the integrand is always positive. If the sum on the right can be shown to be positive, then $d^{N-1}h(\gamma_1)/d\gamma_1^{N-1}$

will be positive, and we can repeat the procedure, after N iterations obtaining $h(\gamma)$. Therefore the sum in (19.54) can be shown to be positive for all $\lambda > 0$ if the sum in (19.55) can be shown to be positive for $n = 3, 4, \ldots$.

Part 6: Finally, we have that the sum

$$\sum_{k=0}^{M-2} \frac{(-1)^k}{(k+2)^p} \binom{M-2}{k}$$

indeed is positive for $p = 1, 2, \ldots$.

Note that (with an additional parameter x_1)

$$(1 - x_1)^{M-2} = \sum_{k=0}^{M-2} (-1)^k \binom{M-2}{k} x_1^k$$

Multiplying by x_1, we have

$$f_1(x_1) = x_1(1 - x_1)^{M-2} = \sum_{k=0}^{M-2} (-1)^k \binom{M-2}{k} x_1^{k+1}$$

which is nonnegative for $0 \leq x_1 \leq 1$. Therefore

$$f_2(x_2) = \int_0^{x_2} f_1(x_1)\, dx_1 = \sum_{k=0}^{M-2} \frac{(-1)^k}{k+2} \binom{M-2}{k} x_2^{k+2} \tag{19.58}$$

is positive for $0 < x_2 \leq 1$, and in particular,

$$f_2(1) = \sum_{k=0}^{M-2} \frac{(-1)^k}{k+2} \binom{M-2}{k} > 0$$

Dividing (19.58) by x_2 and again integrating, we obtain

$$f_3(x_3) = \int_0^{x_3} \frac{f_2(x_2)}{x_2}\, dx_2 = \sum_{k=0}^{M-2} \frac{(-1)^k}{(k+2)^2} \binom{M-2}{k} x_3^{k+2}$$

which is positive for $0 < x_3 \leq 1$, and again in particular,

$$f_3(1) = \sum_{k=0}^{M-2} \frac{(-1)^k}{(k+2)^2} \binom{M-2}{k} > 0$$

Continued iteration gives the desired for every p, and we can conclude that

$$\sum_{k=0}^{M-2} \frac{(-1)^k}{(k+2)^p} \binom{M-2}{k} > 0 \qquad \text{for } p = 1, 2, \ldots$$

We have therefore proved that

$$\frac{\partial^2 P_d(\lambda;\mathbf{\Gamma}_0)}{\partial \lambda_{12}^2} < 0 \qquad \text{for all } \lambda > 0$$

To complete the proof of the second-order conditions we must demonstrate that

$$\frac{\partial^2 P_d(\lambda;\mathbf{\Gamma}_0)}{\partial \beta_{12}^2} < 0 \qquad \text{for all } \lambda > 0$$

However, it can be shown by the same procedure used for the λ_{ij} that

$$\frac{\partial^2 P(s_1|s_1)\Big|_{\mathbf{\Gamma}=\mathbf{\Gamma}_0}}{\partial \lambda_{12}^2} = \frac{\partial^2 P(s_1|s_1)\Big|_{\mathbf{\Gamma}=\mathbf{\Gamma}_0}}{\partial \beta_{12}^2}$$

and

$$\frac{\partial^2 P(s_M|s_M)\Big|_{\mathbf{\Gamma}=\mathbf{\Gamma}_0}}{\partial \lambda_{12}^2} = \frac{\partial^2 P(s_M|s_M)\Big|_{\mathbf{\Gamma}=\mathbf{\Gamma}_0}}{\partial \beta_{12}^2}$$

thus completing the proof of Theorem 19.2.

19.4 GLOBAL OPTIMALITY WHEN $M = 2$

In the special case when $M = 2$ the orthogonal signal set can be shown to be globally optimum.

THEOREM 19.3[1] *When $M = 2$, placing the signals orthogonal to one another is globally optimum for all signal-to-noise ratios.*

Proof: Set $M = 2$ in (19.24) and make the transformation

$$u_1 = \sqrt{|\mathbf{\Gamma}|}\, r_1 \cos\theta_1 \qquad u_2 = \sqrt{|\mathbf{\Gamma}|}\, r_2 \cos\theta_2$$
$$v_1 = \sqrt{|\mathbf{\Gamma}|}\, r_1 \sin\theta_1 \qquad v_2 = \sqrt{|\mathbf{\Gamma}|}\, r_2 \sin\theta_2$$

After expressing the integrations with respect to θ_1 and θ_2 in terms of modified Bessel functions, we obtain

$$P_d(\lambda;\mathbf{\Gamma}) = \tfrac{1}{2} \exp\left(-\tfrac{1}{2}\lambda^2\right)(1 - \lambda_{12}^2 - \beta_{12}^2) \int_0^\infty dr_1\, r_1 \exp -\tfrac{1}{2}r_1^2$$
$$\int_0^\infty dr_2\, r_2 \exp\left(-\tfrac{1}{2}r_2^2\right) I_0\left(\lambda \sqrt{1 - \lambda_{12}^2 - \beta_{12}^2}\, \max_i r_i\right)$$
$$I_0\left(\sqrt{\lambda_{12}^2 + \beta_{12}^2}\, r_1 r_2\right)$$

We now see that the only signal parameter effecting P_d is

$$u^2 = \lambda_{12}^2 + \beta_{12}^2$$

[1] Due to C. Helstrom [19.8].

which, after substitution, becomes

$$P_d(\lambda;\mathbf{\Gamma}) = (1 - u^2) \exp\left(-\tfrac{1}{2}\lambda^2\right) \int_0^\infty dr_1\, r_1 \exp -\tfrac{1}{2}r_1^2$$
$$\int_0^\infty dr_2\, r_2 \exp\left(-\tfrac{1}{2}r_2^2\right) I_0(\lambda \sqrt{1 - u^2}\, r_2) I_0(u r_1 r_2) \quad (19.59)$$

At each signal-to-noise ratio this expression can be shown to take on its maximum value when $u = 0$ (for details see Helstrom [19.8]).

For solutions to the signal-design problem for noncoherent systems in which the bandwidth restriction is such that the orthogonal signal set is no longer admissible, see [19.11]. In particular, when $D = 4$, globally optimal signal structures have been determined when $M = 2, 3, 4, 6, 12$ by Schaffner [19.13].

PROBLEMS

19.1 Show that the noncoherent signal set is also the optimal choice when the model is the same as used in this chapter, with the exception that the amplitude of the transmitted waveform is any random variable unknown to the receiver.

19.2 Consider the M-ary statistical decision problem where the observer has available the vector $\mathbf{Y}$, whose probability density function is one of the following $p_k(\mathbf{Y}|\boldsymbol{\varepsilon})$; $k = 1, \ldots, M$, which have a priori probabilities π_k, $k = 1, \ldots, M$, respectively. The vector $\boldsymbol{\varepsilon}$ represents a set of unknown random variables, such as amplitude, and phase of the kth signal $\mathbf{S}_k(\boldsymbol{\varepsilon})$ which is assumed to have probability density function $q_k(\boldsymbol{\varepsilon})$ under the kth hypothesis.

(*a*) Determine the receiver which minimizes the probability of error.

(*b*) Let H_0 be the hypothesis that no signal is present, and assume that $\mathbf{Y}$ has probability density function $p_0(\mathbf{Y})$ under H_0. H_0 is not considered one of the M choices available to the receiver. Show that the probability of correct decision can be expressed

$$P_d = E(\max_k L_k|H_0)$$

where

$$L_k \triangleq \pi_k \int d\boldsymbol{\varepsilon}\, q_k(\boldsymbol{\varepsilon}) p_k(\mathbf{Y}|\boldsymbol{\varepsilon})/p_0(\mathbf{Y}) \qquad k = 1, \ldots, M$$

(*c*) From the result in (*b*) derive directly the expression for the probability of correct decision for M coherent signals as given by Eq. (14.11).

(*d*) Repeat (*c*) for noncoherent systems and derive the probability of correct decision as given by Eq. (19.24). (See Helstrom [19.12].)

REFERENCES

19.1 Helstrom, C.: "Statistical Theory of Signal Detection," Pergamon Press, New York, 1960.

19.2 Wainstein, L. A., and V. D. Zubakov: "Extraction of Signals from Noise," Prentice-Hall, Englewood Cliffs, N.J., 1962.

19.3 Middleton, D.: "Introduction to Statistical Communication Theory," McGraw-Hill, New York, 1960.

19.4 Balakrishnan, A. V., and I. J. Abrams: Detection Levels and Error Rates in PCM Telemetry Systems, *IRE Intern. Conv. Record,* 1960.

19.5 Nuttall, A. H.: Error Probabilities for Equicorrelated M-ary Signals Under Phase-coherent and Phase-incoherent Reception, *IRE Trans. Inform. Theory,* vol. IT-8, no. 4, July, 1962, pp. 305–315.

19.6 Reiger, S.: Error Rates in Data Transmission, *Proc. IRE,* vol. 46, May, 1958, pp. 919–920.

19.7 Turin, G. L.: The Asymptotic Behavior of Ideal M-ary Systems, *Proc. IRE,* vol. 47, no. 1, January, 1959.

19.8 Helstrom, C. W.: The Resolution of Signals in White Gaussian Noise, *Proc. IRE,* vol. 43, no. 9, September, 1955, pp. 1111–1118.

19.9 Scholtz, R. A., and C. L. Weber: Signal Design for Non-coherent Channels, *IEEE Trans. Inform. Theory,* vol. IT-12, no. 4, October, 1966.

19.10 Lindsey, W. C.: Coded Non-coherent Communications, *IEEE Trans. Space Electron. Telemetry,* vol. SET-II, no. 1, March, 1965.

19.11 Weber, C. L.: A Contribution to the Signal Design Problem for Incoherent Phase Communication Systems, *IEEE Trans. Inform. Theory,* vol. IT-14, March, 1968.

19.12 Helstrom, C. W.: Scholium, *IEEE Trans. Inform. Theory,* vol. IT-14, April, 1968.

19.13 Schaffner, C. A.: "The Global Optimization of Phase-incoherent Signals," doctoral dissertation, California Institute of Technology, Pasadena, Calif., April, 1968.

Appendixes

A. SUMMARY OF CONDITIONAL GAUSSIAN PROBABILITY DENSITY FUNCTIONS

Let $\mathbf{x}$ be an n-dimensional gaussian variate with mean $\mathbf{m}_x$; let $\mathbf{y}$ be an m-dimensional gaussian variate with mean $\mathbf{m}_y$; let

$$\mathbf{z} \triangleq \begin{bmatrix} \mathbf{x} \\ \mathbf{y} \end{bmatrix}$$

with gaussian probability density function

$$G\left(\mathbf{z}; \begin{bmatrix} \mathbf{m}_x \\ \mathbf{m}_y \end{bmatrix}; \mathbf{\Lambda}_z\right)$$

where

$$\mathbf{\Lambda}_z \triangleq \begin{bmatrix} \mathbf{\Lambda}_{xx} & \mathbf{\Lambda}_{xy} \\ \mathbf{\Lambda}_{yx} & \mathbf{\Lambda}_{yy} \end{bmatrix}$$

where

$$\mathbf{\Lambda}_{xx} = \text{covariance matrix of } \mathbf{x}$$
$$\mathbf{\Lambda}_{yy} = \text{covariance matrix of } \mathbf{y}$$
$$\mathbf{\Lambda}_{xy} = \mathbf{\Lambda}^T{}_{yx} = \text{covariance matrix between } \mathbf{x} \text{ and } \mathbf{y}$$

Then the conditional density of $\mathbf{x}$, given $\mathbf{y}$, is

$$f(\mathbf{x}|\mathbf{y}) = G(\mathbf{x};\boldsymbol{\beta};\mathbf{C})$$

where

$$\boldsymbol{\beta} = \mathbf{m}_x + \mathbf{\Lambda}_{xy}\mathbf{\Lambda}_{yy}^{-1}(\mathbf{y} - \mathbf{m}_y)$$

and

$$\mathbf{C} = \mathbf{\Lambda}_{xx} - \mathbf{\Lambda}_{xy}\mathbf{\Lambda}_{yy}^{-1}\mathbf{\Lambda}_{yx}$$

If $\mathbf{y}$ is a singular density, it is sufficient to consider the largest number of components of $\mathbf{y}$ whose density is nonsingular, neglecting the other components.

B. KARHUNEN-LOEVE EXPANSION

A stochastic process $n(t)$ is said to be *continuous in the mean* (of order 2) if and only if

$$\lim_{\Delta \to 0} E(|n(t) - n(t + \Delta)|^2) = 0$$

The covariance function $R(t,s)$ of a stochastic process is continuous in t and s if and only if $n(t)$ is continuous in the mean.

The Karhunen-Loeve expansion states that any zero mean stochastic process which is continuous in the mean in the interval $[0,T]$ and whose covariance function satisfies

$$\iint_0^T |R(t,s)|^2 \, dt \, ds < \infty$$

has the representation

$$n(t) = \underset{K\to\infty}{\text{l.i.m.}} \sum_{k=1}^{K} n_k \varphi_k(t)$$

for all t in the interval $[0,T]$, where l.i.m. is the probability limit in the mean

$$\lim_{K\to\infty} E\left(\left|n(t) - \sum_{k=1}^{K} n_k\varphi_k(t)\right|^2\right) = 0$$

The random coefficients are given by

$$n_k \triangleq [n(t),\varphi_k(t)] \triangleq \int_0^T n(t)\varphi_k(t)\, dt$$

The $\{n_k\}$ are jointly uncorrelated, and

$$E(|n_k|^2) = \lambda_k$$

where λ_k is the kth eigenvalue of $R(t,s)$ with associated eigenfunction $\varphi_k(t)$.

In particular, if the process is gaussian, the random coefficients are mutually independent. See Refs. [A.1, A.2].

C. MODIFIED BESSEL FUNCTION OF THE FIRST KIND

The modified Bessel function of the first kind of order zero is the solution of the differential equation

$$\frac{d^2y}{dx^2} + \frac{1}{x}\frac{dy}{dx} - y = 0$$

It has the integral representation

$$I_0(x) = \frac{1}{2\pi}\int_0^{2\pi} \exp\,(x\cos\theta)\, d\theta$$
$$= \frac{1}{2\pi}\int_0^{2\pi} \exp\,(x\sin\theta)\, d\theta = \frac{1}{2\pi}\int_0^{2\pi} \exp\,[x\cos(\theta+\phi)]\, d\theta$$

Its series expansion (the first two terms of which are used as an approximation for small x) is

$$I_0(x) = \sum_{m=0}^{\infty} \frac{x^{2m}}{2^{2m}(m!)^2}$$

Its asymptotic expansion (the first term of which is used as an approximation for large x) is

$$I_0(x) = \frac{\exp x}{\sqrt{2\pi x}}\left[1 + \frac{1}{8x} + \frac{1^2 3^2}{2(8x)^2} + \cdots\right]$$

The approximation for small x is

$$I_0(x) \approx 1 + \frac{x^2}{4}$$

and the approximation for large x is

$$I_0(x) \approx \frac{\exp x}{\sqrt{2\pi}\,\sqrt{x}}$$

$I_0(x)$ is plotted in Fig. C.1.

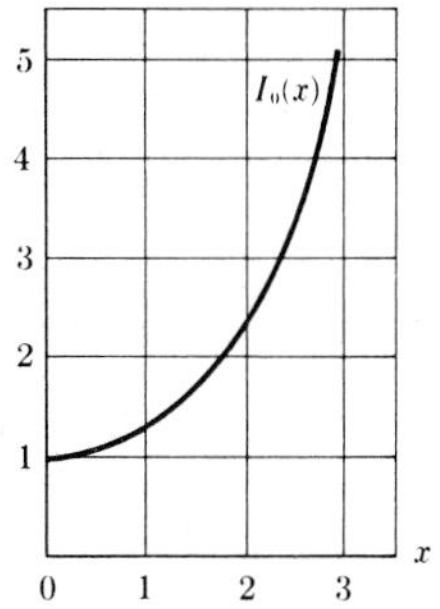

Fig. C.1 Modified Bessel function of the first kind of zero order.

D. MARCUM'S Q FUNCTION

The Q function is defined as

$$Q(\gamma,\beta) \triangleq \int_\beta^\infty x \exp\left(-\frac{x^2+\gamma^2}{2}\right) I_0(\gamma x)\, dx$$

The *generalized Q function* of order M is defined as

$$\begin{aligned} Q_M(\gamma,\beta) &= \int_\beta^\infty x\left(\frac{x}{\gamma}\right)^{M-1} \exp\left(-\frac{x^2+\gamma^2}{2}\right) I_{M-1}(\gamma x)\, dx \\ &= Q(\gamma,\beta) + \exp\left(-\frac{\gamma^2+\beta^2}{2}\right) \sum_{r=1}^{M-1} \left(\frac{\beta}{\gamma}\right)^r I_r(\gamma\beta) \end{aligned}$$

Clearly, when $M = 1$,

$$Q_1(\gamma,\beta) = Q(\gamma,\beta)$$

$I_r(x)$ is the modified Bessel function of the first kind of order r and is tabulated in complete books of tables. The Q function has been tabulated by Marcum in Ref. [A.3]; see also [A.4].

E. SUMMARY OF TETRACHORIC SERIES

THEOREM E.1 *Let F be a bivariate gaussian distribution with zero means, unit variances, and correlation coefficient ρ. Define*

$$d = \iint_{hk}^{\infty} dF$$

Then

$$d = \sum_{r=0}^{\infty} \frac{\rho^r}{r!} \tau_r(h)\tau_r(k)$$

where

$$\tau_r(x) = \left(-\frac{d}{dx}\right)^{r-1} G(x) \qquad G(x) = \frac{1}{\sqrt{2\pi}} \exp -\tfrac{1}{2}x^2$$

and the convergence is uniform for $|\rho| \leq 1$.

This is called the *tetrachoric series* for the bivariate gaussian distribution.

THEOREM E.2 *For the M-dimensional case,*

$$d = \int_{h_1}^{\infty} \cdots \int_{h_M}^{\infty} G(\mathbf{x};\mathbf{0};\boldsymbol{\alpha})\, d\mathbf{x}$$

where

$$\boldsymbol{\alpha} \triangleq \begin{bmatrix} 1 & \ddots & \lambda_{ij} \\ \lambda_{ij} & \ddots & 1 \end{bmatrix}$$

Then

$$d = \sum_{r=0}^{\infty} \Bigg[\underbrace{\sum_{n_{12}} \cdots \sum_{n_{ij}} \cdots \sum_{n_{M-1,M}}}_{\text{such that } \sum\sum_{i>j} n_{ij} = r} \prod\prod_{i>j} \frac{(\lambda_{ij})^{n_{ij}}}{(n_{ij})!} \prod_{k=1}^{M} \left(-\frac{d}{dx}\right)^{q_k - 1} G(h_k) \Bigg]$$

where

$$\sum_{\substack{i=1\\i\neq j}}^{M} n_{ij} = \sum_{\substack{i=1\\i\neq j}}^{M} n_{ji} = q_j$$

See Ref. [A.5] for more complete development of these theorems.

Note the following:

1. $\{n_{ij}\}$ and $\{q_j\}$ depend on the value of r.
2. The $r = 0$ term is

$$\prod_{i=1}^{M} \int_{h_i}^{\infty} G(x)\, dx$$

3. If we wish to expand

$$d' = \int_{-\infty}^{h_1} \cdots \int_{-\infty}^{h_M} G(\mathbf{x};\mathbf{0};\alpha)\, d\mathbf{x}$$

instead of d, the summation $\sum_{r=1}^{\infty}$ remains unchanged, but the $r = 0$ term becomes

$$\prod_{i=1}^{M} \int_{-\infty}^{h_i} G(x)\, dx = \prod_{i=1}^{M} \phi(h_i)$$

F. CHI-SQUARED DISTRIBUTION

The chi-squared distribution is the distribution of

$$y \triangleq \chi^2 \triangleq \sum_{j=1}^{k} \left(\frac{x_j - u_j}{\sigma_j}\right)^2$$

where the $\{x_i\}$ are independently distributed gaussian random variables with means $\{u_i\}$ and variances $\{\sigma_i^2\}$, respectively. That is, χ^2 is the sum of squares of independent gaussian random variables each with zero mean and unit variance.

The characteristic function of the chi-squared distribution is

$$C_y(u) = E(\exp iuy) = (1 - 2iu)^{-k/2}$$

The probability density function, determined from the inverse Fourier transform of $C_y(u)$, is

$$p(y) = \frac{1}{[(k/2 - 1)!2^{k/2}]} (y)^{k/2-1} \exp - \frac{y}{2}$$

This is a particular form of the gamma distribution and is referred to as the chi-squared distribution. The chi-squared distribution and its cumulative distribution have been tabulated for integer values of k.

REFERENCES

A.1 Davenport W. B., and W. G. Root: "An Introduction to the Theory of Random Signals and Noise," McGraw-Hill, New York, 1958.

A.2 Papoulis, A.: "Probability, Random Variables, and Stochastic Processes," McGraw-Hill, New York, 1965.

A.3 Marcum, J. I.: *Rand Report* RM339, January, 1960.

A.4 *IRE Trans. Inform. Theory,* vol. IT-6, no. 2, April, 1960.

A.5 Kendall, M. G.: Proof of Relations Connected with Tetra-Choric Series and Its Generalizations, *Biometrica,* vol. 32, p. 196, 1941.

Index

Abrams, I. J., 257
α space, admissible, 175
Antennas, 4
Antipodal signals, 35
A posteriori probabilities, 108
A posteriori receiver, 36
A priori distribution, 8
  least favorable, 41
Asymptotic approximations, 178
Average risk, definition, 14
  minimum, 23

Baker, C. R., 97, 102
Balakrishnan, A. V., 102, 149, 189, 218, 257
Battin, R. H., 80
Bessel function, modified, first kind, 282
Binary communication, colored noise, 73–90
    coherent, 82–84
    noncoherent, 84–90
    via random vectors, 103
    via stochastic signals, 97–102
Binary communication system, differentially coherent, 71
  frequency-shift-keying, 53
  noncoherent, 63–68
  partially coherent, 70
  phase reference signal, 71
Binary decision function, 20–25
Binary system, coherent, 30–35
  composite hypothesis testing, 52–68
  definition, 13
  multiple signal waveforms, 37
  non-return-to-zero, 20
  on-off modulation, 38
  performance, 29
  return-to-zero, 19
  vector model, 25–30
Biorthogonal signal structure, 185, 243
Blackwell, D., 17
Block encoding, 106

Channel, 3, 4
Characteristic function, 165
Chi-squared distribution, 285
Communication efficiency, 184
Communication system, analog, 6
  binary (*see* Binary communication system)
  digital (*see* Digital communication system)
  M-ary (*see* Digital communication system)
Complex envelope, 61–63
Conditional risk, 15
Continuous method, 32, 97–100
Convex body theory, 152
Convex hull, 153, 167, 171
Correlation detection, 33
Correlation detector, envelope, 57
Correlation receiver, envelope, 88
Cost function, definition, 13
  simple, 13
Cost matrix, 22
  definition, 24
Cuboctahedron, 253

Decision function, 9
  admissible, 16
  complete, 16
  definition, 12
  deterministic, 12
  equalizer, 42
  equivalent, 16
  essentially complete, 16
  minimax, 41
  nonrandomized, 12
  randomized, 13
Decision regions, 17
Decision rule, Bayes, 14, 21
  definition, 14
  minimax, binary, 16, 40–44
    M-ary, 43
  Neyman-Pearson, 46–48

Decision space, 8, 11
Decision theory, sequential, 9
Deflection, 97
Detection, in colored noise, 74–82
  complex envelope representation, 61–63
  correlation, 33
  matched filter, 33
  noise-in-noise, 93–102
    perfect detectability, 103
  nonsingular, 27, 81, 91
  perfect, 80
  singular, 27
  stochastic signal in noise, 93–102
Detection integral equation, 79
Detection probability, 46
Detection theory, radar, 45–50
Detector, quadratic, 95
Differential phase systems, 256
Digital communication system, 4
  binary (*see* Binary communication system)
  M-ary, 106–113
  M-ary coherent, 109–111
    nonwhite noise, 116
    Rayleigh fading channel, 114
  M-ary noncoherent, 111–113
    nonwhite noise, 116
  M-ary partially coherent, 115
Dunbridge, B., 216

Electromagnetic waveguides, 4
Entropy, 13
Envelope, complex, 61–63
Envelope correlation detector, 57
Envelope correlation receiver, 88
Error function, 28
  complementary, 28

False alarm probability, 46
Ferguson, T. S., 10, 17
Filter, causal, 6
Filtering, 5
Fourier transform, 165
Fredholm equation, 79
Frequency shift keying, noncoherent, 267
Friedman, B., 96

Girshick, M. A., 17

Helstrom, C. W., 70, 80, 102, 277
Hermite polynomial, 271
Hilbert space, reproducing kernel, 81
Hypothesis testing, composite, 52

Ideal observer, 24
Ideal receiver, 24
Information, 3
Ionospheric reflection, 4

Jamming, 5

Kadota, T. T., 97, 101, 102
Kailath, T., 81
Karhunan-Loeve expansion, 281

Lagrange variational technique, 193, 244
Laguerre polynomials, 273
Landau, H. J., 216, 253
Laning, J. H., 80
Likelihood ratio, 19
  definition, 23
  generalized, 54

Marcum Q function, 283
Maser, 122
Matched filter, 27, 109
Matched-filter detection, 33
Maximizing the minimum distance, 214, 254
Mean squared error, 5
Mean width, 153
  polygon, 155
  in two dimensions, 154
Mercer's theorem, 78
Message space, 7
Middleton, D., 80
Minimax theorem, 41
Multipath communication systems, 93
Multipath propagation, 5

Natural ordering, 16
Neyman-Pearson lemma, 46–48, 50, 74
Noise, colored, 73
  definition, 3, 4
  thermal, 5

Nuisance parameters, 54
Nuisance vector, 54
Nuttall, A. H., 257

Observation space, 8, 11
  partitioning, 27
Orthogonal noncoherent signal structure, 265
Orthogonal signal set, 160, 184

Phase-locked loop, 4
Polyhedra, regular, 254
  semiregular, 254
  symmetric, 253
Prediction, 5
Predictor, 6
Probability, of correct decision, 18
  of detection, 18
    asymptotic approximation, 178
    gradient, 164
  of error, 5
    definition, 17
Probability density function, gaussian, 281
  rician, 273

Q function, 60, 283

Radar detection, K-pulses, 69
  one pulse, 56–61
Radar detection theory, 45–50
Radar system, 4, 6, 8
  vector model, 48–50
Radar system design philosphy, 46–48
Ragazzini, J. R., 79, 80
Range rate, 6
Rayleigh probability distribution, 58
Receiver, 4
  a posteriori, 36
  Bayes, 14
  best, 14
  envelope correlation, 88
Regular simplex coding, 189–217
Rician probability density function, 60, 273
Risk, average, 14
  conditional, 15
Robinson, R. M., 253
Root, W. L., 92

Sampling, continuous, 7
  discrete, 7
Sampling method, 31
Scatter communication systems, 93
Schaffner, C. A., 228
Scholtz, R. A., 269
Schwartz, M. I., 37
Series expansions, 178
Shannon, C. E., 121
Signal-to-noise ratio, 5
Signal sets, biorthogonal, 243, 249, 250
  equicorrelated, 162
  linearly dependent, 157
  linearly independent, 158
  orthogonal, 160
  regular simplex, 160
Signal space, 7, 8, 11
Slepian, D., 102, 216, 253
Sonar system, passive, 93, 97
Spectral density, rational function, 79
Stochastic process, continuous in the mean, 281
Support function, 153
System performance, effect of dimensionality, 249
Systems, digital, 4
  sonar, active, 4
    passive, 4

Telemetry system, 4
Tetachoric series, 178, 284
Transmitter, 4
Tropospheric propagation, 5
True target range, 18
Twisted cube, 253
Type I error, 46
Type II error, 46

Uncertainty, 13

Viterbi, A. J., 71

Welch, L., 261
Wiener-Hopf equation, 80

Zadeh, L., 79, 80